HANDBOOK

for

SHELL COLLECTORS

REVISED EDITION

ILLUSTRATIONS AND DESCRIPTIONS *of* OVER 2,000
MARINE SPECIES FOREIGN TO THE
UNITED STATES *of* AMERICA

By
WALTER FREEMAN WEBB

LEE PUBLICATIONS WELLESLEY HILLS, MASS.

PRINTED IN THE UNITED STATES OF AMERICA

INTRODUCTION

While there are really rare shells in this book, I had no trouble in securing them for my private collection. People who start now will have more trouble than I did, as all the old dealers in shells have passed on and there has been no one to take their place. To accumulate a real stock of the shells of the world is a life work and always will be.

About 90% of all known forms of mollusca are small and a vast number are minute when fully adult. The shell collector should have a good magnifier and with its aid you can usually determine whether your specimen is of a species fully adult or the young of some other species.

There are over 100,000 species of shells in the world and more being discovered all the time. Only one complete check list has ever been printed as far as I know, and that was in 1887. It made 3 vols. covering 1500 pp. Such a list if printed today, would be vastly larger, but someone with plenty of time and money should bring it out.

In the back of the book you will find an index to the Latin and common names. The Latin names are the only truly universal name and you should learn them on the start. The common names are for those who simply make a small collection and have no real desire to make a study of them. The common name of a shell may differ in every country where found.

This volume with my book on Foreign Land Shells and the one on United States Mollusca, cover many thousand species and contain information which would cost many hundred dollars if you purchased a general library on Conchology.

WALTER FREEMAN WEBB

Revised Edition

The main change in this edition is the addition as an appendix of 15 plates covering 325 species of shells. These shells had come on the market and were for sale by dealers, hence I thought it advisable to illustrate them. While there are just as many shells in the sea as ever, it is a fact that many hundred fine varieties have not yet appeared since the war and are not obtainable at any price. Over 2,000 species are illustrated herein.

WALTER FREEMAN WEBB

JACKET ILLUSTRATION

THE PICTURE on jacket is Conus gloria-maris, Hwass. It has been found in the Philippines and certain South Sea Islands. Considered one of the rarest of marine univalves. This specimen came from the noted collection of Mrs. S. L. Williams of Chicago and now is in the American Museum of Natural History, New York. Mrs. Williams paid $500 for it around 1900. The dealer there secured it from the famous collection of Dr. James Cox of Australia. The data read Marquesas Islands. Although discovered a century and a half ago, only about 7 perfect specimens are known and about 6 others of poorer quality. It is called the Glory of the Sea. The illustration is shown through the courtesy of the American Museum of Natural History, New York.

GASTROPODA

IT IS BELIEVED that three-fourths of the marine mollusca are *Gastropoda* which refers to shells of one piece. The other fourth are mainly *Pelecypoda* or bivalves. There are still three other classes very small by comparison in number of species. *Cephalopoda* which includes squids, etc. *Pteropoda* which cover very small free swimming mollusks which live on the surface of the sea in very warm water, the *Chitinidae* which consist of eight pieces, and the *Scaphopoda* or Dentaliidae. Collectors usually include the recent Brachiapods which are not real shells at all, but are included in the Molluscoidea.

To attempt to give one general fairly complete idea of the Gastropoda would take a couple hundred pages hence these notes have to be brief. The greater percentage of species are coated with an external covering ranging from very thin to very thick.

In most cases the shell is sufficiently large to contain the entire mollusk but you will find as you look over the following pages there are very many exceptions. The spiral growth of shells is as nearly of true mathematical regularity as is possible in an organic body. Typically the operculum is spiral also yet in many cases its growth is annular. It ranges all the way from horny to calcareous.

The mollusk is only attached to the shell at one point on the columella and by means of a columellar muscle, which passing through the foot is attached at the other end to the operculum. The mantle border is the principal agent in the secretion of the shell. With the organic basis of this secretion is mingled carbonate of lime, originating in the cells of the mollusk. The external layer of the shell is a skin or epidermis, having no lime in its composition.

While the growth of the shell is thus provided for by additions to the aperture margin from the mantle border, the whole mantle is equally capable of producing shelly substance; and not only are shells thus thickened from within by the mantle surface, but breaks are repaired with new material by a similar provision; such repaired and interior portions are devoid of epidermis and of color, the pigments being found only in the free border of the mantle. The mollusks are even able to secrete shelly matter to prevent against threatening dangers from the boring of other animals into their shell.

The anterior notch in many shells is occupied by the siphonal tube which is usually placed near the margin of the columella. If greatly prolonged it is called a canal. This siphon has various uses in the life of the mollusk. You will find it largely in carniverous species. There may be a small siphon extending backward and there is a special notch for same as in Cypraea and Conus. It serves as an exit for water.

The relations of the shell to the breathing organs is very intimate, it being essentially a calcified portion of the mantle of which the breathing organ is a very specialized part. The shell is characteristic of the mollusca that they have been commonly called "testacea" (from testa, a shell).

The most important feature of the study of the shell is its importance geologically, as it is almost the only portion of the vast number of fossil species which has been preserved to us. By its study we have been able to compare the fossil and recent species and in many cases predicate the external appearance, the anatomy and the physiology of all known forms of shell life.

All mollusca acquire a rudimentary shell before they are hatched except the Agonauta, and this becomes the nucleus of the adult shell. It is often differently colored and shaped from the rest of the shell and becomes of great importance in the study of species.

Sea weeds filter the salt water and separate lime as well as organic elements. Without this lime, shells would not exist. Where shells are unnaturally thickened there is an over abundance of lime. In those districts wholly destitute of lime, there are no mollusca.

The texture of shells is various and characteristic. Some when broken present a dull lustre like marble or china and are termed *porcellaneous;* others are *pearly* or *nacreous,* some have a fibrous structure, some are *horny* and others *glassy* and *translucent.*

The nacreous shells are formed by alternate layers of very thin membrane and carbonate of lime but this alone does not give the pearly lustre which appears to depend on minute undulations of the layers. Nacreous shells when polished form "mother of pearl." This is the most easily destructible of shell textures, and in some geological formations we find only casts of the nacreous shells, while those of fibrous texture are completely preserved.

The epidermis of shells has life but not sensation, like the human scarf-skin. It protects the shell against the influence of weather and chemical agents. It soon fades or is destroyed after the death of the animal in situations where, while living, it would have undergone no change.

The shell growth is formed by the mantle, as each layer of it was once a portion of the mantle, either in the form of simple membrane or as a layer of cells, and thrown off by the mantle to unite with those previously formed.

The epidermis is formed by the mantle or collar of same. The membraneous and nacreous layers by a thin and transparent portion which contain the visera, hence we find the pearly texture only as a lining inside the shell as in the Nautilus and Turbos.

Many shells cease to grow in winter and these periods of rest are often indicated by interruption of the otherwise regular lines of growth. In shells like the Murex the period of growth is marked by a fringe or a row of spines.

In some shells the mollusk vacates the apical whorls which fall off and then are filled with nacre. The colors of shells are usually confined to the surface beneath the epidermis and are secreted by the border of the mantle which often has similar tints and patterns. The secretion of this color by the mantle depends greatly on the action of light. Shallow water shells as a class are brighter colored than those from deep water.

The operculum of Gasterpods is developed on a particular lobe of the foot and consists of horny layers, sometimes hardened and shelly matter. The spines on shells depend on folds of the mantle-margin, are formed one at a time and usually indicate the periodicity of growth. The varices of Murex are simply the thickened lips of former mouths.

There is nothing in nature more beautiful than the regular geometrical progression of the growth of a shell or the certainty with which each species and genus grows in its normal pattern although these modes vary among themselves so widely. For example we have the simple depressed cone of the Patella, all eperture and no spire, and from it in every graduation, from the Haliotis almost equally depressed and broad, to the long many whorled Terebra or the Vermetus which is a Terebra partially unrolled into a simple long tube, the opposite of Patella.

The whorls of most shells are closely wound around its axis but in the Conus and Cypraea the shell only shows externally its last whorl.

The details of sculpture of a shell as striae, sulcations, ribs, spines, etc., result from similar ornamentation of the mantle. If the spine of a Murex is closely examined, it will be found to have a longitudinal seam upon its front face, showing that it has been formed by a corresponding digitation of the mantle.

If the axis of a shell around which the whorls are coiled is open or hollow, the shell is said to be *perforated* or *umbilicated*, like Solarium. The columella is that portion of the inner wall of the shell which invests the axis. It is in the lower portion of the inner margin of the aperture in most spiral shells and is entirely wanting in such shells as Bulla and Cypraea.

A whorl is a single complete revolution of the spiral cone. Its periphery is an imaginary spiral upon the outer wall, half way between the suture and the base, or line of greatest width. The last whorl of a shell ending with the aperture is called the body whorl; the others are called spire whorls. In the females of some species the whorls enlarge more rapidly than in the males.

The distance between the apex and base of a Gastropod shell is termed its height. The aperture is entire in most of the vegetable feeders but notched or reduced into a canal in the carniverous families. This canal surrounds a siphon which is respiratory in its office and does not necessarily indicate the nature of the food. The margin of the aperture is called Peristome.

A complete description of the many parts and their uses of the mollusk that forms the shell as its home would fill several hundred pages. The collector who wishes to made a study of such characters should purchase works that cover such information.

PLATE 1

1. **Murex brevifrons,** Lam. Central America. A very dark brown shell with prominent sharp fronds, almost similar in form to ramosus. Long narrow canal with long spines on same. 3″ or more.

2. **Murex foliatus,** Gmel. Lower California. A white shell with prominent folds. Canal is completely closed. Rather rare. 2½″.

3. **Murex sauliae,** Sow. Moluccas. The ground color is white, with numerous circular bands of chestnut. Tip is pink, and the edges of fronds are frequently so marked. Rather rare. 2½″.

4. **Murex trunculus,** L. Naples. A rounded rough shell, with numerous dark bands. Interior of rich purple. It is one of the famous dye shells of antiquity, and there are many mounds of shells in the islands of the Aegean Sea which have lain there for 25 centuries or more. 2″.

5. **Murex zealandicus,** Quoy. New Zealand. A small white shell, with numerous long prongs, which seem to be arranged along the edge of the aperture showing various periods of growth. 2″.

6. **Murex trialatus,** Sow. San Diego and southward. A small shell with three prominent wings, which gives it a triangular appearance. The shell seems to be uncolored, and is very often completely covered with bryozoan growths, making its cleaning for the cabinet a real work of art, if ever accomplished. 1½″.

7. **Murex maurus,** Brod. Phillipines. Similar in form to **Sauliae,** it is not a common shell. The fronds are low and short, the tiny circular ridges are tipped with brown, and I fail to find any other prominent characteristics. 2½″.

8. **Murex cabritti,** Bern. West Indies. A shell of the form of cut, with very low varices, circular lines of brown, nodules at regular intervals on last whorl. A shell from deep water rarely seen. 2″.

9. **Murex rectirostris,** Sow. Panama. A small flesh-colored shell, three prominent varices, with one or two points on each, narrow nearly closed canal. 2½″.

10. **Murex tribulus,** L. Loo Choo Ids. There are a number of forms of Spiney Murex like this cut and they are often very hard to distinguish. The three prominent varices have spines of different lengths. The body of the shell has many circular small ridges and the canal is rather long and nearly closed. 3″.

11. **Murex adunco spinosus,** BK. Philippines. A white shell of the general form of the preceding, the short spines are curved and sharp. The canal about half the length af the shell. 2¼″.

The Family Muricicinae are commonly called rock shells. They are mostly tropical and subtropical and are found in shallow water to 50 fathoms or more. There are more than 150 fossil species commencing with the Eocene. Usually the aperture ends in a canal, which is generally partly closed.

The common Murex erinaceus feeds largely on the oyster beds around Europe and is often very destructive of same. The Murex will clasp an oyster and applies its rostrum to the surface of the shell invariably near the beak. The regular movement of the body to the right and left for about three or four hours, enables it to pierce the shell through a round hole which it has drilled. This exposes the vicera and quickly kills the mollusk. The bored oyster, exhausted, opens its valves when many other forms of sea life hasten to eat a full meal.

Many ancient people obtained their purple dye from species of Murex. The small shells were bruised in mortars, the animals of the larger ones taken out. Many heaps of Murex trunculus are still to be seen along the Tyrian shore. Along the coast of Morea many such heaps of Murex trunculus have been found.

✦

According to Aristotle, the ancients used to pound up the smaller shells of Murex and Thais to secure the Tyrian Purple used as dye. They found difficulty of removing the mollusc and save the gland dyes at same time. They were particular to pound up the shells alive for if they die they spit out the purple. The old plan of catching them was with bait only, the result of which was when the line was hauled in, many were lost by dropping off the bait, and this they were sure to do if gorged. They then devise the idea of hanging a small basket on the line below the bait and if the Murex fell off, it tumbled into the basket and was caught.

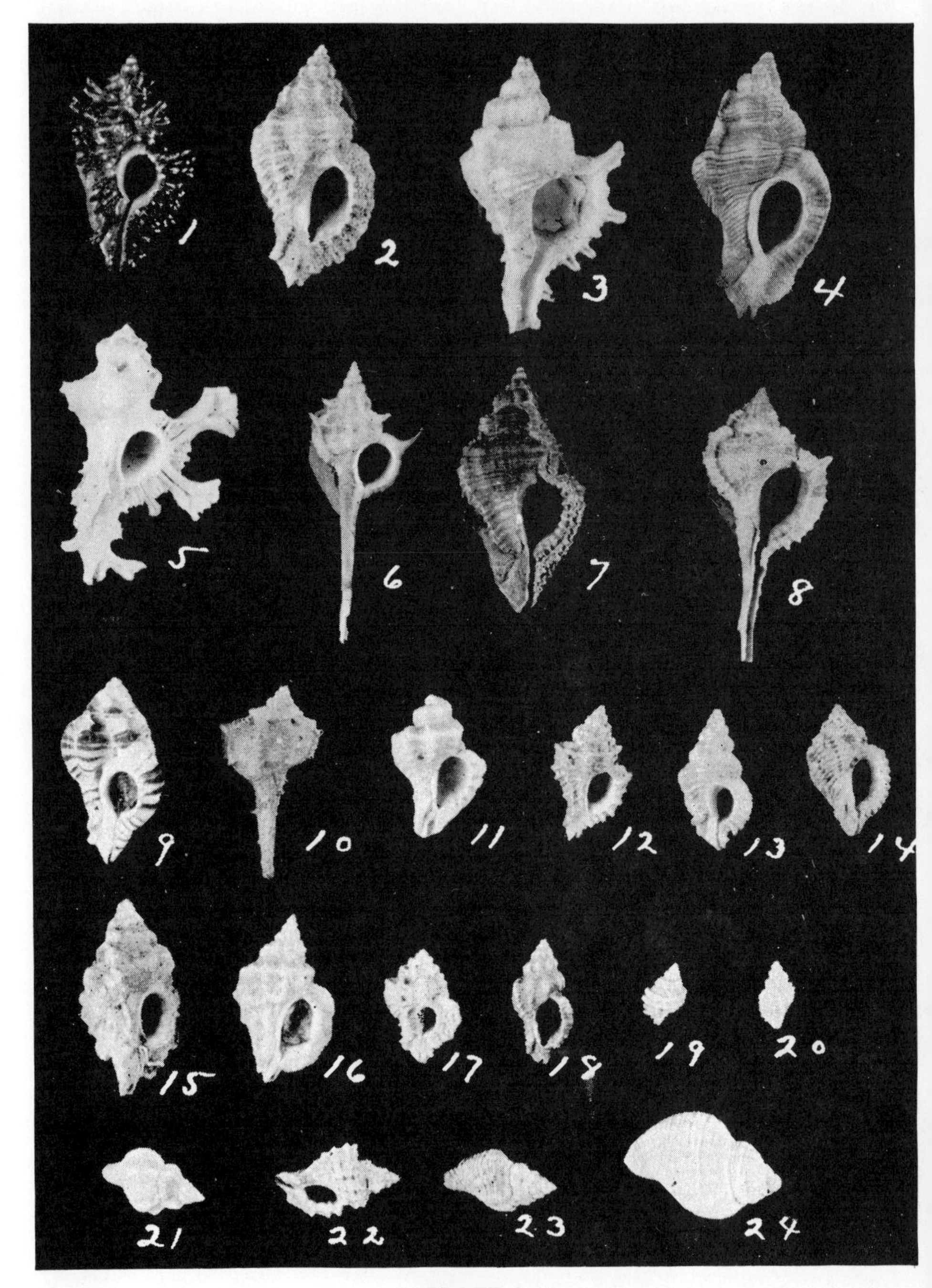

PLATE 2

1. **Murex rufus,** Lam. Key Largo, Florida. A small black shell of splendid form. The varices are finely frilled, even on the upper whorls. 1 to 1½".

2. **Murex erinaceus,** Lam. Mediterranean Sea. The shell is uncolored with fairly smooth varices and prominent circular lines. Lip is wide, canal closed. 1¼".

3. **Murex despectus,** AAd New Caledonia. A small white shell with no unusual characteristics. There are several small blunt spines on edge of lip. 1½".

4. **Murex festivus,** Hinds. California coast. A small shell with frills on edge of lip and similar frills at former lines of growth. The body is covered with fine circular lines. Canal completely closed. 1½" to 2".

5. **Murex anatomica,** Perry. Japan. A pure white shell with remarkable wide spine on edge of lip. The whole shell appears disjointed. The canal is closed and upper whorls are knobby. 1½".

6. **Murex sobrinus,** AAd Formosa. A very trim and neat little shell with one prominent spine on edge of lip and few others on varices. There are faint bands of brown. About 2".

7. **Murex capucinus,** Lam. Moluccas. The shell is a deep brownish-black, the lip of aperture is slightly frilled, and the entire shell ringed with small ridges. 1½".

8. **Murex funiculatus,** Rve. China. The small shell has only a suggestion of spines, completely covered with small nodules. Upper whorls of a pinkish cast and lighter below. 1½".

9. **Murex gemma,** Sow. California coast. A small shell with several frilled ridges. Aperture round and short, nearly closed canal. Color is white with prominent brown bands. 1".

10. **Murex brevispina,** Lam. Arabia. A small shell with tiny spines, round open aperture and slender canal. Light brown color. 1½".

11. **Murex erinaceus cingulifera,** Lam. Mediterranean Sea. It is rather smaller than the type with several blunt fronds and circular ribs. 1".

12. **Murex erinaceus tarentina,** Lam. Mediterranean Sea. More slender than the type, with finer ribs and wide flaring aperture. Almost uncolored.

13. **Murex nodulifera,** Sow. Philippines. A rather tall slender small shell uncolored, with deep ridges, round aperture and short curved canal.

14. **Murex circumtextus,** Stearns. California coast. A small ridged shell with many varices. Aperture is over half the length of the shell. There is one small band of brown.

15. **Murex fournieri,** Crosse. Loo Choo Ids. Shell is uncolored, three prominent ridges, aperture oblong, canal closed.

16. **Murex crassilabrum,** Gray Chili. A small uncolored shell with many deep ridges. The aperture has a flaring lip, with very thin horny operculum. The whole shell has a checkered appearance.

17. **Murex incisus,** Brod. California coast. A very small shell with several prominent ridges. Aperture is small, round with about closed canal.

18. **Murex edwardsi,** Payr. Malta. A rather elongated deeply ridged small shell with oblong aperture and very short canal. Uncolored.

19. **Murex scrobiculate,** Dkr. South Africa. A very small light colored shell, with deep circular ridges.

20. **Murex purpuroides,** Dkr. Natal, Africa. Another very small and slender light colored shell with prominent ridges. There are many such small forms of this genus, some of which gradually merge into other allied genera.

21. **Urosalpinx mexicana,** Rve. Mexican coast. A more slender shell than the variety cineria of our coast, with tiny ridges and light color. Closely allied to its northern relative.

22. **Eupleaura nitida,** Brod. Panama. The small shell is rather flat with rows of ridges both ways. A neat little shell as are all of this genus.

23. **Urosalpinx birilifi,** Lisch. Japan Sea. A small dark round shell with fine ridges and no prominent knobs. Aperture wide extended to end of canal.

24. **Urosalpinx Rushii,** Pils. Uruguay. A strong round solid shell of a very light brown appearance and many tiny circular ridges. Aperture large, white.

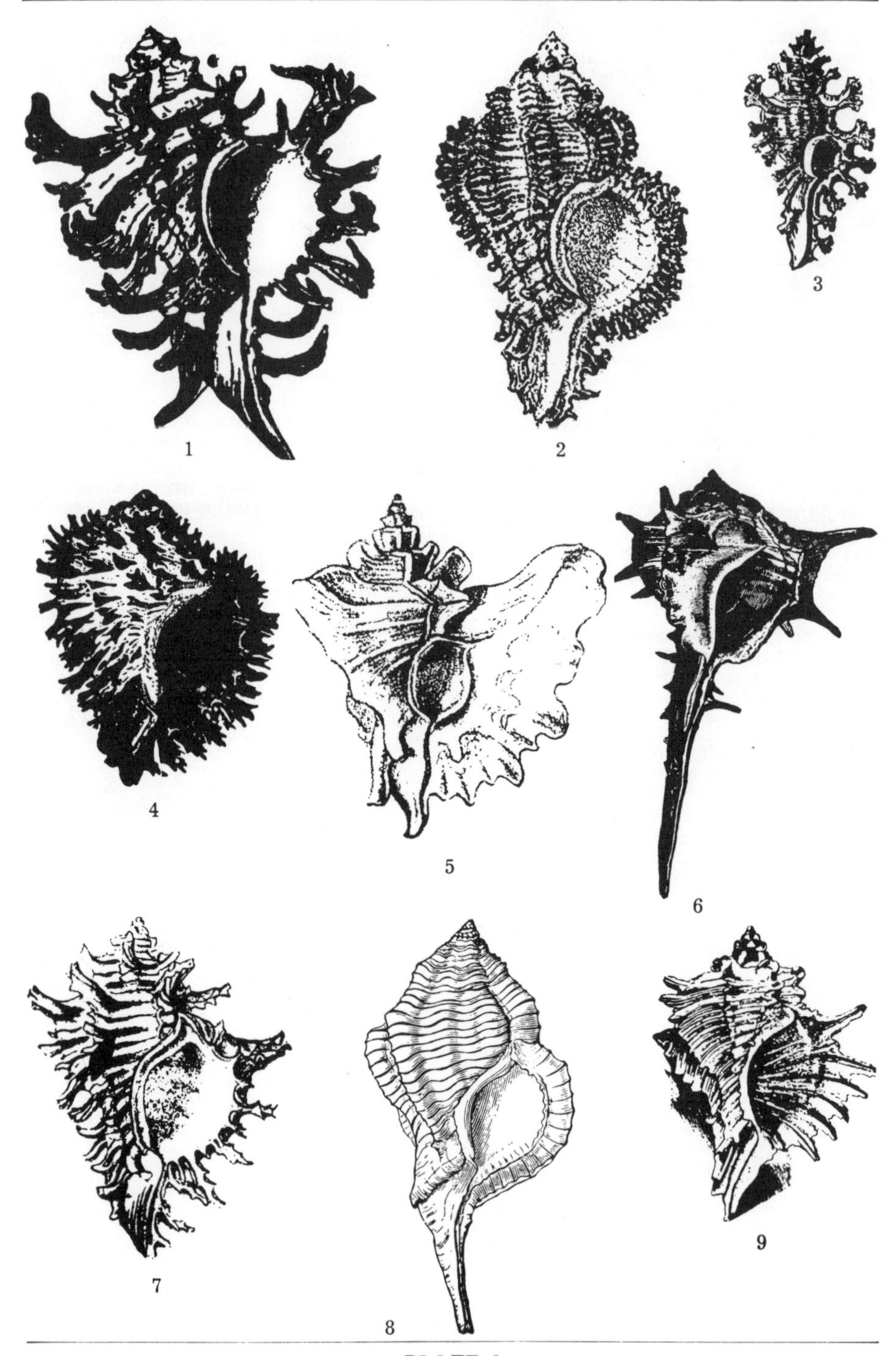

PLATE 3

1. **Murex endivia,** Lam. Philippines. A handsome brown and white marked shell with curved spines or none at all, which is fairly common on the shores of Cebu and other similar situations. Usually about 3″.

2. **Murex stainforthi,** Rve. Stainforth's Murex. Northwest Australia. Has several rows of blackish ridges which will distinguish it from other forms. There are few shell collectors where it is found hence not very common in collections. 2 to 3″.

3. **Murex palmarosea,** Lam. Indian and Pacific Oceans. Never very common anywhere. It is a wonderfully fine brownish species and from some localities the edges of the fronds are pink. There are other forms of similar style and rarer. There never seems to be enough shells of this species to supply the demand. 3″.

4. **Murex radix nigritus,** Phil. Black Murex: West Mexico and Panama. There is another similar form with only a few rows of varices and is found to run much larger in size than this form. Usual size of rich black shells is 3 to 4″. Small 2″. Specimens have very sharp slender spines, which thicken up with age.

5. **Murex aduncus,** Sow. Winged Murex. Japan. A small shell seldom over 2″ with very prominent thin wings. Much desired. Usually a pale brown color.

6. **Murex cornutus,** Lam. African Horned Murex. West Africa. A dark brownish shell attaining 6″ or more with fine curved horns. Fairly common if there were any collectors in the territory where they are found. But there is never enough of them on the market to supply the demand.

7. **Murex princeps,** Brod. Princess Murex. Gulf of California to Panama. A very fine species ranging from 2 to 4″ with several rows of varices. The body whorls are well marked with rich deep reddish-brown bands. Not rare but the larger and older specimens are usually much eroded by enemies of all such mollusca. The medium sized shells are best.

8. **Murex elegans,** Beck. Lined Murex, West Indies. A handsome smooth species of 3″ which must be uncommon as it is so rarely seen in collections in fine condition. If you make trips to the Bahamas, as so very many do now days, look for it and find out if possible where it is most often found. It much resembles the variety **motacella** from Senegal.

9. **Murex saxatilis,** L. African Murex. West Africa. One of the large fine forms of light brown color, ranging from 4 to 6″. Must be fairly common as I can remember back 40 years ago when it was one of the most common species on the market but of recent years very few are seen. It has three bands which show inside the white aperture.

In the Journal of Malcology, Vol. 8, Page 341, the Rev. A. H. Cooke describes a mound located at Sidon. It stands immediately to the south of the present town, a short distance outside the walls and forms part of a low cliff near the shore. It is about 60 by 20 feet but it may extend into the hill at one side. An Arab burial ground occupied the upper part of the cliff. The shells were exclusively of Murex truncułus. On examining the specimens I found that they had evidently all been broken at exactly the same point, I suppose to extract the dye. If a specimen is held with the spire uppermost and the aperture toward the observer and slightly inclining to his right, a large hole is observed in the ultimate and penultimate whorls, laying bare but not fracturing the axis of the spire. I imagine this hole was made by some punch or stamp. Occasionally it misses the mark, and one or more of the upper whorls are fractured. The hole in all cases would be near where the dye gland is located. There are very many such mounds in the Mediterranean region. Archaeologists claim some of these mounds date back to 1600 BC.

✦

The Amphipeperas (Ovula) belong to the Cypraea and they are all shiny of various colors. The largest form is A. ovum or Egg shell and the strangest is A. volva, the Weaver Shuttle shell. Of the latter I have seen 5″ specimens but 3″ is more common and just as attractive. Some of the small forms live on the colored Gorgonias and Sea Feathers and in such cases they assume the color of the host. There are nearly 100 species and a few are found fossil in the Tertiary.

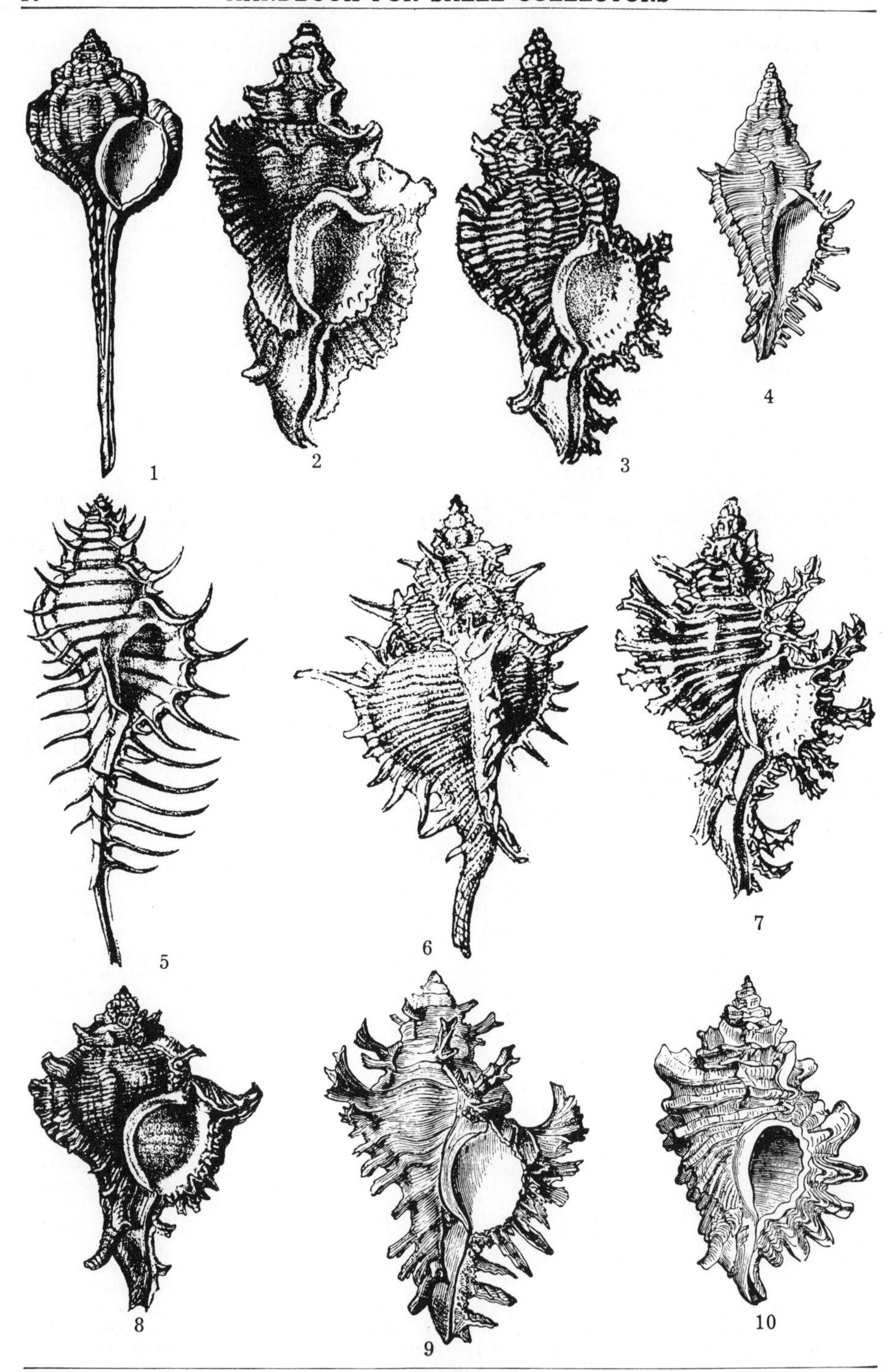

PLATE 4

1. **Murex haustellum,** L. Snipe-bill Murex. Philippines. This is one of the very odd and curious forms of this great genus. There are over 400 varieties of Murex in the world and if you could see them all or even most of them, you would find some ever more curious than this one. It is 4 to 5″, has high ridge back of aperture and spiral brown lines.

2. **Murex phlorator,** Ad. and Rve. Japan. This is an odd and uncommon winged shell which is not as large as the cut would indicate. Usually runs about 1¼″ but may come a little larger.

3. **Murex torrefactus,** Sow. Philippines. There are a number of species of this type and they are usually some shade of brown. Most of them are fairly common in their range. They often require a great deal of work to prepare them for the cabinet, as all sorts of marine life love to use them as host. 3 to 4″.

4. **Murex hexagonus,** Lam. Panama. This is a species of about 1 inch but it is a beauty. Most specimens I have seen contains more spines than are shown on this cut. There are similar forms, most of which live on rocky shores.

5. **Murex occa,** Sow. China. One of the unique forms of the spiney Murex as the lined body and peculiar curved spine, make it easily recognized. Attains 4″ and has faint brown bands. Not at all common in my lifetime but no telling what the future will bring forth.

6. **Murex pliciferous,** Sow. Japan. A pure white species that ranges from 3 to 4″ with short spines. Most of the Murex have a horny operculum. This specimen has one that perfectly fits the aperture. So many collectors fail to preserve this important part of the shell.

7. **Murex elongatus,** Lam. Fringed White Murex. China Seas. Usually attains 4 to 5″ but may be found much larger as it resembles the very common White Murex. Brownish-white with frilled spines.

8. **Murex anguliferus,** Lam. Angular Murex. Red Sea. Usually 2 to 3″ of a brownish color, with short stubby spines or none at all. There are a number of similar types.

9. **Murex adustus,** Lam. Black Asiatic Murex. Philippines. A small black shell with frilled edges that is fairly common in Sulu Sea and thereabouts. It must be thoroughly cleaned to bring out the rich color. 2 to 3″.

10. **Murex erinaceous,** illustrated and described on Plate 2.

Strombus goliath, Ch. is the largest Conch shell in the world. It is similar in form to gigas, but the lip is higher and rolls over towards the back and the shell is really larger than any gigas I have seen. I have only had two specimens in my lifetime and they came with collections purchased. Paetel in his list of shells of the world gives Nicoya as the locality where found, which is on west coast of Costa Rica, but I think that is an error as I have always thought they would be found around the "nose" of Brazil in the Atlantic. The shell readily brings a huge price and I hope some traveler to Brazil will make inquiries and see if any are being collected now that we are living in the air age.

✦

In the Annual Report of the Smithsonian Institution for 1916 is an article by Dr. Paul Bartsch on Pirates of the Deep. It is interesting to the shell collector as it explodes many myths which still exist. You will see illustrations taken from old works of an Octopus capturing a crab, a Paper Nautilus and its shell sailing along blissfully over the ocean, an Octopus attacking a ship winding its 30-foot arms around the masts, Dore's illustration of Gilliatt's fight with an Octopus, sailors' encounters with an Octopus, Sea Serpents, fight between a sperm whale and a giant squid and another with a whale and an Octopus, most of which just never happened. Stories of how the Nautilus by expelling the water from its body and retaining the air rises to the top of the ocean and sails around, which is not so. But there is a fine illustration of the trap used by the Filipinos in catching the Nautilus which is accurate. As the squids and octopus belong to the Mollusca, most collectors have small specimens in preservative in their collections.

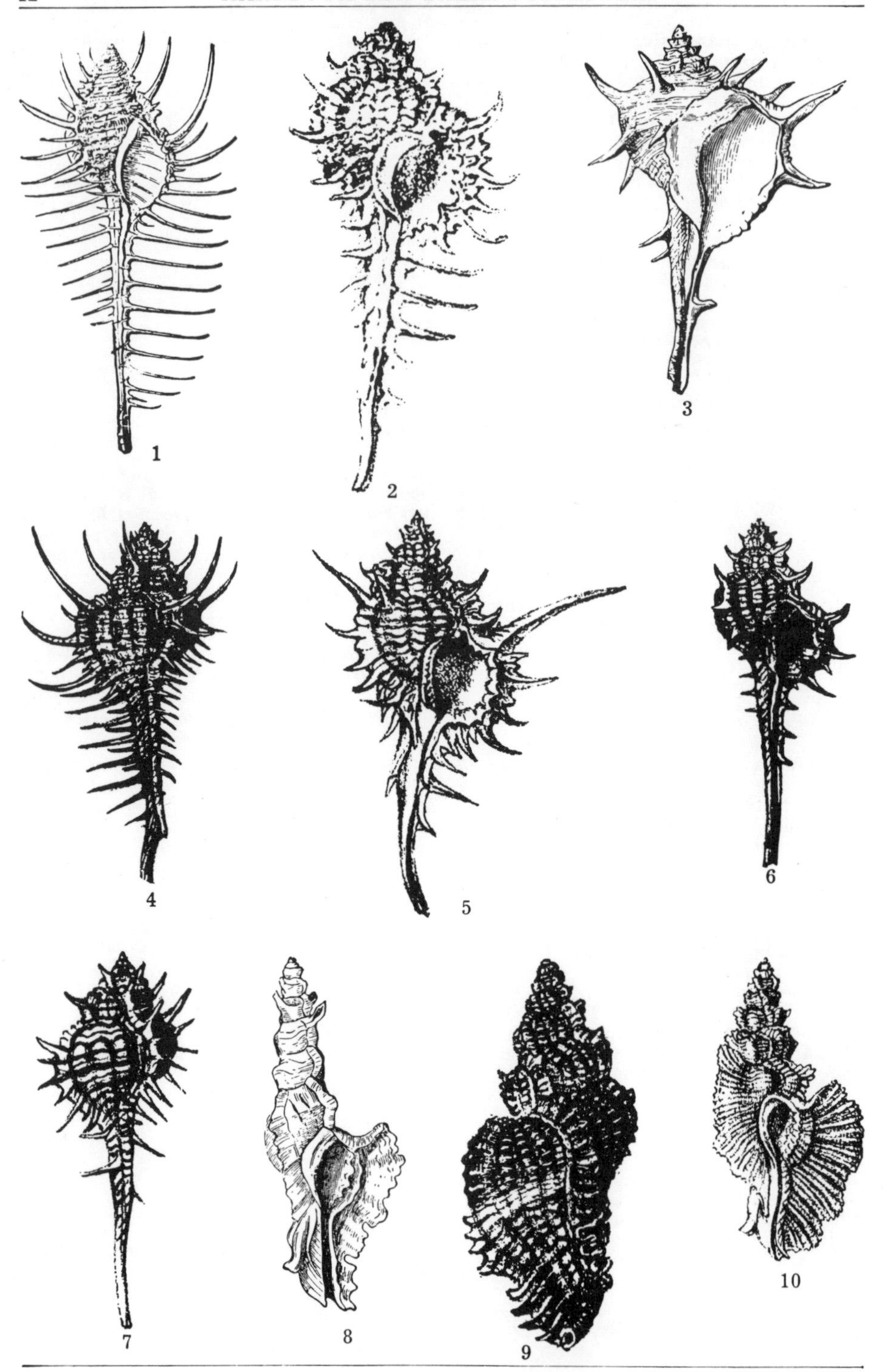

PLATE 5

1. **Murex tenuispina,** Lam. Venus Comb. Philippines to Japan. It is not a rare shell these days, but not always easy to procure with all spines perfect. One of the daintiest of all Murices and much admired. Whitish 4 to 5″.

2. **Murex plicatus,** Sow. Plicate Murex. Gulf of California. A handsome stubby species 2½″ to 3″, fairly common, and while fully ridged it is smooth to the touch. There are a number of similar forms.

3. **Murex brandaris,** L. Branded Murex. Naples. A very common species from a territory that has been inhabited for more than two thousand years by people who loved shells and who, in many cases, diefied them on their coins and pottery. 2 to 3″.

4. **Murex nigrospinosus,** Rve. Black-tipped Murex. Japan. Of a corneous color, usually attains 3 to 4″ and the many sharp curved spines are tipped with dark color. This is the main distinguishing feature except its spiney pattern.

5. **Murex spinosus,** A.Ad. Banded Spiney Murex. Red Sea. This is one of several spiney forms which are not always very easy to classify. Usually each tropic sea has certain forms more common than elsewhere so that locality is often very important. The elongated spine on upper part of the aperture is a fairly good distinguishing mark. 3 to 4″.

6. **Murex ternispina,** Lam. Chinese murex. China coast. One of the many smaller forms of spiney Murex that range 3 to 4″. The little prongs being short and sharp. Fairly common with bands of brown in fresh specimens.

7. **Murex martinianus,** Rve. Martins Murex. Japan. Quite similar to the preceding species with short sharp spines and usually about the same size. A draw of all the forms of Spiney Murex make an elaborate display. 3″.

8. **Murex clavus,** Kien. Spike Murex. Philippines. A very rare shell that has brought as much as $40 in recent years for a 4″ specimen. It is white, slender, three-sided with flaring aperture. I have had them from West coast of Luzon.

9. **Murex triqueter,** Born. Philippines. Differs from most all other forms of the genus, as it is covered with smooth ridges. Ranges about 2″. There are several smaller forms down to half inch of similar structure.

10. **Murex pinnatus,** Wood. China coast. A small triangular, 2″ pure white form of remarkable dainty beauty. I know of no other species at all like it. Not at all rare but very few seem to come on the market, so that it is not an easy shell to procure.

The shell Pterocera bryonia is the largest of the genus. During the late war one commander of a ship in the Pacific sent me a kodochrome of a shell the natives found that was 13 inches long and colored in the aperture. He felt sure it was very rare but the fact was he was not in the proper locality to find them by the hundred. At Funafuti they live in gravelly flat on the west side of the lagoon, at which the water was waist deep at low tide. In such situation many of the shells had lost their fingers and were much distorted. On the opercula of many specimens were found fine specimens of Hipponyx australis. The native value the Pterocera mollusk as food both raw and roasted. Shells that are very rare in one locality may be very common in another.

✦

The Island of Mauritius lies 20 degrees south latitude and about 5 degrees east of Madagascar in the Indian Ocean. I have for a number of years received literature that has been published on the shells found there and I quote from one of these bulletins as follows: The Marine species and varieties listed cover 2096, bivalves 279 or 2375 species from that one island and small islands included nearby. Here are the totals on a few of the genera. Terebra 70 varieties, Conus 170, Oliva 75, Harpa all known varieties but one, Voluta only 2, Mitra 265, the largest list I have ever heard of from one small territory, Murex 54, Drupa 46, Triton 48, Ranella 30, Cypraea 125 which covers about half all known forms in the world, Strombus 26, Pecten 41 and included is gibbus and nodosus both found here in Florida. How did these two common kinds ever get over in the Indian Ocean. What a territory for a man to spend one or two years with plenty of funds for hiring good help. The population is largely Portuguese who are always fine seaman. I tried for years to get a satisfactory collector there but my efforts were always a failure. Hope some of my friends will be more successful.

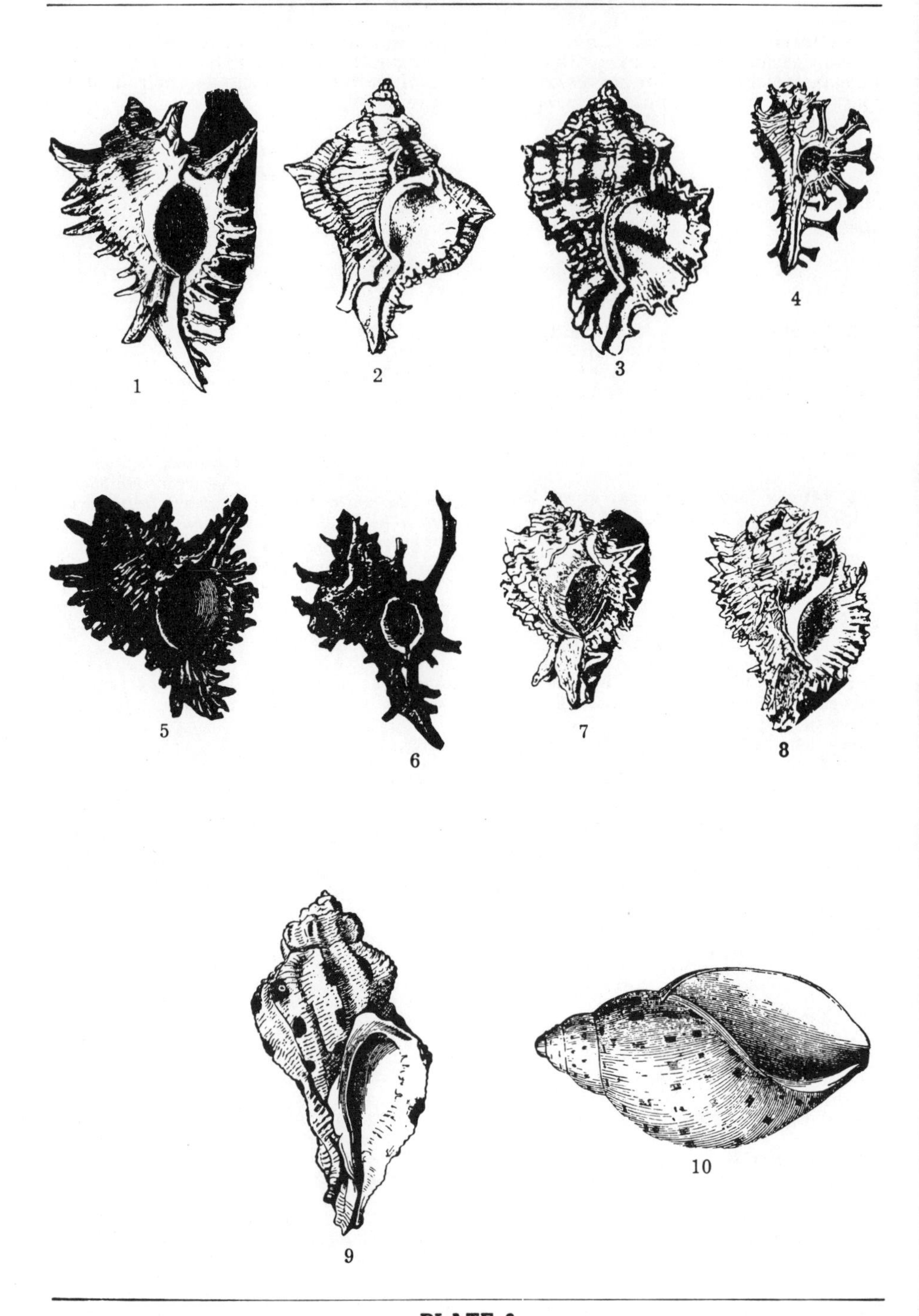

PLATE 6

1. **Murex ramosus.** Lam. White Murex. Red Sea and nearby territory. It must be a very common species as forty years ago it was imported into this country in vast quantities and in all sizes from 3″ to mammoth 8″ specimens. It is pure white, although some specimens show traces of brown and I have seen fresh specimens with reddish apertures.

2. **Murex trunculus,** the Banded Murex is also illustrated and described on Plate 1.

3. **Murex brassica,** Lam. Banded Pink Murex. Gulf of California. Fairly common in this territory where fine large round shells range from 3 to 4″ and I have seen perfect 8″ specimens which are very rare. Aperture is pink with brown bands on the whorls. Found on the wide mud flats of that region and specially on rocky shores.

4. **Murex anatomica,** also illustrated and described on Plate 2.

5. **Murex megacerus,** Sow. Found over much of Oceanica. It is a solid dark colored shell of about 3″ but may come larger. There are other solid shells similar in same territory.

6. **Murex axicornis,** Lam. Small Horned Murex. Moluccas. The horns or spines of this little fellow are unusually long for the size of the shell and I consider it one of the daintiest species obtainable. There are so very many choice forms found in and around the hot Dutch East Indies. Of a reddish-brown color, about 2″.

7. **Murex regius,** Wood. Rose Murex. Panama. A fine large species with rose colored aperture. It used to be a fairly common commercial shell and today is one of the most common forms of the Panama region also one of the largest. Usually 3 to 5″. It requires plenty of hard work to remove all the many forms of sea growth that are quite sure to infest the little spaces between the nobby spines.

8. **Murex bicolor,** Val. Pink Murex. Panama. This species and the Rose abovementioned are both found at Panama but of late years it is not real common there. It has been found more common at points farther north on Mexican coast. The aperture is deep pink color in fresh specimens and the body whorl has the usual number of spiney knobs. 3 to 4″.

9. **Murex salebrosa,** King. Mazatlan. to Panama. The Murex are divided into 19 sections or subgenera and this species comes under Vitularia. By arranging the family into sections as shown in recent monographs you will see a gradual graduation into allied genera. It lives under rocks and not very common anywhere. 2½″.

10. **Halia priamus,** Meusch. Cadiz, Spain. A light brownish smooth shell. Slightly mottled, found in very deep water. It is quite rare. There seems to be no known affinity with other genera of marine shells. In the usual classification it is placed next to Cancellaria. 3″.

The Great Barrier Coral Reefs stretching along the east coast of Australia for some 1500 miles, extending into the ocean for sixty miles are great collecting grounds. Ships can only approach the shore where channels have been kept open. When the tide goes out, immense stretches of coral fields are exposed. You will see a wonderful lot of large photos of this territory in the American Museum at New York. In the past few years a fine book has been published on The Wonders of the Barrier Reef. It is illustrated with many colored photos, made under the sea. The text is well written and very readable, containing much information never published before and some good pictures of shell life; I usually carry it in stock.

✦

Mollusks living on the ocean floor at great depths where the pressure is several tons to the square inch are composed of tissue so constituted as to permit the free permeation of the water through every part in order that the pressure may be equalized. How this is possible without putting an end to all organic functions is one of the great mysteries of abyssal life.

✦

The operculum of abyssal mollusks is very much smaller than similar forms from shallow water and many genera such as Mangilia of which there are large numbers in the deep sea, the operculum is entirely absent. They apparently do not need the protective devices that are so common in shallow water shells.

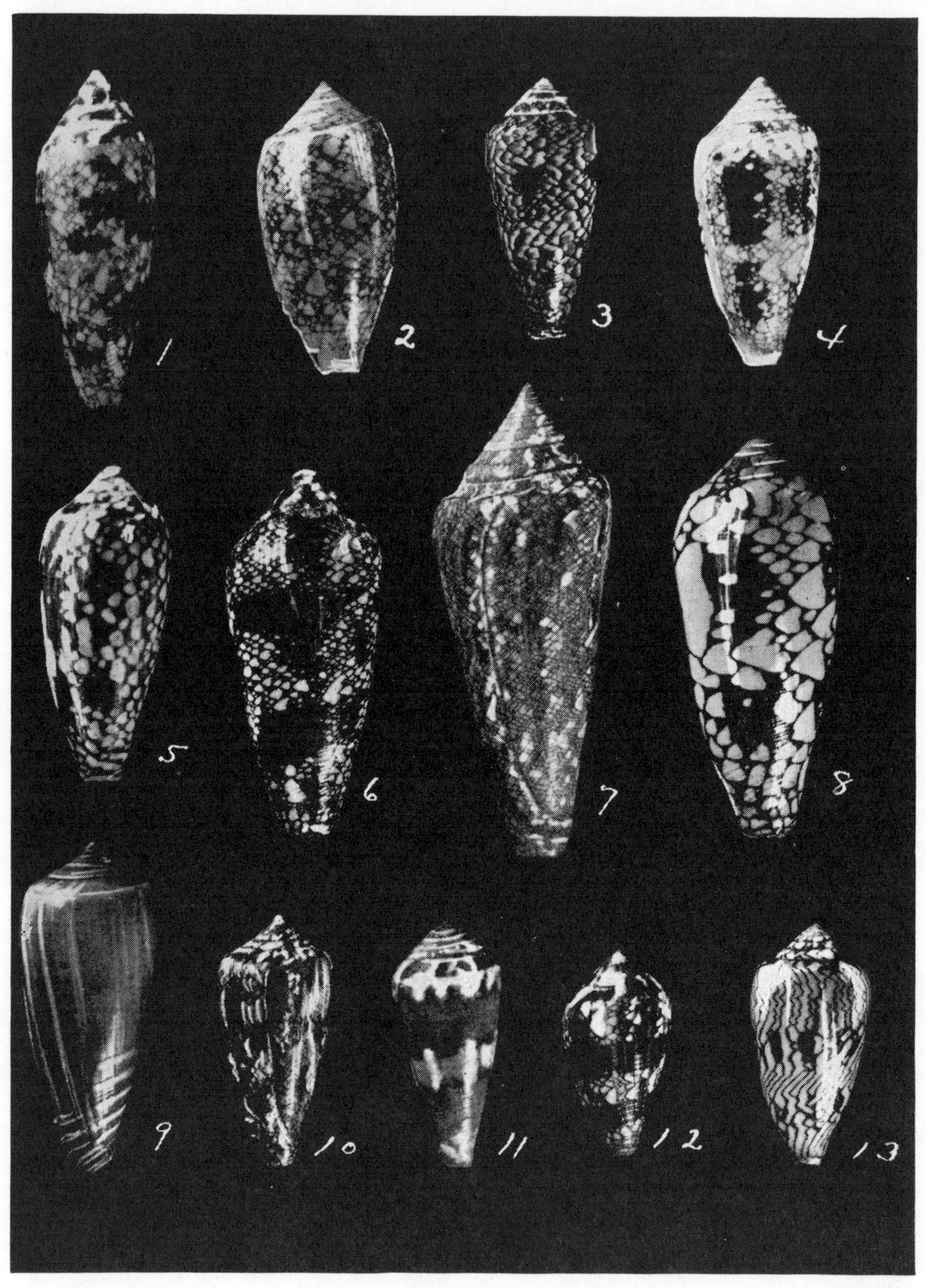

PLATE 7

1. **Conus aureus.** Hwass. Moluccas. The ground color is reddish-chestnut with wavy vertical lines of darker color. The tent-like markings cover much of the surface. Spire rounded and elevated. There are numerous extremely fine tent marks. 3″.

2. **Conus textile scriptus,** Sow. Mauritius. A rounded bulbous shape shell with very large, medium and minute tent-like markings. Sharp elevated pointed spire 2¼″.

3. **Conus praelatus,** Brug. Mauritius. The ground color is a rich deep brown with occasional patches showing nearly white surface the whole covered with blue and white tent-like markings. A very distinct type of coloring. 2″.

4. **Conus legatus,** Lam. Mauritius. The ground color is flesh which is completely covered with rather small and fine tent-like markings and these in turn covered with large splashes of reddish-chestnut. Spire moderately elevated and rounded. ½″.

5. **Conus colubrinus,** Lam. Mauritius. The ground color is light yellow which is completely covered with white tent-like markings of fairly large size. Spire rounded and elevated. 2½″.

6. **Conus magnificus,** Rve. Moluccas. The ground color is a rich dark brown with the small white tent-like markings only partially covering same. In fact the tents are arranged in waves leaving wide spaces of ground color. Spire rounded and elevated. 3″.

7. **Conus gloria-maria,** Hwass. Moluccas. The ground color is a very light shade of chestnut which is completely covered with hundreds of minute white tent-like markings. Lip curved and sharp, being deflected back at top a full half inch. Spire elevated to point and tent markings cover same. 3½″.

8. **Conus episcopus,** Brug. Red Sea. The ground color is a rich chestnut with very large bold white tent markings. Spire much rounded and slightly elevated. 3″.

9. **Conus radiatus,** Gmel. Fiji Ids. The shell is of uniform fawn color with small white band at top of whorl. There are six circular small ridges at base of shell. Spire moderately elevated to point. Interior white. 2½″.

10. **Conus natalensis,** Sow. South Africa. The ground color is light flesh and the entire surface covered with very minute tent-like marks like a fine net. There is two wide faint bands of slightly darker color. Spire moderately elevated and covered with same markings. 2″.

11. **Conus pertusus festivus,** Chan. Mauritius. A reddish-orange shell with two bands of splashes of flesh color. Top rounded and few faint ridges at base. 2″.

12. **Conus retifer,** Mke. Oshima, Japan. A short stubby pyramidal shell quite distinct from all others of its class. The ground color is reddish-chestnut with elongated stripes of blackish-brown and two bands of white tent-like markings. Spire rounded and elevated. 1½″.

13. **Conus textile eutrios,** Sow. Mauritius. Same general form as type but much smaller. It has zigzag elongated thin line of light brown on a white surface. There were two circular bands of reddish-chestnut. A rather distinct color pattern and differs from all other varieties of this specie. 1¾″.

The Family Conidae is a very large one and includes usually only the one genus. They inhabit mainly the equatorial seas and the mollusk has a mouth of peculiar structure. Their favorite hiding place is in the holes in rocks and thousands of miles of shallow coral reefs. They bore holes in the shell of other mollusks and suck out the juices, when other marine life finish the job. They are found most common in the Asiatic seas, the great island of Mauritius claims around 170 species. Around 50 species are found in Western Hemisphere. Usually live in shallow water but are found to 30 or 40 fathoms.

✦

More than 400 species are known and around 100 fossil species from the Cretaceous. From the following descriptions of the many species illustrated you will get an idea of the variety of form, some with long pointed spire and others almost flat on top, etc. They have been usually divided into 14 sub-genera and if your collection is arranged by these sections, you will find them more interesting.

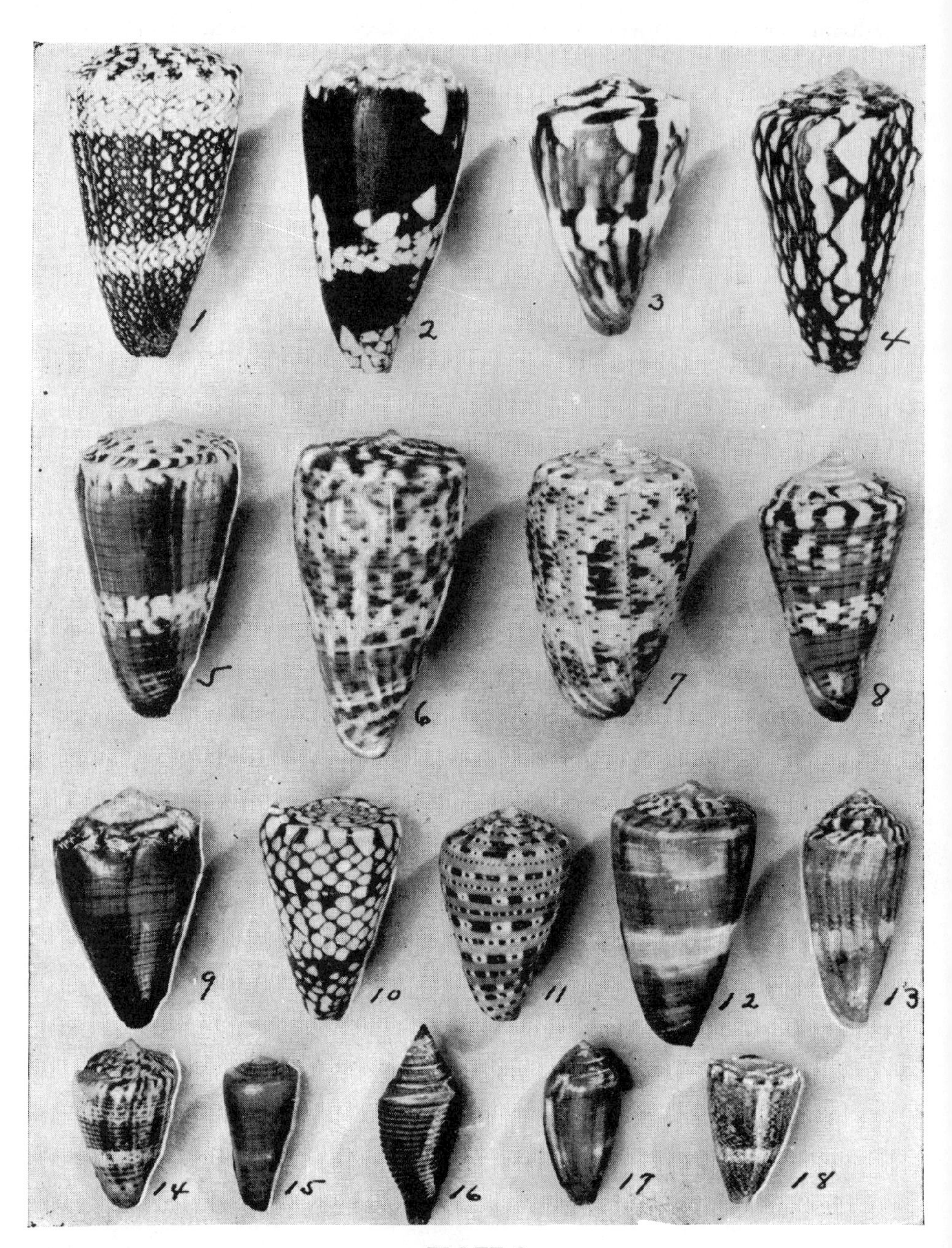

PLATE 8

1. **Conus vidua,** Rve. Copue Id. Tent-like markings on white background. Two broad bands of black. Very distinct. 2½″.

2. **Conus nocturnus Deburghiae,** Sow. Moluccas. Dark blackish-brown, with white tent markings. 2½″.

3. **Conus sumatrensis,** Hwass. Berbera, East Africa. Diagonal glossy markings of orange and chestnut, white markings thru center and on top. 2″.

4. **Conus crosseanus,** Fisch. New Caledonia. Ground color rich brown with wide white tent markings. Very distinct. 2½″.

5. **Conus vitulinus,** Hwass. P. I. Light brown with one white marking, same on top, six rows dots near base. 2¼″.

6. **Conus promethius,** Brug. Africa. Yellowish, shading in form of bands, six rows dots. Largest form of genus. Grows to 8″ or more.

7. **Conus obesus (Ceylonicus,)** Chem. Ceylon. White and pink back ground with blotches of reddish-brown. 2¼″.

8. **Conus splendidulus,** Sow. Madagascar. Conical, light brown, white markings and brown lines. 2¼″.

9. **Conus brunneus,** Gray. Panama. Brown throughout, with faint circular dark brown lines. 1¾″.

10. **Conus marchionatus,** Hinds. Marquesa. Light brown with uniform prominent white tent markings. 1¾″.

11. **Conus genuanus,** L. Ascension Id. Olive with brown lines and white marks and other lines of white and brown. 1½″.

12. **Conus pulchellus,** Swain. Madagascar. Pink and russet with white band. Pink lip. 2″.

13. **Conus granulatus,** L. Antiqua. Rich red granulated surface, circular ridges and faint white band. 1¾″.

14. **Conus classiarius,** Hwass. Red Sea. White with numerous brown lines and white band. 1″.

15. **Conus seychellensis,** Nevill. Seychelles Ids. Reddish-brown throughout. 1″.

16. **Conus coromandelicus,** Smith. Persian Gulf. 170 fath. Conical, elevated spire, brown, fine circular ridges. 1¼″.

17. **Conus glans tenuistriatus,** Sow. Philippines. Light brown and pink. 1″.

18. **Conus praetextus,** Rve. Marquesas. Richly mottled with brown and white. 1″.

Panama shore lines present a typical tropical beach. There is a good tidefall and a variety of stations. Where it is rocky you will find just above tide water Truncatellas, Melampus, Littorinas and Siphonaria. They cling to the rock after the tide goes down and are able to live there until it comes up again. Along the mangrove shore among the roots are Littorinas, Ostraea, and crawling in the mud Cerithiums, Cyrenas and Arcas. Where the tide has receded leaving little tide pools are Cerithium, Thais, Columbellas, Nassas and Crepidula. At low water mark of the average tides you will find Caecum Vitrinella, and other very small forms while under the edge of rocks you will find several forms of Cypraea, Cantharus, many more Columbellas and more small forms. Where the rock is muddy you will find Conus, Turritellas and Latirus. In the cracks between rocks are Chitons but some forms are all over the rocks, where the surf is great. At low water mark are Acanthina, Leucozonia and Vermetus, and burrowing, are Pholas, some forms of Mytilus. Under edge of huge rocks are Venus, Doliums, Murex and on the operculums of the large shells you are sure to find nice Crepidulas. On the liquid mud flats are Marginellas, Nassas, Columbellas gliding about, and in some pools Olivellas by the hundreds, with several species of Naticas. Further down in the shallow water are such bivalves as Chione, Callistas, Tellinas, etc. Just locate the usual habitat of a genera and then you find the shells you wish.

If you live near fish docks much valuable material could be had by making arrangements with the owners to empty the paunches of large fish like Haddock into buckets of clean water. This debris will often contain choice specimens of bivalves like Leda, Nucula, Nuculina, Pectens and in the Australian region very rare Voluta have been found in this manner. A visit to the great fish markets where thousands of fish are processed daily will bring good results.

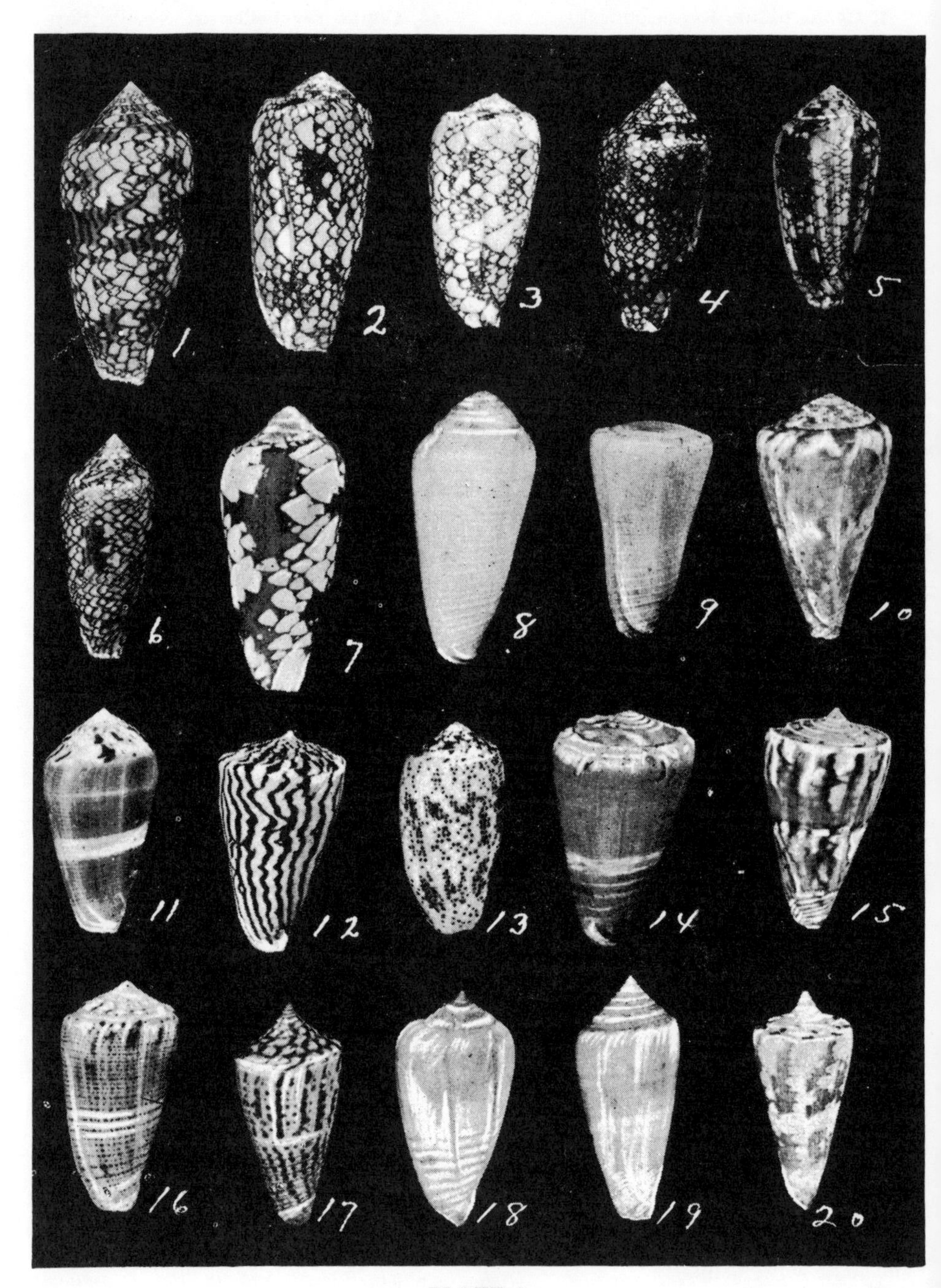

PLATE 9

1. **Conus textile vicaria,** Lam. Mauritius. The shell has a ground color of orange, spire elevated, the body completely covered with tent-like markings of white, arranged in such form there appears to be two wide bands of body color. 2½″.

2. **Conus omaria,** Brug. Amboina. A rather slender rounded shell with background of brown, completely covered with tent-like markings of white. Spire slightly elevated. Most of these tent shells are covered with a yellowish periostracum when freshly collected. 2¼″.

3. **Conus omaria pennaceus,** Born. Mauritius. Ground color chestnut, with numerous tent-like markings of white. Spire only slightly elevated. 2″.

4. **Conus tigrinus,** Sow. Mauritius. Shell rather conical, spire well elevated. Ground color dark reddish-chestnut almost completely covered with tent-like markings, some of which are extremely small and range up to a quarter inch in size. Interior of lip pink. 2″.

5. **Conus textile verriculum,** Rve. Mauritius. A white shell completely covered with minute tent-like markings and many others of larger size arranged in rows which look as if corresponded with grown lines. 1¾″.

6. **Conus archaepiscopus,** Brug. Mauritius. Ground color reddish-chestnut, completely covered with white tent-like markings. Spire well elevated. There is also perpendicular markings of brown. 1¾″.

7. **Conus canonicus,** Hwass. Moluccas. Ground color chestnut, the whole shell covered with large prominent tent-like markings. Spire moderately elevated. 2¼″.

8. **Conus terebra,** Born. Mauritius. Shell white; completely covered with tiny circular ridges. Tip of base lavender. Closely allied if not identical with terrebellum. Mart. 2¼″.

9. **Conus emaciatus,** Rve. Viti Ids. The shell is light yellow with purple tip at base. Top perfectly flat, the lower half of shell with numerous spiral lines. 1¾″.

10. **Conus nemocanus,** Hwass. Aden, Arabia. The general color pattern is light yellow, with moderate spire, body whorl streaked with chestnut markings, few spiral lines at base. 2″.

11. **Conus fulmen,** Rve. Loo Choo Ids. A drab shell with one white band in middle and numerous dark patches on the moderately elevated spire. Few very fine ridges at base. 2″.

12. **Conus princeps lineolatus,** Val. Panama. The true type of princeps is uncolored and this variety has same reddish-chestnut base color with numerous wide irregular perpendicular lines of brown. Spire nearly flat with fine tip. 2 to 3″.

13. **Conus stercus-muscarum,** L. Moluccas. A white shell with numerous small black dots, usually arranged in waves. Spire moderately elevated, with same dotting, and aperture is orange. 1½ to 2″.

14. **Conus flavidus,** Lam. Oceanica. Uniformly yellowish-brown with dark tip at base, almost flat spire, faint nobs around top of whorl and dark, purplish interior. 2″.

15. **Conus centurio,** Born. West Indies. A beautiful pyriform shell with almost flat top, each whorl outlined with fine ridge and minute tip. The base color is white with perpendicular shadings of light brown, and one darker row of blotches in middle of whorl. About 2″.

16. **Conus magus ustulatus,** Rve. Philippines. Ground color yellowish-white with many circular rows of tiny dots and three white bands in middle of whorl, which includes two rows of tiny dots. Top only slightly elevated. 2″.

17. **Conus capitaneus,** L. Philippines. Rather pyriform, much elevated, fine spire, ground color yellowish-white, completely covered with rows of light brown spots. Fine ridge at base. 2½″.

18. **Conus radiatus parius,** Rve. Philippines. A yellowish shell throughout with circular ridges extending upwards from base to half length of whorl. This is the most prominent characteristic. 1¾″.

19. **Conus rutile,** Mke. Australia. A yellowish-white shell with two prominent wide bands of deeper yellow, elevated spire with fine tip. Numerous spiral lines at base. 2″.

20. **Conus generalis maldivus,** Hwass. Moluccas. A pyramidal shell of glistening white, with numerous irregular yellow splashes of color. Top mostly flat, the final 4 or 5 whorls forming a sharp tip. 1½ to 2″.

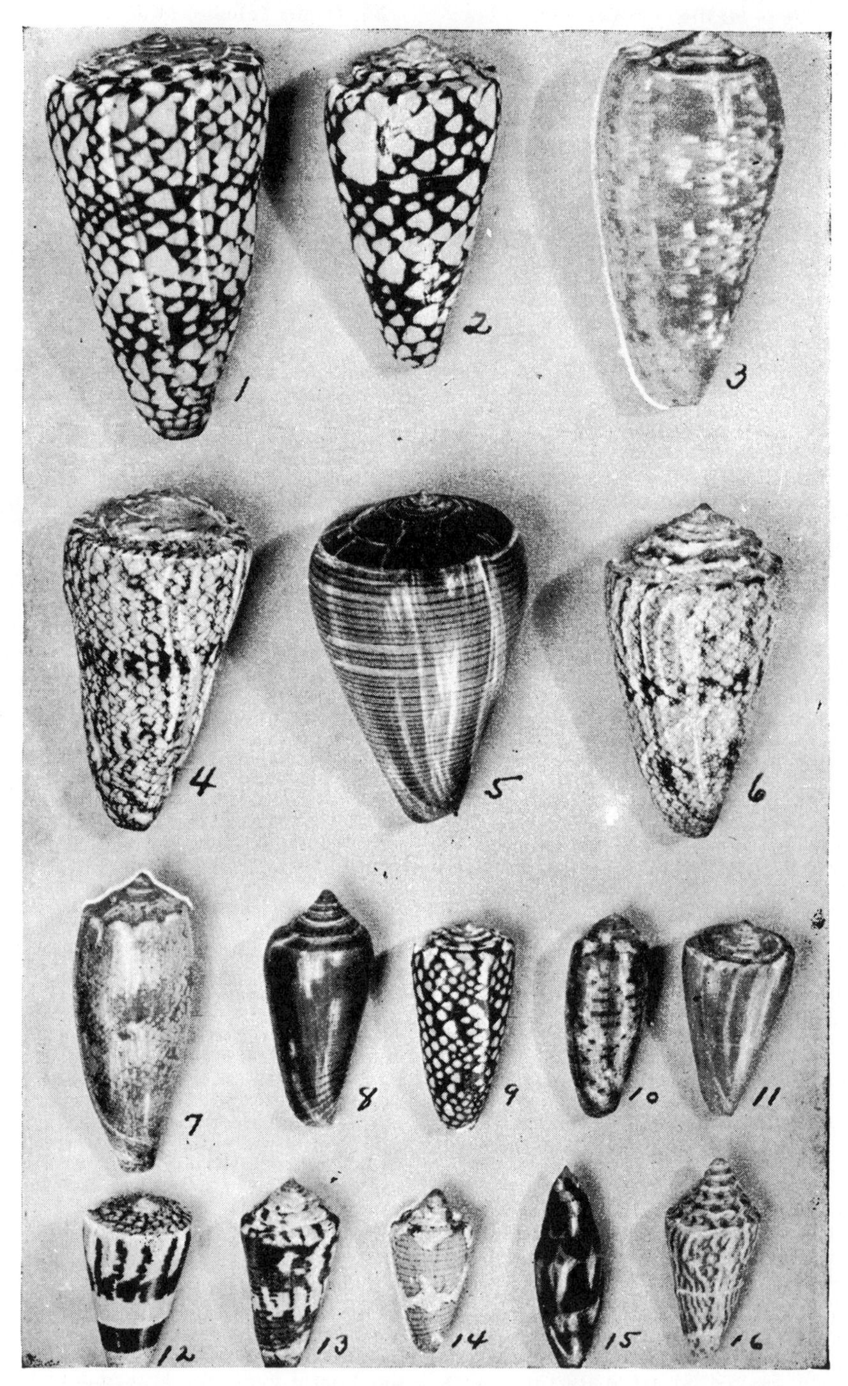

PLATE 10

1. **Conus marmoreus bandanus,** Hwass. New Caledonia. Brilliant white tent markings on black background. 3½".

2. **Conus marmoreus pseudomarmoreus,** Desh. New Caledonia. Similar to preceding. 3".

3. **Conus cervus,** Lam. Moluccas. White tent blotches on yellow background. Very rare. 3¼".

4. **Conus arenosus nicobaricus,** Brug. Nicobar Ids. Fine brown tent markings, darker band through center. 3".

5. **Conus figulinus,** L. Amboina. Uniformly light brown with darker lines of brown. Top deeper brown. Fine apex. 3".

6. **Conus arachnoides,** Gmel. Mauritius. Fine tent markings of brown and two darker bands. 3".

7. **Conus geographus intermedius,** Rve. Philippines. Pink with brown tent marks like geographus. 2½".

8. **Conus ochroleucus,** Gmel. New Guinea. Uniformly yellowish-brown. Fine circular ridges gradually deeper near base. 2".

9. **Conus nobilis,** L. Philippines. White tent marks on reddish-brown background. Very distinct. 2".

10. **Conus circumcisus,** Born. Moluccas. Pinkish-brown with circular interrupted lines, brown and white. 2".

11. **Conus daucus,** Hwass. Mauritius. Reddish-yellow throughout. 2".

12. **Conus aplustre,** Rve. Australia. Finely dotted with brown, one interrupted band and one lower brown band. 1½".

13. **Conus luctificus,** Rve. Oran, Algeria. Brown with a band of white blotches. 2".

14. **Conus cedonulli,** Klein. Mauritius. Yellow with large irregular blotches of white and fine dots. 1½".

15. **Conus cylindraceus,** B & S. Society Ids. Elongated pure brown with white streaks. 2".

16. **Conus scalaris,** Val. Panama. Irregular yellow perpendicular markings on white. Elongated scalare spire. 2".

Always look for eggs of the Gasterpod mollusks. The Whelks deposit their eggs in ovicapsules generally anchored to some stationary body. The prickly coils of eggs of Fulgars are common along most U. S. A. shore lines and the smaller cases that are similar are frequently Pyrula. The Buccinum deposits its capsules in a heap like grains of corn but so arranged as to admit the sea water to every part. The capsules of Thais stand on end like little vases in groups upon the rock, often under what is termed bladder weed. Chrysodomus heaps its capsules in a cylindrical tower often six inches high. Volutas lay a few large eggs of capsules which contain several embryo. These can be preserved in alcohol. If the embryos are well developed they make a very interesting specimen. The Melongenas usually lay their capsules in a form of a ribbon 3 or 4 inches long. A collection of these various capsules make a fine display.

✦

The distribution of shell life in the sea has always been a fascinating study. The U. S. A. Steamer Blake spent part or all of ten years making a survey of the Gulf of Mexico region and the southeastern shores of the U. S. generally. The volumes published contain the following data: Number of genera 180. Number of species 709 of which 403 were Littoral, 376 were Archibenthal, 129 were Abyssal. Species common to two areas 250. Species common to all areas 49. I suspect a study of other regions would show quite a similar result.

✦

For many years Mr. Charles Hedley was a correspondent and friend. He was the leading conchologist of Australia. He told me once that the coast of Queensland contained 1800 species of marine shells or more and it is likely the other coasts are as rich. The people of that great ocean continent take a great interest in Conchology and there are many active workers in the science with hosts of private collectors. The great bulk of the population live not far from the sea and it is quite natural that they should have a deep interest in ocean life.

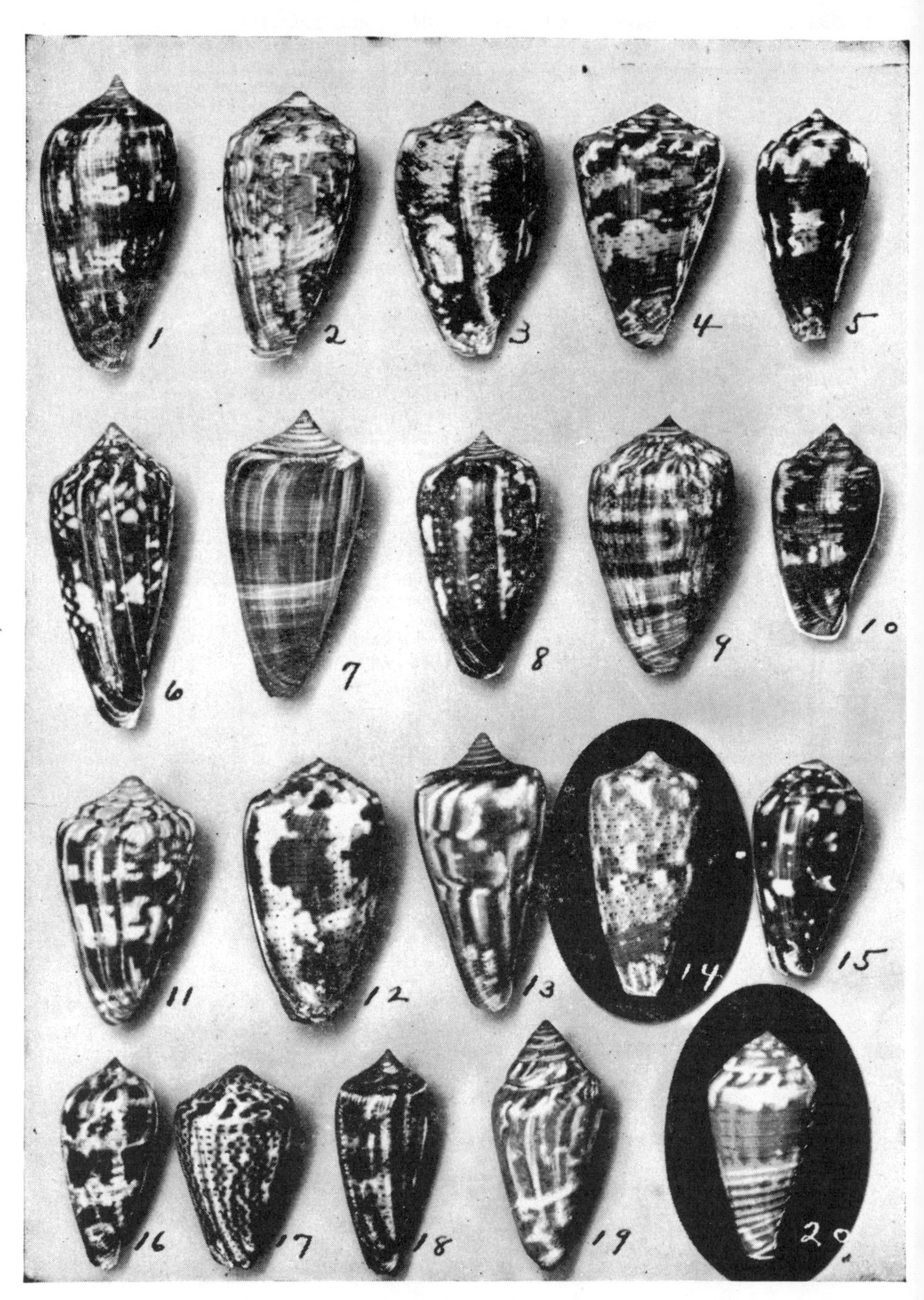

PLATE 11

1. **Conus cinereus,** Hwass. Moluccas. A rich chestnut-brown shell with patches of flesh white in form of two interrupted bands. Spire rounded, elevated to sharp point. A rich deep colored shell. 2″.

2. **Conus anemone,** Lam. Victoria, Australia. A rather thin light shell which is completely covered with irregular and zigzag markings of chestnut on flesh background. There is a wide white indistinct band through middle of whorl. Spire moderately elevated. 1¾″.

3. **Conus anemone maculata,** Sow. Northwest Australia. Similar to type but more bulbous with wider aperture. The markings are heavy and dark with perpendicular white patches. Rather light and thin. 1¾″.

4. **Conus venulatus,** Hwass. Cape Verde Islands. The shell is rather pyramidal and completely blotched with bluish-white and brown shades much resembling a small **purpuratus.** Top almost flat with tiny spire—interior flesh color. 1½″.

5. **Conus boeticus,** Rve. Philippines. A slender pyramidal blackish-chestnut shell, with few white blotches arranged in form of two bands. The shell is completely covered with fine ridges and dots, more prominent near base. Spire elongated and pointed. 1½″.

6. **Conus lienerdi,** Bern. New Caledonia. A rich shining olive-brown with numerous white tent-like marks mostly arranged in two bands. Very fine ridges at base. Spire elevated and pointed. A rather dark shell with few white markings. 2″.

7. **Conus lignarius,** Rve. Philippines. Uniformly light brown with very faint streaks of white. Numerous fine ridges at base. Spire almost flat, few ridges and small point. 1¾″.

8. **Conus cinereus bernardi,** Kien. Moluccas. Two shades of brown with several white streaks, some in form of tent markings. Prominent ridges at base. Spire rounded and slightly elevated to sharp point. Interior flesh color. 1¾″.

9. **Conus lachrymosus,** Rve. Australia. A shell of the general form of **anemone maculata** on this plate, but the surface is mottled with fine brown markings and five narrow circular bands of darker color. Spire moderately elevated to point. Interior flesh. 1¾″.

10. **Conus anemone nova-hollandiae,** A.Ad. New Holland. Rather more slender than type, of a very deep mahogany brown with few splashes of lighter color and completely covered with fine circular ridges. Spire ridged and elevated. 1½″.

11. **Conus venulatus nivosus,** Lam. Teneriffe. A white shell mottled with reddish-chestnut, the white markings mostly arranged in form of two bands. Spire moderately elevated. 1½″.

12. **Conus catus nigropunctatus,** Sow. Red Sea. It is white mostly covered with olive-brown marks. These in turn are covered with tiny circular lines of darker color. Spire ridged and moderately elevated. 1½″.

13. **Conus virgatus,** Rve. Panama. A pyramidal shell of flesh color with irregular perpendicular stripes of russet. Flat top with elevated center point. 2″.

14. **Conus nimbosus,** Hwass. Ceylon. Rather cylindrical and covered with fine circular ridges, which are mottled with spots of reddish-chestnut. The whole markings have a wavy appearance. Spire nearly flat. A rare shell. 1½″.

15. **Conus lautus,** Rve. South Africa. The shell is a rich reddish-chestnut color with irregular splashes of flesh-white mostly arranged in form of bands. Top rather rounded and moderately elevated. 1½″.

16. **Conus cinereus,** Hwass. Philippines. A shiny drab colored shell with oblong patches of brown arranged in the form of bands. Spire moderately elevated and pointed. 1⅜″.

17. **Conus puncticulatus,** Hwass. West Indies. A rather pyramidal thick shell completely covered with circular rows of brown spots on a bluish-white background, some forming wavy lines. Spire elevated. 1½″.

18. **Conus cinereus gabrielli,** Kien. Moluccas. A rather cylindrical shell with a background of rich olive. There are circular rows of dots and dashes with lighter spaces between. Spire moderately elevated to point. A very smooth shining shell. 1½″.

19. **Conus mediterraneus oblonga,** Buc. Medit. Sea. Similar in form to type, of a bluish shade with occasional streaks of brown. The general effect being of a grayish pattern. Lip is curved and spire round and elevated . 1½″.

20. **Conus pertusus,** Hwass, Mauritius. A shell that is rich orange-pink with moderately elevated spire and prominent circular ridges throughout. One band through middle, of flesh-white. Interior pink. 1½″.

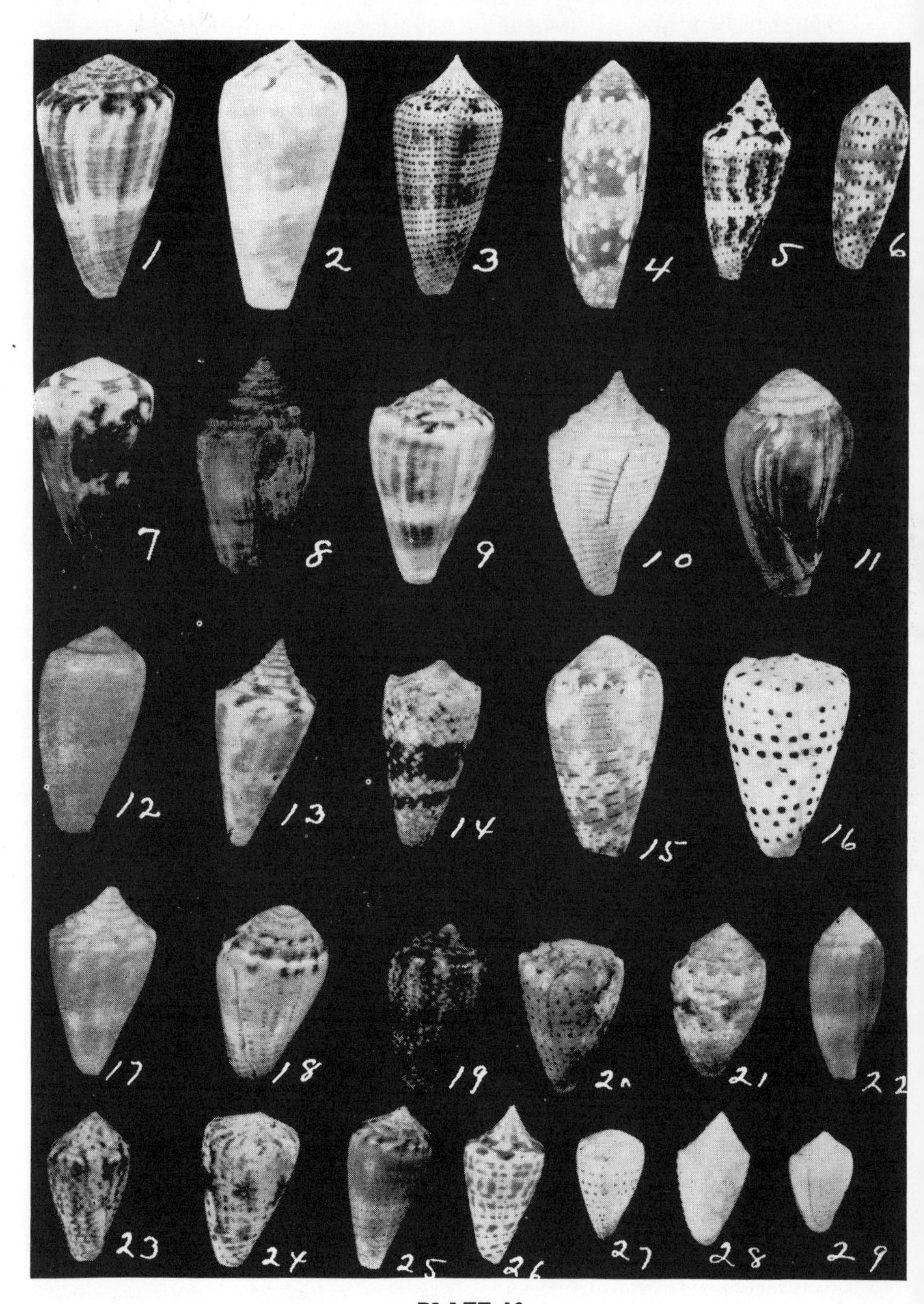

PLATE 12

1. **Conus gladiator,** Brod. Panama. The shell is white, completely covered with perpendicular chestnut streaks. Spire moderately elevated and deeply shaded. Numerous circular lines at base. 1½".

2. **Conus ustulatus,** Rve. Solomon Ids. A white shell with yellow shading in the form of two wide bands. Spire moderately elevated with brown markings. Fine spiral lines at base. 1¾".

3. **Conus interruptus,** Brod. West Mexico. A flesh colored shell completely covered with tiny dots arranged in circular lines. Spire well elevated to sharp point. Interior of shell purple. 1½".

4. **Conus clavus,** L. Philippines. A slender yellow shell with two or more bands of very fine tent markings, and numerous larger similar markings arranged irregularly. Spire rounded and elevated. 1½".

5. **Conus interruptus mahoganyi,** Rve. Panama. Similar in form to type, but spire is more elevated, the shell is covered with tiny ridges and irregular lines of dark blackish-brown. 1½".

6. **Conus nussatella,** L. Mauritius. A slender elongated shell, with circular lines of prominent dots and numerous splashes of yellow over entire shell. 1½".

7. **Conus tahitensis rattus,** Hwass. Australia. A dark mahogany colored shell, the top of which is white, with few blotches and a narrow band of white through middle of shell. Interior purplish. 1 to 1½".

8. **Conus mediterraneus,** Hwass. Mediterranean Sea. A dark shell completely covered with blotches of drab, spire elevated. There are numerous faint circular lines. 1 to 1¼".

9. **Conus mus,** Hwass. Gulf of Mexico. The ground color is faintly bluish, spire slightly elevated. There are splashes of brown throughout on the lower end arranged as a band. 1 to 1¼".

10. **Conus cancellatus,** Hwass. Hong Kong, China. An entirely white shell with much elevated spire to point. The upper whorls show as ridges. Body whorl completely circled with fine ridges. Faint splashes of brown throughout. 1½".

11. **Conus californicus,** Hinds. California coast. The shell is of a drab color, completely covered with russet periostracum. Rounded top and moderately elevated spire. 1½".

12. **Conus rosaceus,** Chem. South Africa. A rather thin shell of a faintly pink color with narrow lighter band in middle. 1¼".

13. **Conus floridanus,** Gabb. Gulf of Mexico. A rather conical shell with sharply elevated pointed spire. The ground color is white streaked with reddish-brown, also an occasional row of circular dots 1½".

14. **Conus areneosus,** Brug. Ceylon. A rosy colored shell completely covered with tent-like markings with one wide and one narrow band of rich reddish-chestnut with fewer markings. 1½ to 2".

15. **Conus magus Metcalfei,** Rve. North Australia. There are many color varieties of this widely distributed shell. This one has numerous narrow circular bands of light brown and grayish spaces between. Also clouded with white. 1½ to 2".

16. **Conus eburneus,** Hwass. Ceylon. Looks like a real small **litteraus** with its white background and numerous circular rows of black spots. Some of the rows are more prominent than others. A strong solid little shell of 1½".

17. **Conus punctatus,** Sow. West Africa. A yellowish shell which much resembles **floridanus** on the opposite side of the Atlantic. I am rather inclined to believe they are closely allied. 1½".

18. **Conus catus,** Hwass. Java. A comparatively common shell of about one inch usually bluish cast with numerous fine lines of brown. Top rounded and some elevated. Quite variable and hard to separate from other similar forms.

19. **Conus abbreviatus,** Nutt. Hawaii. A small brown shell finely nodulated throughout. Top slightly elevated. Body with numerous circular rows of tiny nodules. Generally mottled brownish and white. 1".

20. **Conus abbreviatus,** Nutt. Another variety lighter colored with more distinct rows of dots.

21. **Conus aristophanes,** Duc. Galapagos. A small bulbous shell with fine circular rows of nodulated ridges and streaked with gray and brown markings. 1".

22. **Conus glans,** Hwass. Australia. A peculiar shaped shell of a brown and purplish cast, prominent spire, narrowed at base and completely covered with very fine circular ridges. 1 to 1¼".

23. **Conus pygmaeus,** Rve. Antilles. A small shell with granulated surface, and circular rows of small knobs, streaked perpendicularly with brown. 1".

24. **Conus erythraensis,** Beck. Red Sea. A small whitish shell with rows of brown dots and splashes of similar color. Rows of tiny circular ridges near base. 1".

25. **Conus pauperculus,** Sow. Korea. The upper part of the last whorl is olive and lower half is much lighter with rows of circular dots. Top slightly elevated and well marked with brown. 1¼".

26. **Conus spurius,** Gmel. West Africa. White with yellowish rows of dots similar to **proteus** of Florida coast. 1½".

27. **Conus musicus,** Hwass. Philippines. Also illustrated on Plate 14 showing specimens better decorated with dots. Fairly common.

28. **Conus verrucosus,** Brug. West Africa. A small white shell with elevated spire completely covered with circular rows of small knobs.

29. **Conus columba,** Brug. Mauntus. A small white shell quite similar to preceding except it is smooth.

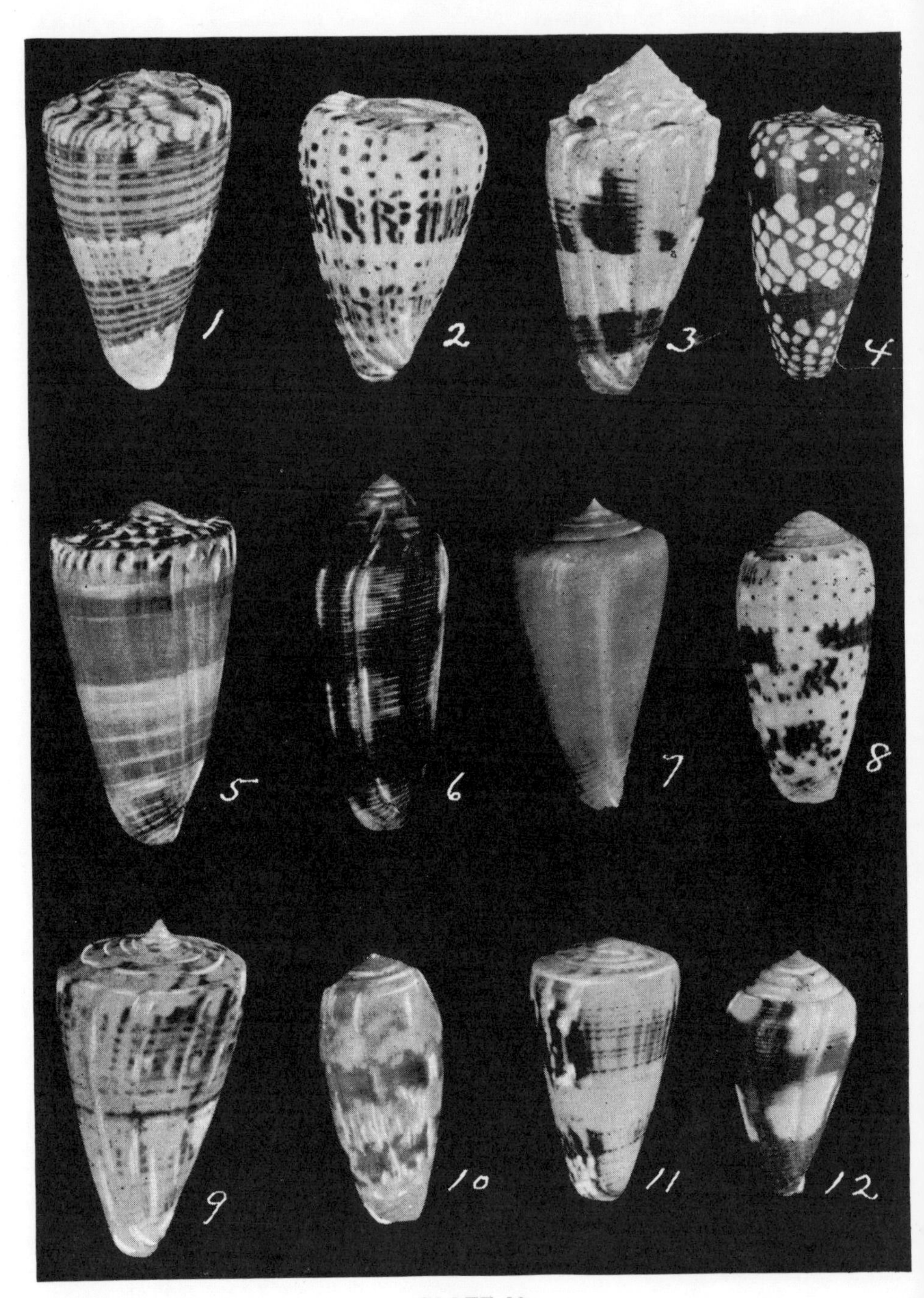

PLATE 13

1. **Conus trigonus,** Rve. N. W. Australia. Light and dark glossy yellow bands with one wide white band and white base. 2″.

2. **Conus crassus,** Sow. Red Sea. Finely ornamented with rows of dots and bands of brown lines, with white space in middle, and fine row of dots. ¾″.

3. **Conus Hwassi,** A.Ad. Mauritius. Drab white with blotches of brown in form of lines. Elongated nobby spire. 2¼″.

4. **Conus cordigerus,** Sow. Philippines. Close to Nobilis L. Base color is reddish-brown, ornamented with white tent markings in form of three bands and similar markings on almost flat top. 1¾″.

5. **Conus planorbis,** Born. Loo Choo Ids. Uniformly reddish-yellow, with bands of lighter color, rich dark markings on top, base pink inside. 2″.

6. **Conus tendineus,** Hwass. Mauritius. A slender elongated shell completely covered with fine circular ridges. Color dark brown with white blotches. 2¼″.

7. **Conus concolor,** Sow. China. Uniformly light yellowish and white fine circular lines near base. 2″.

8. **Conus vicarius,** Lam. Mauritius. Of a rosy pink with blotches of reddish-brown and numerous small dots of same. About two inch. Very rare.

9. **Conus malaccanus,** Hwass. Malacca. Ground color white with numerous elongated and circular markings of yellow and brown. Top almost flat with tiny apex of 5 whorls. 2″.

10. **Conus julii,** Lineard. Mauritius. A small white shell with numerous reddish-yellow markings very similar to **bullatus.** 1¾″.

11. **Conus Chenui,** Crosse. New Caledonia. Ground color flesh-white with two prominent bands of yellowish-brown, top smooth. The bands are formed of fine lines and perpendicular stripes. 1½″.

12. **Conus scabriusculus,** Chem. New Caledonia. A white shell completely covered with fine circular ridges. White spots and russet-red markings. 1½″.

During and since the last war I had many letters from G. I. boys about cats eyes. I was for a time puzzled as to what they had as the real cats-eyes of the 19th century was a brilliant dark pearly specimen of a little over half an inch which made very showy necklaces when pierced. Most of the specimens were Japanese made. But it seems the operculum of the many kinds of Turbo are called cats-eyes and seem to be quite generally called such over the Pacific area. Some are brilliant deep polished green color, others are pure white and others have a granular surface. Of course in collecting, the operculum should be put back in the shell after boiling to remove the soft parts of the mollusk. Hardly a week goes by but some one wants to know where he can market his bag full of cats-eyes he has brought home from the war. I usually refer him to some wholesale firm manufacturing jewelry from shells.

✦

It is surprising what a vast number of kinds of sea shells live under the edge of rocks, in crevices. In such situations you are sure to find many kinds of Cypraea and Trivia if you are working a territory where they are common but the larger forms may be found on the sand. One collector who dived for shells in Hawaii waters wrote that whenever he found any Cypraea arabica reticulata they were always in pairs. He wanted to know if they were males. It was too deep for me, but he insisted that where he found them at all, there was always two fine specimens near together.

✦

The Family Siphonariidae I like to include with my marine shells as they live on the rocks on the ocean shore and have the power to stick to their home when the tide goes out. They look like Limpets except that one side swells out more than the other and yet they are placed between the Chilinas and the Gadinia, the latter of which is definitely fresh water. The largest species I have ever had of the Siphonaria is SIPHO *from Panama which attains most 2″. Most of the handsome shells are about one inch and really brilliant on the inside. There are fully 100 species or more but only few fossil from the Miocene. Are world wide in distribution.*

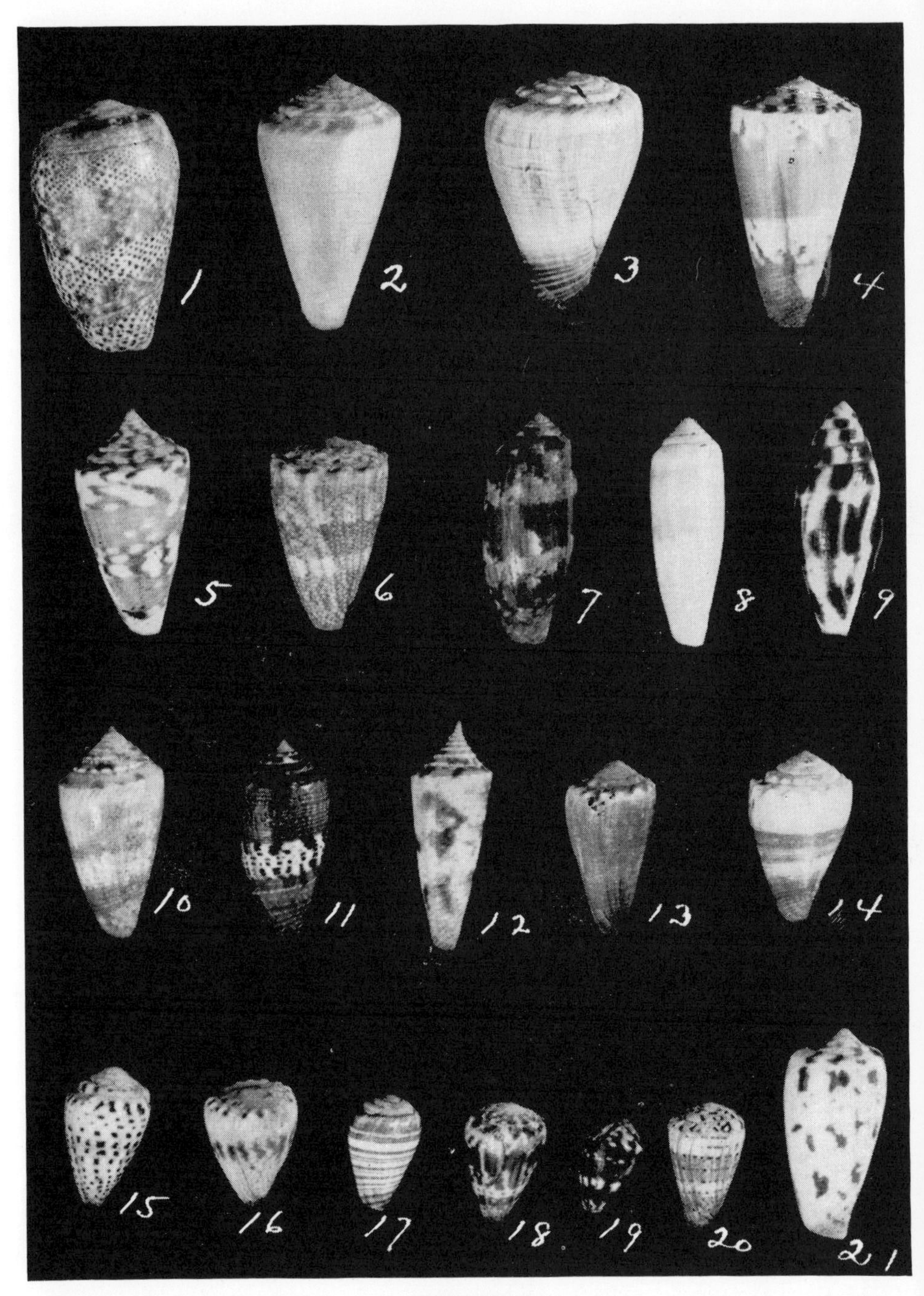

PLATE 14

1. **Conus rhododendron,** Couth. Australia. Of a rich pink color with three bands of fine brownish dots. Shell solid, lip thick, top slightly oval. Very rare. 1½".

2. **Conus ambiguus,** Rve. West Africa. White with very faint tinge of yellow. Top is slightly elevated and distinctly spotted. Light brown. 1½".

3. **Conus roseus,** Lam. Persian Gulf. A thick solid shell of rosy-brown color, one whitish indistinct band on lower half, and darker blotch at base. 1½".

4. **Conus senator,** L. Philippines. A yellow shell with white band on lower half and similar narrow band at top, many dark blotches on top. 1½".

5. **Conus eximius,** Rve. Moluccas. A short pyramidal shell of yellowish color, with numerous irregular white patches. Elevated spire also blotched. 1¼".

6. **Conus eucaustus,** Kien. Marquesas Ids. A grayish-brown shell completely covered with white dots, and two whitish bands. Top ridged. 1¼".

7. **Conus geographus obscurus,** Rve. Society Ids. A thin shell of the form of **tulipa, geographus,** etc. A rich dark brown with two irregular lighter bands. 1½".

8. **Conus tenellus,** Chem. New Caledonia. A slender elongated shell with finely reticulated surface and three faint yellowish bands. 1½".

9. **Conus mitratus,** Hwass. New Caledonia. A very peculiar shaped shell for this genus, as the spire is about one-third of the shell. Elongated oval, the whole shell striped with rich brown shades. Lip always curved. 1½".

10. **Conus mindanus,** Brug. Seychelles. A pyriform shell with elongated spire, finely mottled with glossy russet and white deep circular ridges near base. ¼".

11. **Conus coccineus,** Gmel. New Caledonia. The shell has a finely granulated surface. Color red with white band, mottled with brownish marks. Spire elongated. Always a rich colored shell. 1¼".

12. **Conus traversianus,** Smith. Aden, Arabia. A very thin elongated shell with elevated spire terminating in sharp point. Fine circular ridges on lower half of shell. Mottled with brown and white blotches. 1½".

13. **Conus oblitus,** Rve. Philippines. An olive-brown shell with tiny nobs at top of last whorl, interior lavender. 1".

14. **Conus tenuisulcatus,** Sow. Mauritius. A chunked small shell with nobby top. Light olive, with bands of brown, interior purplish. 1".

15. **Conus pusillus,** Chem. Seychelles. A small white shell completely covered with brown dots arranged in circular rows.

16. **Conus sponsalus,** Chem. New Caledonia. A small drab shell with a band of diagonal bars of red, top flat. Tip of base dark.

17. **Conus fuscolineatus,** Sow. Sierra Leonne, Africa. A small shell much resembling some forms of Columbella. Completely covered with circular brown and white stripes, top rounded.

18. **Conus ceylonensis,** Hwass. Fiji Ids. A small shell with nobs on top of last whorl. Mottled with spots and blotches of russet and white.

19. **Conus hieroglyphicus,** Ducl. Curacao. A very small dark shell, brown with white spots and spotted top.

20. **Conus musicus,** Hwass. variety. Aden, Arabia. A small shell with bluish and white bands which are ornamented with circular rows of dots.

21. **Conus andamanensis,** Smith. Andaman Ids. A rather thin flesh colored shell with blotches of russet, spire slightly elevated. 1½".

How Cameos Are Made

Take Cassis madagascarensis as a sample. The shell is first cut in pieces of the required size by means of diamond dust and the slitting mill or sometimes with a blade of steel fed with emery and water. It is then carefully shaped into a square, oval or other form. It is then cemented to a block of wood, which serves as a handle to be grasped by the artist, while tracing out with a pencil, the figure to be cut on the shell. The operations that follow are too lengthy to be given here.

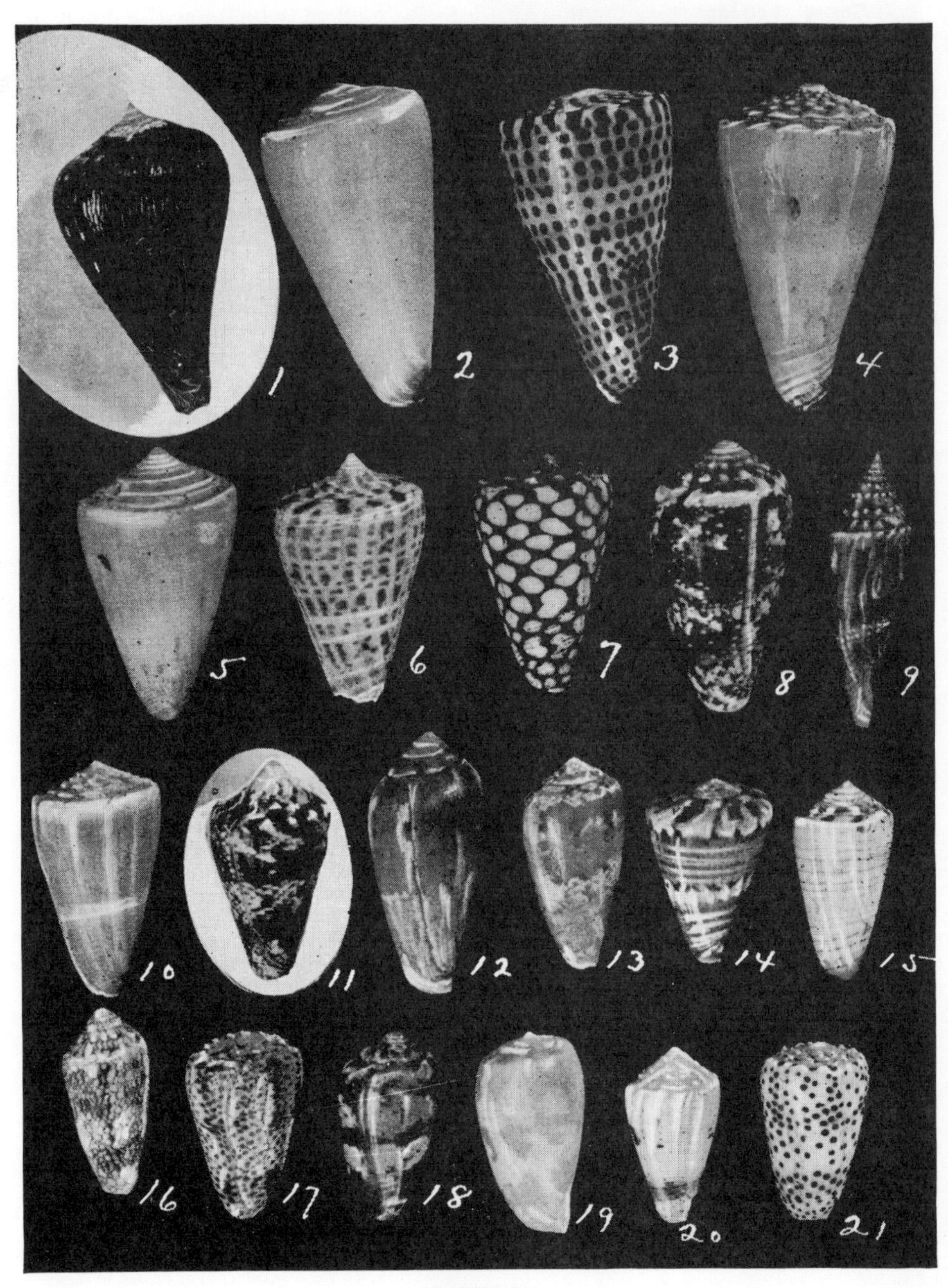

PLATE 15

1. **Conus pyriformis,** Rve. Panama. Rather pyriform, of a delicate shade of pink or flesh color, when the rich glossy brown periostracum is removed. Not very common. 2½".

2. **Conus virgo,** L. Pacific and Indian Oceans generally. Flesh color, almost flat top, the base always of a purple color. 2½ to 3½".

3. **Conus litteratus,** L. China. Top almost flat. Ground color white completely covered with regular rows of dots and splashes of black. There is frequently bands of yellowish ground color. 2½ to 4".

4. **Conus mutabilis,** Ch. China. Moderately white. Top with rows of nobs on each whorl. There are two wide bands of yellowish color. Tip of base brown, with several circular ridges. 3".

5. **Conus quercinus,** Hwass. Red Sea and Indian Ocean generally. A yellow shell with slightly elevated tip, the entire surface covered with minute circular lines. 2 to 3".

6. **Conus proteus,** Hwass. Gulf of Mexico. The entire shell is covered with circular rows of dots and dashes of Chestnut. 2" and occasionally larger.

7. **Conus marmoreus,** L. China. The ground color is almost black and the shell is completely covered with patches of tent or irregular white. One of the commonest world species. 2 to 3, 4".

8. **Conus testudinarius,** Mart. West Indies. The spire is slightly elevated and whole surface mottled with flakes of white, the ground color being chestnut-brown. There are often circular lines of brown. 2".

9. **Conus orbignyi,** Aud. Loo Choo Ids. A very thin and slender shell with elevated pointed spire. The tip is slightly curved. There are wavy marks of brown perpendicular and tiny circular lines. 2½".

10. **Conus lividus,** Hwass. Philippines. The top is mostly flat and slightly nobbed. There is a white ring near top and a narrow white band lower down. General color olive with tip at base dark brown. 1¾".

11. **Conus nebulosus,** Sol. East Florida coast and West Indies. Top with somewhat elevated spire. General color a rich glossy reddish-brown with blotches of white, mainly at top and bottom with a middle band of same. 1½ to 2".

12. **Conus elongatus,** Chem. South Africa. The specimen figured is a very light shade of brown, spire slightly elevated. Most specimens in collections from this territory are not of fine quality being usually shore collected. 2".

13. **Conus lamarcki inflata,** Sow. South Africa. A rather thin light shell of a yellowish color with numerous scattered markings of white. 2".

14. **Conus capitaneus variety,** Lam. Mauritius. A yellowish shell, top almost flat. There is a white strip near top of whorl and another half way down and there are numerous fine bands of brown. 1½ to 2".

15. **Conus lineolatus,** Hwass. Mauritius. A yellowish shell with numerous fine lines of brown and elevated spire. 1½".

16. **Conus pyramidalis,** Lam. Mauritius. A tent cone. The shell is completely covered with small tent-like markings. 1½ to 2".

17. **Conus vautieri,** Kien. Marquesas. A white rugged shell completely covered with fine yellowish dots and some more prominent splashes of same color. Top rounded and almost flat. 1½".

18. **Conus tinianus,** Hwass. South Africa. The shell is of a purple color with numerous splashes of chestnut, top rounded and somewhat elevated, aperture purple. 1¾".

19. **Conus conspersus,** Rve. Java. A white shell with yellowish markings. Top almost flat with tiny apex. Aperture wide, lip thin, base of shell with numerous fine ridges. 1½".

20. **Conus varius,** L. Philippines. The shell is flesh-white, top elevated, and nobby markings of chestnut in two bands. Usually rough surface. 1½".

21. **Conus pulicarius,** Hwass. Tahiti. A white shell well covered with black dots, often arranged in two main bands. The top is rounded, almost flat with nobby surface. 1½".

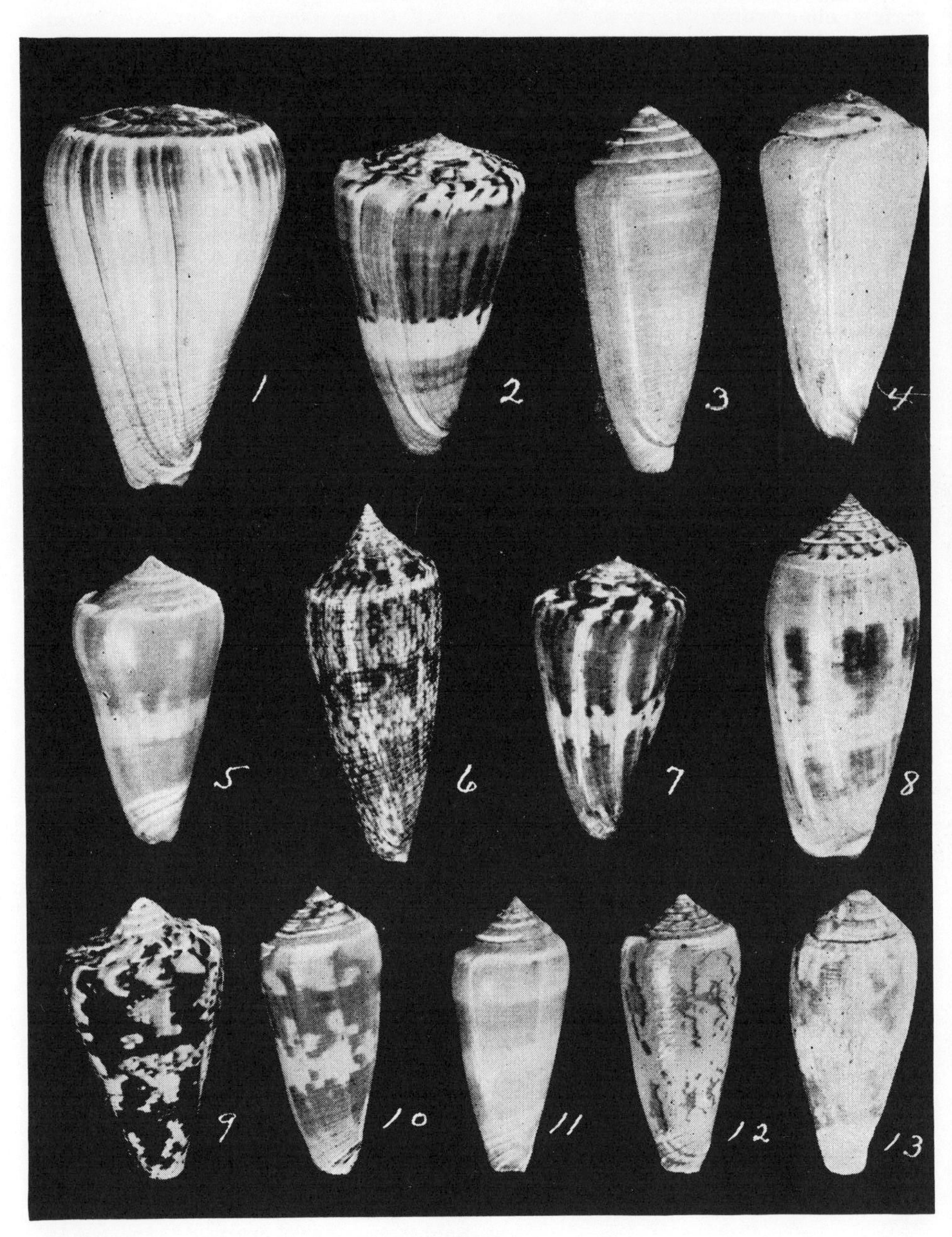

PLATE 16

1. **Conus loroisi,** Kien. Amboina. Uniformly olive-brown. Large thick, and heavy. 3″.

2. **Conus coffea,** Gmel. Red Sea. Light brown with one wide white band. Top striped. 2½″.

3. **Conus terebellum,** Mart. Viti Ids. Flesh-white with fine circular lines. 3″.

4. **Conus terebellum thomasi,** Sow. Red Sea. Flesh-white with faint wide lavender bands and deeper color at base. 2¾″.

5. **Conus consors,** Sow. New Caledonia. A rich orange-yellow with lighter yellow band and faint light blotches. 2″.

6. **Conus australis,** Chem. Australia. Conical with deep circular ridges, mottled with brown. Pointed spire. 3″.

7. **Conus laevigatus,** Sow. Mauritius (close to Sumatrensis Hwass). Perpendicular wavy brown and white with white band. 2″.

8. **Conus terminus,** Lam. Ceylon. Light pink with two wide bands of light brown blotches. 3″.

9. **Conus portoricanus,** Rve. West Africa. Rich blotches of pinkish-white on brown background, fine lines. 2″.

10. **Conus consors daulli,** Cross. Ceylon. More elongated than the type with white blotches on yellow background. 2¼″.

11. **Conus anceps,** A.Ad. Moluccas. Faint yellow bands on white. Elevated spire. 2″.

12. **Conus epistomioides,** Wkf. West Africa. Faint yellow streaks on flesh color background. 2″.

13. **Conus mozambicensis,** Brug. Mozambique. Faint yellow blotches on white circular lines near base. 2¼″.

In tropical regions the coral reefs, whose tops are often accessible at low water, have a rich and varied fauna of their own. The stems and aerial roots of the mangrove are favorite haunts of Littorinas, tree-oysters, Cerithidea, and the like, many of which remain for hours out of the water. In the mud at their base Arca, Saxicava and many others are almost inextricably mixed with the tubes of Vermetus, Petaloconchus and various worms. When the roots extend into clear water they are a favorite haunt of the salt water species of Neritina and Nerita. Under the overhang of rocks and the sides of boulders which stand between tide marks the amphibious Siphonaria and Gadinia may be found associated with chitons and true limpets. Among the pebbles at low water mark may be found hosts of Turbinidae, like Uvanilla and Pachypoma.

✦

The Astartidae consists almost entirely of the genus Astarte. The shells are mostly round and flat and inhabit such locations as Behrings Straits and other northern points, also are found in the deep sea. Mostly of a brown color, They are usually very common in real cold water. Between two and three hundred species have been found fossil from the Carboniferous and later. Collectors should never overlook these shells from polar waters as most of them are in a class by themselves.

✦

In the narrow ditches commonly cut in New England for the quicker drainage of salt marshes, as the tide recedes, Littorinella is often abundant with Bittium upon vegetation. In half submerged beds of peat Petricola and Zirphaea live in borings which they enlarge as they grow, while the piles of old jetties or the softened wood of wrecks when split open reveal the borings of Teredo, Xylotrya and Martesia, often containing the author of the damage. The canals of sponges, of the brittle sort known as "bread sponges" often contain small gastropods which take refuge there; and such sponges are common in the oyster beds and in the pools on stony beaches just below the low water mark, especially on our southern and southeastern shores. The crannies of old weed-grown seawalls are a good collecting ground and in such places on floats of old piling some minute species usually occur which may be vainly sought elsewhere.

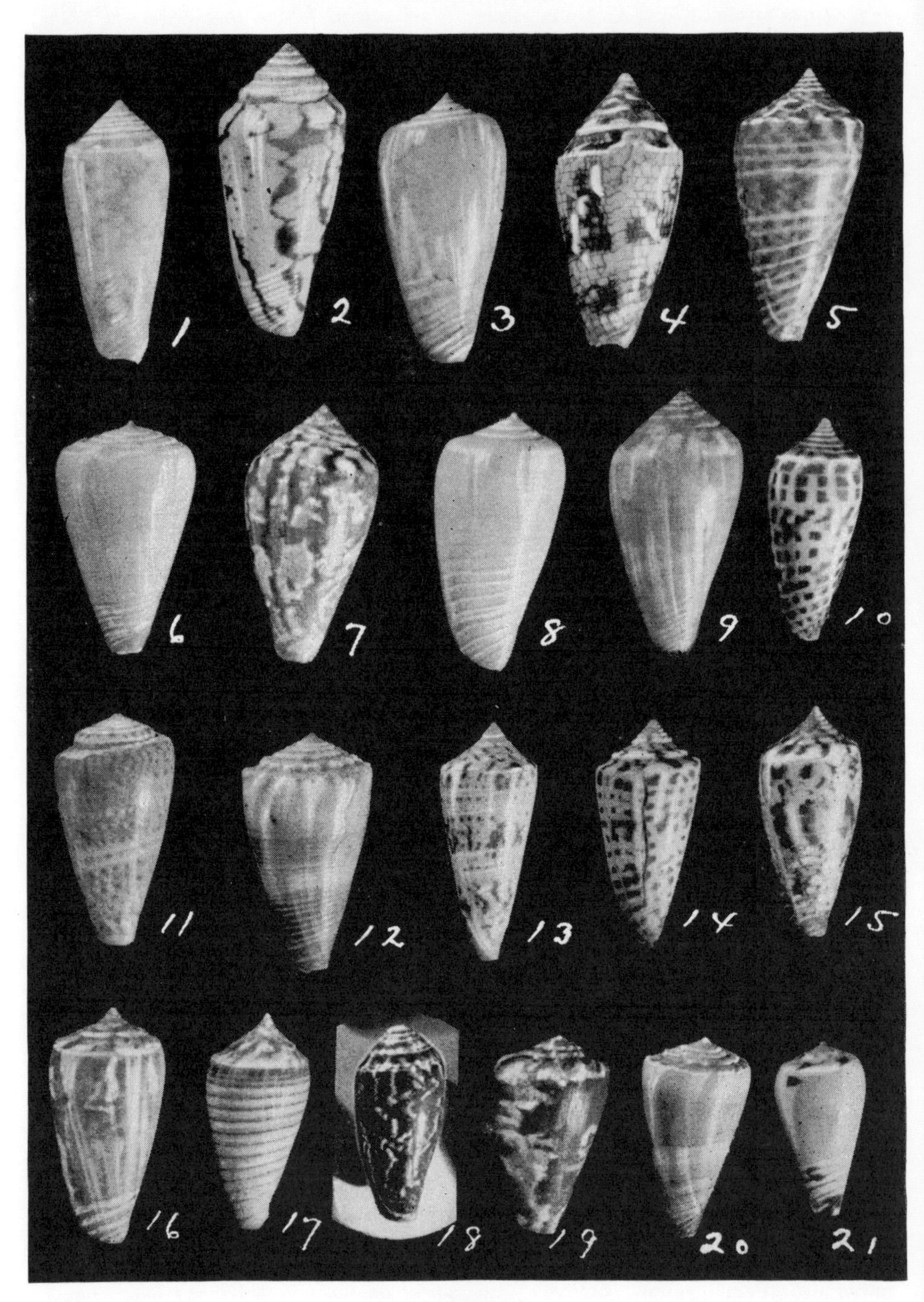

PLATE 17

1. **Conus timorensis,** Hwass. Mauritius. A creamy-white shell covered with faint splashes of pink. Lip curved, spire moderately pointed. 1¾".

2. **Conus simplex,** Rve. South Africa. Of a glossy creamy-white with irregular perpendicular wavy lines of brown usually about six in number. Aperture flesh. Spire deeply ridged and elevated. 1¾".

3. **Conus spectrum daphne,** Boiv. Moluccas. Very light yellow with a faint white band through middle of whorl. There are several faint ridges at base. Top rounded almost flat with tiny point. 1¾".

4. **Conus lucidus,** Mawe. West coast Central America. The shell is rather pyramidal in form, thick and heavy with lattice lines of brown and numerous blotches of chestnut. The design is rather unique. Spire rounded and sharply elevated. 1 to 1¾".

5. **Conus planiliratus,** Sow. Persian Gulf. The ground color is a faint bluish-white with circular rows of dots of light brown. There are two wide rows of chevron markings of same color. Spire moderately elevated to sharp point. Interior flesh. 1¾".

6. **Conus suturalis,** Rve. Moluccas. Uniformly glossy flesh color with two wide very faint bands of light yellow. Few fine ridges at base. Top almost flat with deep ridges, the fine central apex just showing. 1½".

7. **Conus hybridus,** Kien. West Africa. A pyramidal shell of bluish flesh-color with perpendicular markings of russet-brown and bluish-white spots. Moderately rounded elevated spire to point. A rather thin shell. 1¾".

8. **Conus spectrum lacteus,** Lam. Moluccas. A pure white shell with 12 or more circular ridges. Spire almost flat with tiny ridges. Tip pink. 1¾".

9. **Conus guinaicus,** Hwass. West Africa. Of a uniform drab color with one faint white circular line. Spire rounded and elevated. 1¾".

10. **Conus fuscomaculatus,** Smith. West Africa. A rather slender cylindrical shell with moderately pointed spire. There are about 8 rows of russet spots that are almost square, and one row through side of chevron markings. 1½".

11. **Conus infrenatus,** Rve. South Africa. A pyramidal flesh-colored shell with circular rows of dots and one band in middle of white with a row of dots in center of same. The spire is almost flat with one deep ridge. 1½".

12. **Conus balteatus,** Sow. New Caledonia. The top third of the body whorl is of shiny flesh-color, bottom part lined with faint brown, getting darker at base. Top nodulated, slightly elevated with rich red tiny apical point. 1½".

13. **Conus dispar,** Sow. West Coast of South America. A rather slender cylindrical shell with pointed top. Flesh-white with two bands of irregular chestnut blotches and four circular rows of dots of same. 1¾".

14. **Conus inscriptus,** Rve. Aden, Arabia. The shell is pyramidal with elevated ridged spire. Flesh-white, with circular rows of light brown dots and square blotches. Interior flesh. 1½".

15. **Conus collisus,** Rve. Moluccas. Moderately pyramidal with pointed spire. There are perpendicular interrupted lines of russet, and circular blotches of same color. Few fine ridges at base. Lip rather thin. 1½".

16. **Conus stigmaticus,** A.Ad. Moluccas. The shell is flesh-white with perpendicular streaks of light brown and darker wide band through middle. Few lines at base. Top almost flat with fine point. 1½".

17. **Conus mucronatus,** Rve. New Britain. The shell is white with regular circular rows of light brown lines. Spire moderately elevated. 1½".

18. **Conus monachus,** L. Asiatic Sea. A shiny mottled blackish-brown shell with rounded top, moderately elevated and pink tip. There are a few circular lines at base. These is perpendicular lines of white and irregular shaped blotches. 1¼".

19. **Conus jukesi,** Rve. Port Jackson, Australia. A rather thin pyramidal shell covered with heavy brown blotches and whiter shadings. Top almost flat. Interior dark. 1¼".

20. **Conus sugillatus,** Rve. China. Top almost flat and slightly knobby. There is a white band at top of whorl and another slightly below middle. Rest of shell is light brown getting darker at base. There are several rows of sharp ridges at base. 1¼".

21. **Conus pica,** A & R. Sumatra. A small bluish-white shell with only 2 or 3 brown blotches on body whorl and more on the almost flat top. Few fine lines at base. 1".

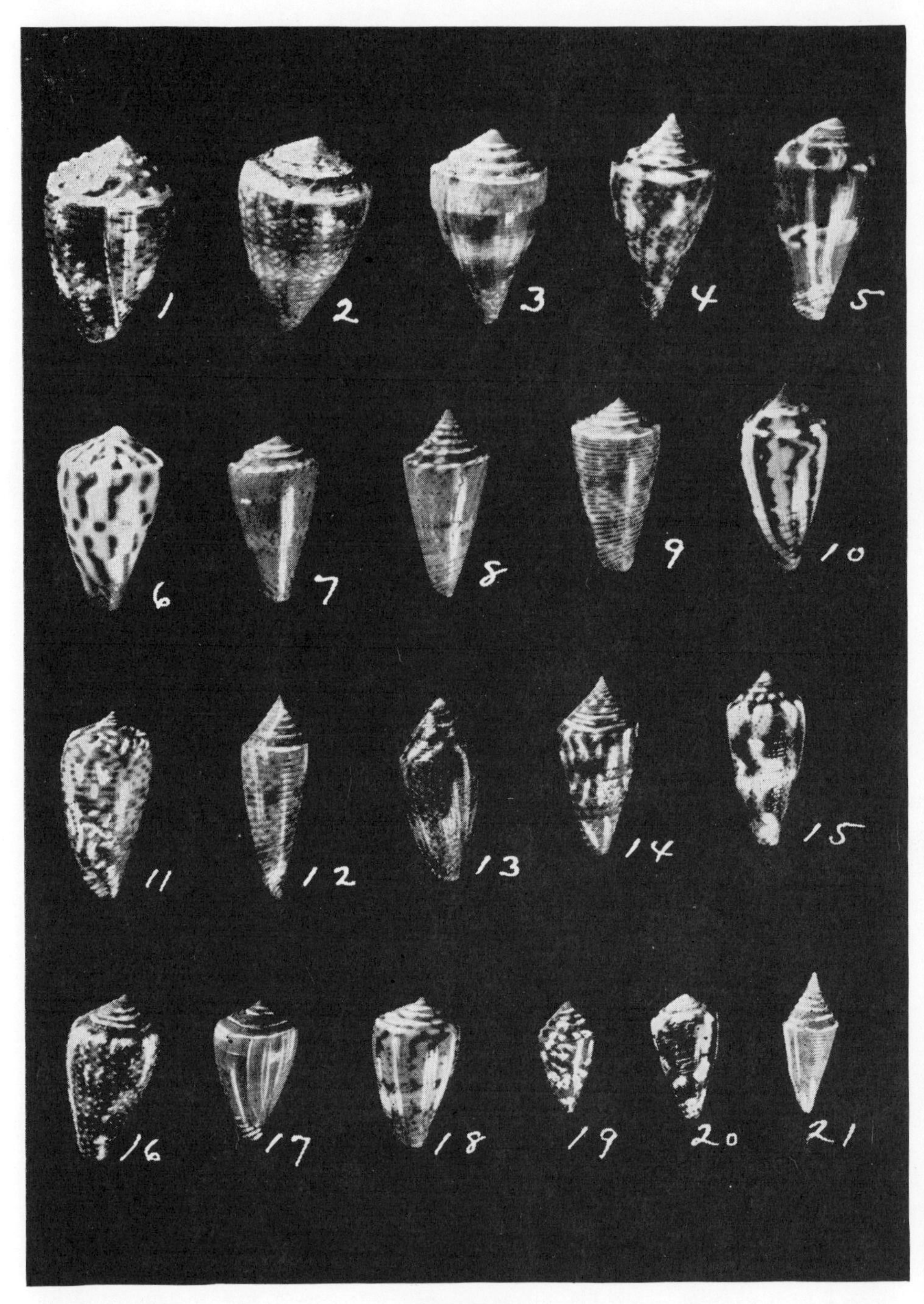

PLATE 18

1. **Conus coronatus,** Dill. New Caledonia. A round solid robust shell with circular dark interrupted lines on a mottled olive surface. Spire slightly elevated with circular row of nodules and pink apex. Interior dark olive. 1¼″.

2. **Conus fulgetrum,** Sow. Loo Choo Ids. Similar in shape to the preceding, of a reddish-chestnut color with numerous fine ridges and flecks of white. Spire just moderately elevated. 1¼″.

3. **Conus propinquus,** Smith. Mauritius. Similar shape of preceding with two wide light brown bands, on a grayish background. Top moderately elevated, aperture slightly purplish inside. 1″.

4. **Conus lentiginosus,** Rve. Bombay. A pyramidal shell with ridged sharp elevated spire, terminating in sharp point. Surface mottled with light brown and flesh. Lip curved outward. 1¼″.

5. **Conus algoensis,** Sow. South Africa. Of a light brown color with white spotted band through middle. Top rounded elevated. Interior flesh. 1¼″.

6. **Conus bairstowi,** Sow. South Africa. A flesh-colored shell with regular perpendicular markings of light brown. Spire moderately elevated. 1″.

7. **Conus pictus,** Rve. South Africa. Of a light reddish-chestnut with two faint bands of spotted pink. Top almost flat and ridged. Interior flesh. 1″.

8. **Conus L'Argillierti,** Kien. West Indies. A sharply pyramidal shell with elevated sharp spire. Light shining chestnut color with faint circular rows of dots. Interior white. 1″.

9. **Conus proximus,** Sow. Moluccas (Moluccensis Ch.). The small shining shell is completely covered with fine interrupted circular lines of reddish-chestnut with some flesh-white spots. Spire moderately elevated to point. Interior flesh. 1″.

10. **Conus zebra,** Lam. Philippines. The rich perpendicular stripes are reddish-brown on a flesh background. Spire rounded elevated. Interior flesh. 1¼″.

11. **Conus broderipi,** Rve. Moluccas. The light brown dots and splashes are arranged in circular rows, spire moderately flat, aperture wide, flesh color. Fine circular lines at base. 1¼″.

12. **Conus acutiformis,** Ad. & Rve. China Sea. A slender pyramidal shell with tall sharp spire. It is completely covered with fine circular ridges and faint spotting of brown. Lip curved outward. 1¼″.

13. **Conus adansoni,** Lam. (Jamaicensis) West Africa. A dark bluish-olive shell with rounded elevated spire. Faint markings of white dots and one distinct lighter band below middle. 1″.

14. **Conus adustius,** Sow. Arabia. A small conical light brownish shell with sharp conical spire. Prominent circular nobby ridges on bottom of last whorl, smooth above. Very few white dots. 1″.

15. **Conus cleryi,** Rve. Venezuela. A small finely nodulated shell with regular prominent perpendicular splashes of reddish-brown on a flesh background. Moderately elevated spire. 1″.

16. **Conus pustulatus,** Kien. Philippines. A small shell completely covered with small nodules. There are several brownish markings on a flesh background. 1″.

17. **Conus miser,** Boiv. Cape Verde Id. A small uniformly light brown shell with moderately elevated spire, fine lines around base. 1″.

18. **Conus pusio,** Lam. Medit. Sea. A small light purplish shell with few markings of reddish-brown. Fine lines at base, slightly elevated spire.

19. **Conus borbonicus,** H. Ad. Bourbon Id. A very small shell mottled with reddish-chestnut and elevated spire.

20. **Conus cabriti,** Bern. New Caledonia. A small very dark blackish-brown shell with few flesh-white markings. Looks like a very small edition of **nocturnus.**

21. **Conus milesi,** Smith. Gulf of Oman. A very small conical shell with very sharp spire. It is completely covered with tiny circular ridges and faint brownish markings. Lip curved outward.

The Capulidae consists of the genus Capulus covering 10 or more species and a squarish form called Amathina. The type is C. Ungaricus common around England. But other forms have been found in all seas and about 20 fossil species in the Silurian. Some collectors call them Bonnet-limpets. Are mostly of a horn color.

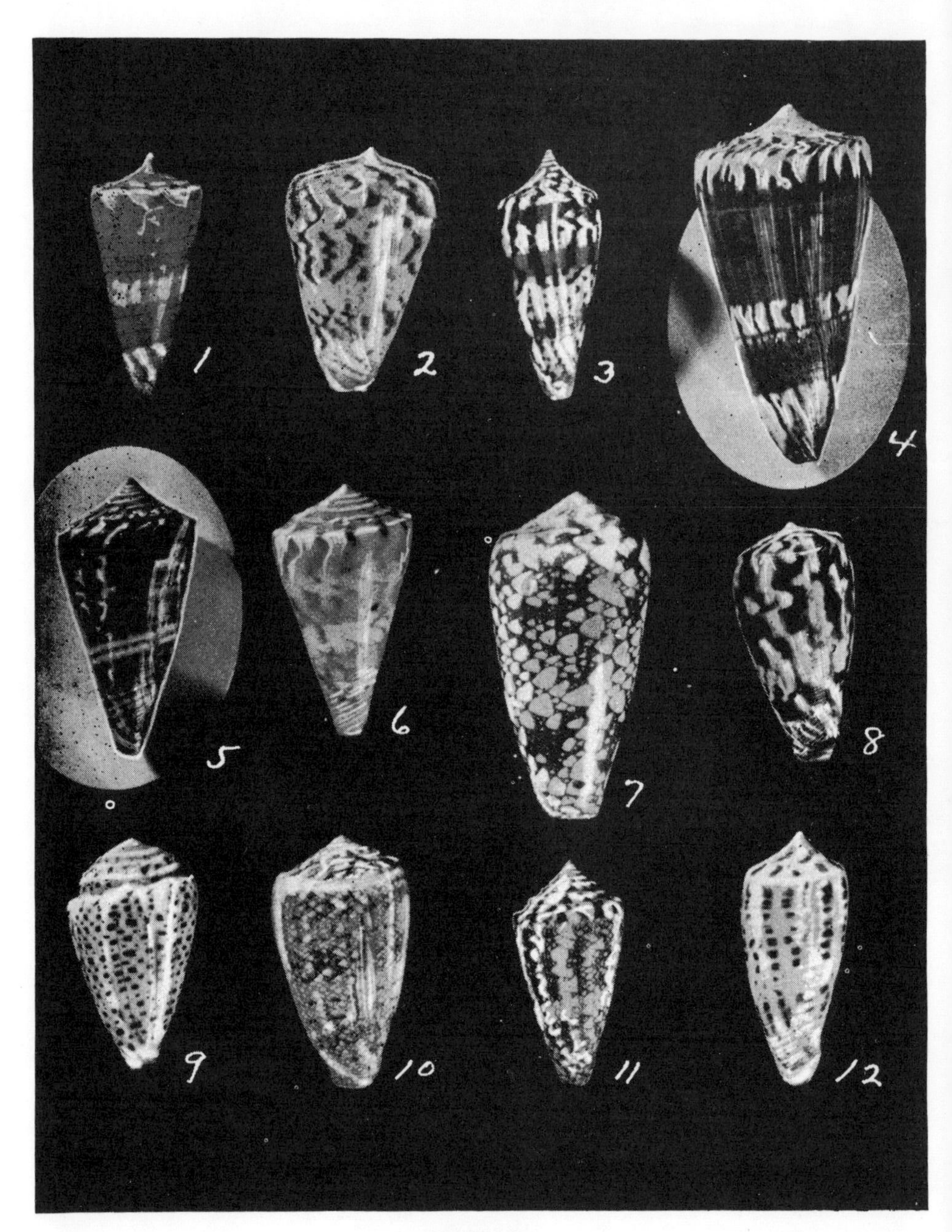

PLATE 19

1. **Conus spiroglossus,** Desh. Reunion Id. The ground color is yellowish-brown with white band in middle, top and base but not continuous, being broken up into splashes. Top flat and sharp point. Lip strong, white inside. 2″.

2. **Conus characteristicus,** Chem. Mauritius. White with irregular parallel elongated strips of chestnut. Top is almost flat with tiny point. The shell is rather heavy for its size. 2″.

3. **Conus janus,** Brug. Mauritius. It has a white ground color with irregular rich brown lines and dots. Three distinct bands of light brown. Spire elevated to point with 7 or 8 whorls. 2″.

4. **Conus generalis,** L. Moluccas. A rich reddish-brown with three white bands striped with same color. Top nearly flat with small point. Very variable. 3″.

5. **Conus regularis,** Sow. Mazatlan. A deep rich reddish-brown with one small lighter band and tiny row of dots in middle of same. Spire slightly ridged and pointed. There are numerous curved blotches. Lip curved outward. 2″.

6. **Conus lorenziarius,** Ch. Brazil. The shell is light yellow with two white bands, and numerous chevron-like markings of white. Spire is ridged and moderately elevated. Lip curved, and shell pyramidal. 2″.

7. **Conus omaria pennaceus,** Born. Mauritius. The ground color is light brown completely covered with large and small white tent-like markings. Spire moderately elevated. 2½″.

8. **Conus spectrum stillatus,** Rve. N. W. Australia. The ground color is white with wide irregular perpendicular bars of reddish-chestnut. Top nearly flat, white point and numerous fine circular lines. Aperture wide, lip thin. 2″.

9. **Conus taeniatus,** Hwass. China Sea. The shell is strong and white with numerous circular rows of dots and lines of black, of a faint bluish cast. Top ridged and elevated. Lip thick and heavy. ¾″.

10. **Conus victoriae,** Rve. No. Australia. The shell is rounded, fat, with ground color of yellowish-brown, and wide faint bands of darker brown, all covered with very fine bluish-white tent-like markings. The design is extremely intricate and hard to describe. 2″.

11. **Conus acuminatus,** Brug. Red Sea. The shell is white, conical with regular elongated stripes of fine white tent-like markings. Spire slightly ridged and elevated to point. 1¾″.

12. **Conus stramineus,** Lam. Moluccas. Of a flesh color with row of elongated interrupted strips of reddish-brown. Spire with fine ridges and moderately elevated. Inside flesh color. 1¾″.

In the Abyssal region alone should we expect to find that any considerable proportion of the fauna has lost all its littoral characteristics, assume characters in keeping with its environment and become disseminated over the ocean bottom throughout a large part of its extent. These expectations in the main are fairly satisfied by the facts as far as the latter are positively ascertained.

In order that their existence may be maintained the Abyssal mollusks require oxygen to aerate their circulation, food to eat, and a foothold upon which they may establish themselves. It is necessary that the conditions should be such as will not prevent the development of the eggs by which successive generations are propagated, and that they do permit it may be assumed from the very fact that mollusks in large numbers have been shown beyond all question to exist on the oceanic floor wherever this has been explored.

In general it seems as if we might safely assume that the composition of Abyssal sea water shows no very important differences from that of other sea water, and that the animals existing in it are not exposed to any peculiar influences arising from this source alone

This cannot be said of the physical conditions. Everyone knows how oppressive to the bather is the weight of the sea water at only a few feet below the surface, and how difficult it is to dive, still more to remain on the bottom, if only for a few seconds. But it is difficult to convey any adequate idea of the pressure at such a depth as 2,000 fathoms, or about two miles below the surface.

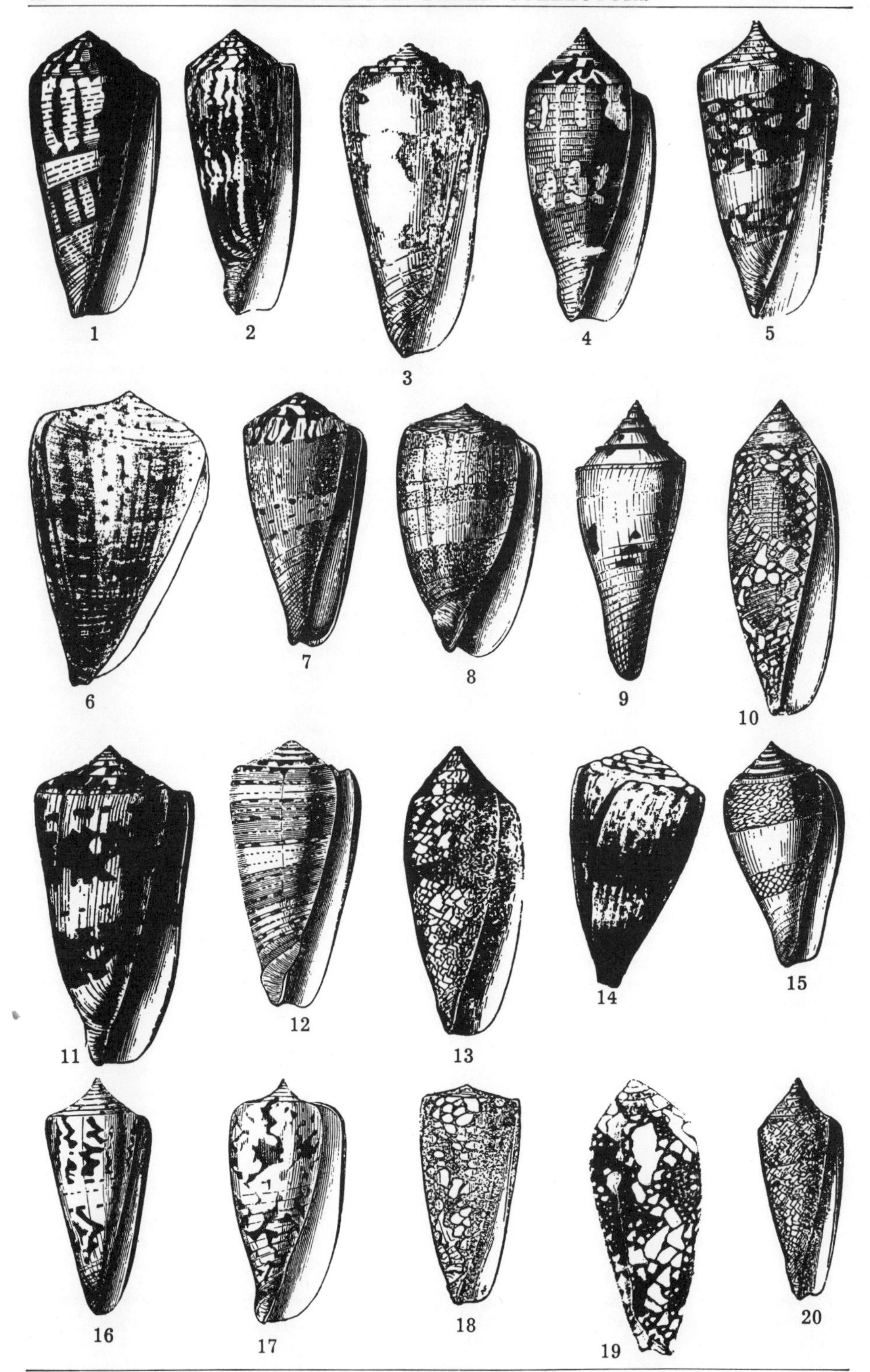

PLATE 20

1. **Conus magus,** L. Philippines. A rather common species ornamented with irregular chestnut markings on white. As it is found over a wide territory, about ten varieties have been named and some of them are very distinct from the type. 2 to 2½″.

2. **Conus magus raphanus,** Hwass. Philippines. This is one of the many varieties mentioned above. The color pattern is quite different and can be easily separated from other forms. 2 to 2½″.

3. **Conus gubernator,** Hwass. Mauritius. A fine light colored species 3 to 4″ with light chestnut markings on a creamy white background.

4. **Conus achatinus,** Chem. The Agate Cone. Philippines. Light russet lines and wave markings of chestnut on white. Some specimens are all grayish-brown with light dots. Very variable shell. 2½″.

5. **Conus ammiralis,** L. Philippines. A very handsome species usually covered with white tent-like markings and two bands. In the old days of the early 19th century it brought fabulous prices, as did many other species of this genus. 2 to 2½″.

6. **Conus betulinus,** L. Singapore. A fine very heavy species of yellowish color, with well spaced rows of dark dots. Ranges up to 5″. The smaller specimens are usually the finest colored.

7. **Conus mustellinus,** Hwass. Philippines. This species is richly ornamented with small spots and dashes of dark drab on light buff background. A fine 3″ shell that is quite distinct.

8. **Conus arenatus,** Hwass. The Dotted Cone. Ceylon. This species is ornamented with hundreds of small dots, often arranged in waves. It is a handsome small chubby species, common to the Indian and Pacific Oceans. 1½″.

9. **Conus sieboldi,** Rve. Japan. Very distinct, rather slender and thin. Has only a few russet markings on white. The elevated spire is typical. Rather scarce but more common of recent years. 2″.

10. **Conus auratus,** Lam. Ceylon. A noble shell ranging from 3″ up with chestnut tent-like markings on white background. Somewhat resembles aulicus but never approaches that species in size.

11. **Conus nocturnus,** Hwass. Moluccas. The white blotches often tent-shaped on a black background, marks this as a dark shell. It is rather rare and few are ever seen on the market. There is a very fine variety called **deburghae,** from Java. Both are about 2½″ and rare.

12. **Conus aurisiacus,** L. Moluccas. A rather rare shell with markings of bands of pink on white, ornamented with chestnut dots and dashes. Only occasionally offered and always at a high price.

13. **Conus aulicus,** L. South Seas. One of the finest and largest of the genus attaining 5 to 6″. It is entirely covered with tent-like markings.

14. **Conus miles,** L. Philippines. A very common species with a broad brownish band on a lighter background. Also dark band at base. It has been sold in a commercial way for generations. 3″.

15. **Conus mercator,** L. Senegal. A very distinct small species which is banded with waves of zigzag chestnut marks on a light buff ground. It is a rather rare shell and not often offered for sale. 2″.

16. **Conus monile,** Hwass. South Seas. A very handsome widely distributed species. It has chestnut dots and splashes on creamy white with often light russet background. 2″.

17. **Conus spectrum,** L. Mauritius. A distinct marked small shell. It has a wide aperture, rather thin edge. Light chestnut markings on white. 2″.

18. **Conus crocatus,** Lam. Mauritius. This species has zigzag markings of dark brown on a lighter russel background with tent-like blotches of white. The cut does this rare shell scant justice as I have seen. 2½″. Specimens bring $50.00.

19. **Conus rubiginosus,** Hwass. Mauritius. The white tent-like markings are on a russet background. There are other similar species which are hard to separate. A very handsome 2½″ shell.

20. **Conus elisae,** Kien. Madagascar. One of the fine and rare small tent cones not often seen in cabinets. They inhabit coral reefs and good specimens are never seen on shore lines. It is a gem shell if you are successful in securing one. There is a species called Dalli found in deep water off Mexican coast to Panama which much resembles this form and quite as beautiful. 2″.

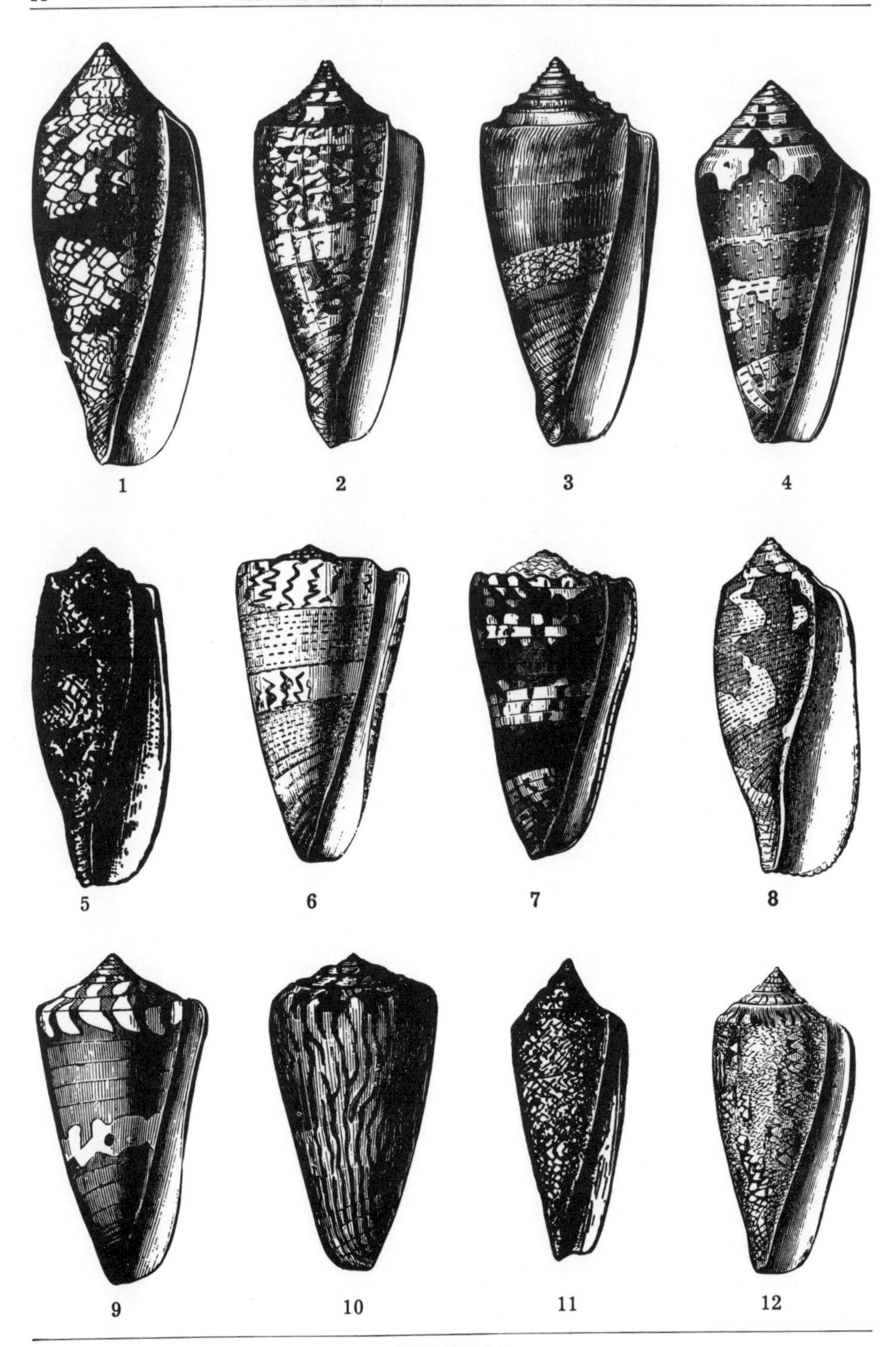

PLATE 21

1. **Conus textile,** L. Tent Cone, Philippines. A handsome shell entirely covered with wavy tent-like lines of dark chestnut on russet. The most widely distributed of the many species of Tent cones found through the Pacific and Indian Oceans. 3 to 4″.

2 and 3. **Conus amadis,** Chem. Ceylon, Australia, etc. The chestnut markings almost cover the background of white. A fairly common species well distributed over a wide territory and various color forms will be found, likely due to different ecological conditions. 3″.

4. **Conus aurantius,** Hwass. Philippines. This is one of the grand Conus and a good series of color forms is very rare. Specimens seen in many collections give a very faint idea of the great variety of color. It has very irregular markings of deep russet. My Philippine collectors have never sent this shell to me. 3″.

5. **Conus geographus,** L. Ceylon to Philippines. A large 4 to 5″ species with wide aperture. The body of the shell is thin as compared with other Conus and is mottled with reddish-brown. An outstanding species that should not be very hard to secure.

6. **Conus thalassiarchus,** Gray, Japan. This shell has faint irregular markings on a white background. It is a rare shell and always has the appearance of being over cleaned, but it just comes that way. 3″.

7. **Conus zonatus,** Hwass. Andaman Ids. A brilliant dark shell ornamented with black markings on a white background. It is a rare species which may cost you a ten spot or more for a choice one. 3″.

8. **Conus tulipa,** L. Philippines. A very distinct shell in both form and color. The wide aperture is somewhat similar to **geographus** but it is usually thicker and smaller. It is of pink color on white background with fine lines and dashes. 3″.

9. **Conus lithoglyphus,** Meusch. Ceylon. The pure white markings on a reddish-russet background, makes this a very striking species. Always attracts attention by its bright colors. 2″.

10. **Conus princeps lineolatus,** Val. Panama. There is a better illustration on Plate 9. Various collectors have sent me choice specimens from Guayamas, West Mexico, where it is fairly common.

11. **Conus telatus,** Rve. Mauritius. A very handsome shell of 2″ and somewhat rare. It has longitudinal zigzag markings on a lighter bluish-black background. White tent-like blotches. The island of Mauritius, I believe, has more varieties of Conus to its credit than any other place in the ocean world.

12. **Conus abbas,** Brug. Ceylon. Entirely ornamented with light and dark chestnut. The tent-like markings show the white background. Very choice and rare. 2″.

I often wonder if famous collecting grounds of the 19th century have not changed. One of the first collections of marine shells I ever bought had many shells marked Egmont Key, Fla. So one winter when I was living in St. Petersburg I got a boat and visited this famous spot. The beaches were as bare of shells as a billiard ball. I asked an old-timer if he knew why I found no shells and mentioned it as a famous spot many years ago. He said when he was a boy there was a stream of very warm water constantly bathed the shores of the key and that likely brought so many shells. Now that stream was several miles out and had not touched the shore for years.

✦

Montijo Bay which lies about 200 miles north of Panama on West Central American coast is usually a good collecting spot. One of the largest Cocoanut plantations known is located here. Most of the shore line is mangrove swamps, but you are likely to find Mitra lens, always a prize, and plenty of Cardita latiscostata, Cypraea cervinetta and other good things and plenty of very common forms of shell life. On the mud flats are endless quantity of such as Strigilla sincera, Mactra angulosa, and in patches of rocks, you will likely find Pleurotomas, Cancellarias, Epitomium, Marginellas, Nassas and many others in superb condition.

✦

A hammer, or small pick and hammer combined, is frequently used to crack rocks or coral for pholads, or rake the gravel from hidden gasterpods. To dig out living bivalves, a spade and a good deal of energy are required. Such collecting is best made the subject of a special excursion.

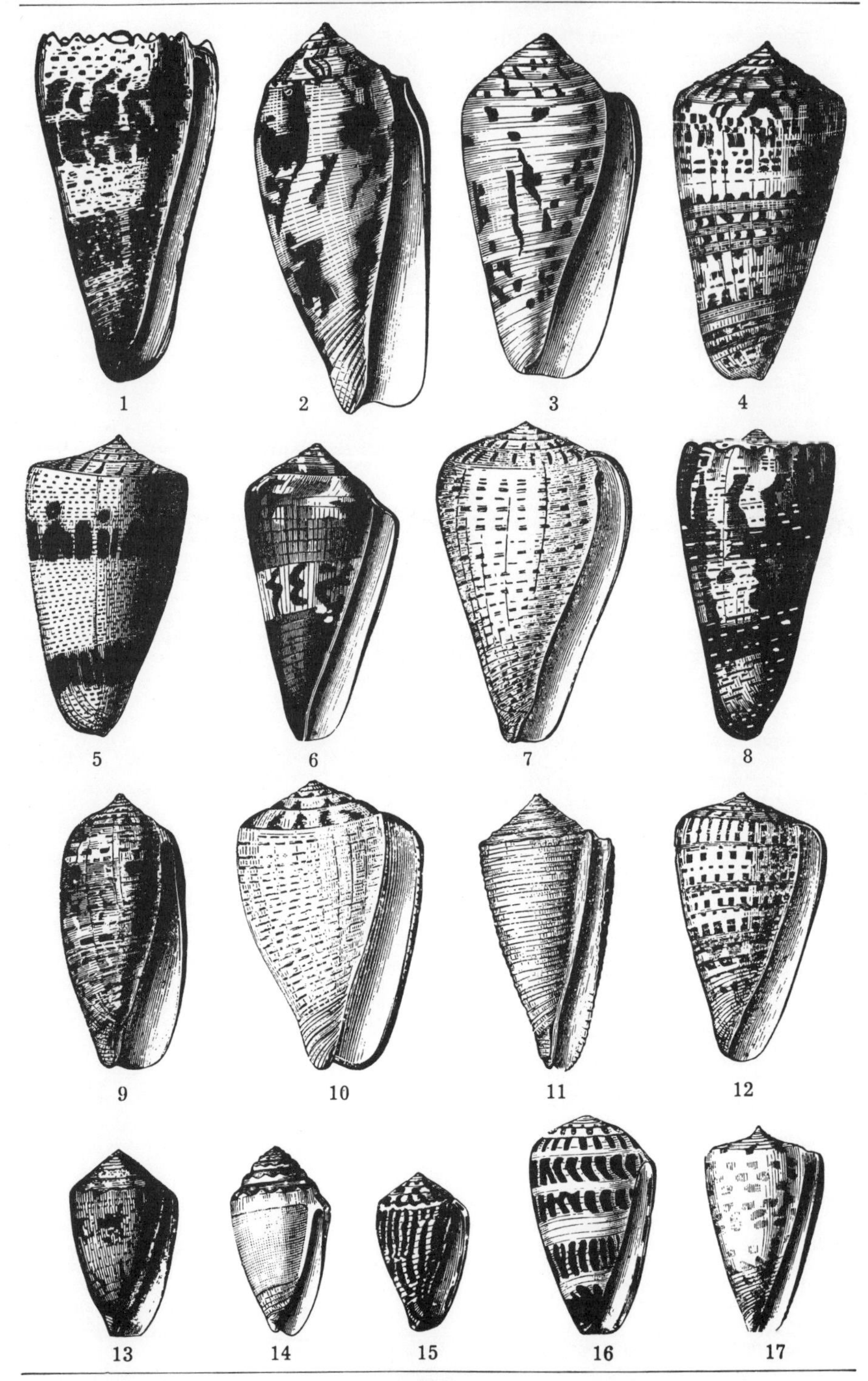

PLATE 22

1. **Conus imperialis,** L. Imperial Cone. Philippines. This species is found over a wide territory. It has yellowish-chestnut markings with black dashes on a white ground color. It is fairly common and one of the finest species. 2½″.

2. **Conus striatus,** L. Striated Cone. Philippines. It has light russet lines, prominent splashes of color on a creamy-white background. A series of shells will run from light to very dark color. 3 to 4″.

3. **Conus floccatus,** King. Itull Id. A very rare shell that is light purplish with longitudinal lines and revolving bands of chestnut. Only occasionally seen in cabinets. 2½″.

4. **Conus siamensis,** Brug. Siam. A fine large solid shell. It has light russet markings in great profusion which almost completely obscure the white background. 2 to 4″.

5. **Conus augur,** Hwass. Moluccas. This very distinct species is ornamented with fine russet dots which merge into two very distinct bands, like the milky way. 2″.

6. **Conus vexillum,** Gmel. Java to Philippines. A fine large species 3½ to 4″. The even yellowish color only shows a few patches of white with a regular design of yellow and white on apex.

7. **Conus suratensis,** Brug. China. This is a fine dotted shell 3″ or more. It has chestnut markings on a yellowish-white background. The pattern of coloring is very distinct and beautiful.

8. **Conus imperialis fuscatus,** Born. Zanzibar. A splendid variety of the Imperial Cone. It has light russet markings on a grayish-white background.

9. **Conus bullatus,** L. New Caledonia. A medium size shell of 2 to 3″ with large aperture which is typical of the species. It has reddish-russet markings on white background. Usual high natural polish, not seen in many other species of this genus.

10. **Conus glacus,** L. Moluccas. A very fine and distinct shell to which the cut hardly does justice. It has russet markings on a light brown background and cannot be confused with any other species. Dark blotches on apex. 2½″.

11. **Conus sulcatus,** Hwass. China. A white shell completely covered with bold circular ridges. Differs from all other species. 2½″.

12. **Conus papilionaceus,** Hwass. Gambia. A fine strong robust shell with russet markings in regular pattern on a white background. A really beautiful shell. 3″.

13. **Conus minimus,** L. Mediterranean Sea. There are many Conus of about this size and they seem to breed in great confusion. A dark mottled shell with irregular splashes of color. 1″ or a trifle larger.

14. **Conus pontificalis,** Lam. Australia. A small 1″ species of a drab color with fine lines and a very distinct apical structure not seen in other species.

15. **Conus hebraeus vermiculatus,** Lam. New Caledonia. This little fellow I call a variety of Hebraeus, as the pattern and size is the same except that the lines are thinner and many more of them. Dark lines on white background. 1″.

16. **Conus hebraeus,** L. Philippines. A neat little species similar to preceding with wider and heavier lines. 1¼″.

17. **Conus tessellatus,** Hwass. Ceylon. A brilliant shell adorned with light reddish markings on creamy-white background. 1½ to 2½″.

My collectors often send me a choice lot of shells from Guayamas, West Mexico, and a little description of the town and environs may be interesting. The little town faces west on a perfectly land locked bay. Steep rocky hills rise abruptly back of it. The entire bay is surrounded by sandy mud flats with an occasional small patch of rocks, which is always a fine collecting spot. In this bay you are likely to find the larger Pleurotoma including P. nobilis the largest form in that region.

✦

The Neritopsidae consists of one species N. radula but more than 20 forms have been found in the Triassic and later. It is unusual to find such a solitary shell. Those I have had were pure white and similar in form to some of the Neritina but of a much more calcareous nature.

PLATE 23

1. **Voluta deshayesi,** Rve. New Caledonia. A shell which resembles many of the forms of vespertilio, with pointed knobs on top of each whorl, four folds in umbilical region. The ground color is pale yellow, with irregular markings on last whorl arranged in two rows of bright russet-red. A very striking shell, with the same shades of color as are found on some of the larger mitra. 4".

2. **Voluta junonia,** Chem. Gulf of Mexico. Has been found from off Cape Lookout, North Carolina, to Florida Keys and in Gulf of Mexico. Most of the specimens in collections have come from West Florida region in 10 to 20 fathoms. The ground color is faint yellowish, which is completely covered with rows of spots often arranged with perfect regularity. Edge of lip usually sharp. 3 to 5".

3. **Voluta bednalli,** Brazier. Torres Strait, No. Australia. A shell that somewhat resembles Gatliffi but is entirely distinct. The ground color is faint yellow. There are three wide lines of dark brown on last whorl, which are crossed longitudinally with similar wide wavy lines, forming a very unusual pattern. There are few shells in the world with such remarkable design. Very rare. 4".

4. **Voluta nodiplicata,** Cox. West Australia, in deep water. The shell is about twice as thick as the usual shell of its size, in this respect resembling some of the Melo. The upper part of each whorl has a fine row of sharp nodules. Ground color yellowish, with faint wide bands of light brown. About 4".

5. **Voluta norrissi,** Gray. North Australia. As in many species of this genus there is a row of small sharp points on top of last two whorls. The shell is brown, with two darker bands on last whorl with faint splashes of white between. 3½".

6. **Voluta reticulata,** Rve. New Holland and West Australia. A shell with brilliant natural polish, completely covered with small tent-like markings of white on a yellowish background. Like a few of the other shells of the genus the last whorl is fully 5/6 of its size. There are four wide bands of darker brown with lighter spaces between. Interior a rich brown. 3½ to 4½".

The Family Volutidae consist of around 150 species and with the genus Cymbium, which include the great Melo shells complete the group. They inhabit the Indian Ocean, Japan, Alaska, Australia, Eastern Polynesia and the Atlantic coasts of Western Hemisphere. A few have been found in South Africa. Australia is the main metropolis of the group. The triangle which includes Japan, Ceylon and Australia would include likely 80% of the species. Most of the species have no operculum but Musica has one. Most of the family are oviparous. I had sent me eggs (so-called) of V. brasiliana from Uruguay. The transparent capsules were perfectly round and could be easily preserved in a solution of formaldrahyde. Where the capsules were well developed one could easily see several small Voluta. The capsules or sacs were about 2 inches in diameter. There are more than 250 species of fossil forms known, starting at the cretaceous.

I consider this genus one of the hardest to complete and most expensive. I have never seen or heard of any museum or private party who succeeded in securing a fine specimen of all of them. Although the group has been well known for two centuries the average collection rarely contains more than 75 kinds and many of these cost $10 to $50 or more each, if in choice condition. Still there is a very good variety that can usually be obtained at modest prices. V. vespertilio and scapha are the only two that have been brought into our country in commercial quantities and must be very common where found.

"In the good old days" the genus was divided into 16 sub-genera but today many of these sections have been raised to generic rank so that few collectors recognize the names when they see them.

The embryonal whorl at the top of many species is of special interest as it is frequently of the size and shape as it emerged from the egg sac and started to develop into a full grown shell.

The Bullidae contains the genera Bulla, which are the largest shells in the family followed by Haminea, Akera (A most remarkable form which whorls open at top) Cylindrobulla and Volvatella, one species from New Caledonia of which there is nothing similar in all the world.

PLATE 24

1. **Voluta volvacea,** Lam. Tasmania. A thick heavy shell rather more chunked and rounded than Scapha. It is gray with faint brown markings. 3½″.

2. **Voluta hebrae,** L. Brazil. The shell is likely a hybrid or variety of this species. It has numerous wavy lines of russet and brown but there are bands somewhat similar to musica. 3½″.

3. **Voluta swainsoni,** Marwick. Stewart Island, N. Z. An elongated fairly smooth shell, the upper whorls being ridged. Flesh color with faint markings. Lip flaring and rounded. 4½″.

4. **Voluta depressa,** Suter. North Island, N. Z. It is much the form of **Jaculoides** with similar row of knobs. The base color is light brown with prominent rows of very dark blackish markings, some of zigzag pattern. 3½″.

5. **Voluta vespertilio serpentina,** Lam. Polynesia. The sharp pointed knobs and zigzag pattern distinguishes it from the usual type of vespertilio. I had many from the Philippines also. 2″.

6. **Voluta nucleus,** Lam. North Australia. One of the smallest species of the genus, seldom attaining over one inch. It has brownish markings on a flesh background.

7. **Voluta nivosa,** Lam. King Sound, North Australia. It is of a deep gray color ornamented with white dots and two bands of brown markings. The upper whorls are finely marked with brown lines. 2 to 3″.

8. **Voluta volva,** Chem. North Australia. A glistening uniform flesh color, the only markings being on the top whorl of fine lines. 3½″.

One of my friends collecting at Puntarenas on the Gulf of Nicoya, West Central America found Potamides pacificus. Many shells come into shallow water to spawn on the sand bars. You will also find here Conus mahoganyi which is really a very dark variety of interruptus, and likely plenty of Conus comptus. As the tides go out you will likely find Galeodes patula digging in and an occasional Oliva angulata the largest form in the Western Hemisphere but more common at Panama further south. He also found seven color forms of the very common Olivella volutella where further north only gray and brown forms were found.

✦

On the Pacific coast a small rod of ⅜th inch iron, hammered to a chisel shape at one end and to a point at the other, is very useful in detaching Haliotis, or large chitons, from the rock. If the blow is sharp and unexpected the mollusk will usually fall without injury, and in many places a small dip net will be of use for securing it. But if the mollusk is irritated or disturbed before the collector strikes in earnest, it is better to pass on, for the creatures when aroused will hardly be detached without injury, so effective is their hold on the rock. The same is true to a minor degree of the limpets.

✦

If chitons are collected they should be placed on a narrow strip of smooth wood like a ruler, well wetted with salt water, before they have time to curl up. By putting them opposite one another and tightly winding soft twine, lint or lamp-wicking around both chitons and stick, they will be kept in a normal posture until the tissues are relaxed and they can then be preserved in spirits or cleaned. If this precaution is not taken they are apt to curl up in a shape which renders them almost useless for dissection or for cabinet specimens and they break rather than flatten out. A large number may be set on a single stick. If, however, they curl before they can set, it is best to put them in a pan of salt water when they will, if alive, eventually resume a normal position.

✦

The Cancellariidae are composed of the one genus Cancellaria which is divided into 10 subgenera. The group as a whole are remarkable. They vary greatly in form, the largest form is 2½″ and most of the species are one inch or smaller. There are nearly 100 species and about 60 fossil forms commencing with the Upper Cretaceous. The one species we have in Florida is fairly common and it may be some of the others are so, but the fact is, I have always found them very difficult to obtain, rarely having more than a very few on hand at a time. You will be very lucky if you ever secure half of the known species.

PLATE 25

1. **Voluta damoni,** Gray. Broome, N. W. Australia. A grand shell hard to describe. The spire consisting of five whorls terminating in a point is rich russet-red with tiny strips on last apical whorl. The main body of shell is a rich glossy white with irregular tent-like markings with two wide bands of similar markings. 4".

2. **Voluta musica tiarella,** Lam. West Africa. The markings are similar to the Music Volute but more of them and finer. The notes of the music band are not so well developed but there are hundreds of fine dots. The two distinct bands are separated with a band of dots. Edge of aperture with numerous black spots. 2½".

3. **Voluta musica,** Lam. West Indies. I illustrate this shell to show the huge size that some specimens attain. This must be due to unusual good food conditions. The color is typical and the black spots on the lip of which are 10, are very prominent. 4".

Smaller illustration on Plate 29.

4. **Voluta musica carneolata,** Lam. Barbadoes. This fine specimen is slender and elongated, with rich russet markings. No black spots on lip. Each whorl to tip has the regular row of knobs. 3¼".

5. **Voluta magellanica,** Chem. Uruguay, 90 fathoms. The shell is very thin, elongated wide aperture, zigzag markings of brown on last whorl. Main color faint yellow. 5 to 6".

6. **Voluta jaculoides,** Powell. New Zealand in 50 fathoms. An elongated flesh-colored shell with prominent row of knobs on last whorl and smaller ones on upper whorls. The specimen shown is prominently marked with brown blotches but some are free of same. 3½" to 4¼".

7. **Voluta pallida,** Gray. Torres Straits, North Australia. A neat little glossy shell with regular lines, some wavy longitudinal. 1¾".

8. **Voluta lyrata,** Sow. Moluccas. The shell has regular deep perpendicular ridges which are ornamented with brown dots. General color, light brown and white. 1¾".

9. **Voluta zebra lineata,** Leach. Queensland. Similar to **zebra** but the rich brown perpendicular lines are very deep color. Lip brown and white. 2¼".

The mollusk of the beach differ widely in different latitudes. In the north Littorina, Purpura, and various limpets frequent the rocky and stony shores, while near low water on the under surface of flat stones and under overhanging ledges several species of Chiton find a congenial home. Among the barnacles and under the profuse fronds of the bladder weed small Rissoids and the urnlike ovicapsules of Purpura are common. On the California coast Haliotis, Acmaea and many Chitons abound in such localities. As we go south the fauna of stony beaches becomes richer, and a vast number of small shell-bearing and naked mollusks inhabit them. Where there is a mixture of stones and sand, large sea-anemones or Actinia live between the pebbles, often covered with fragments of shell and bits of gravel, amongst which careful examination reveals many small shells sticking to the adhesive surface of the Polyps. These must be secured by means of the forceps, as they are not easily detached.

On the sandy beaches will be found a special fauna, without taking into account the species thrown up by the waves from deeper water. Natica is one of the most common gastropods, and living specimens may readily be detected by the little mound of sand which they push before them as they plow their way just below the surface.

✦

A small sieve, with meshes of 1/16th of an inch, is often useful for sifting the sand out of drift material which collects at high water and along the ripple marks. This can be put in a bag or bottle and picked over at home. If is often very rich in small species. An old table knife is useful for detaching limpets or chitons from smooth rocks. The dip net, if the frame is solid and the meshes small, may be used to dredge out small bivalves from the loose sand near low-water mark or from the soft mud of marshy shores.

PLATE 26

1. **Voluta vespertilio var. mitis,** Lam. New Holland. The parent species is very variable, being found rather smooth and with prongs on top of last whorl. This is a well marked variety but always with slightly different pattern. 3 to 4″.

2. **Voluta viresceus Sol** (polygonales Lam.) Gulf of Mexico. This rare shell was described in 1811 by Lamarck. It is from deep water and has been reported from coast of Texas in the Gulf of Mexico to coast of Columbia. There is a row of pointed knobs on upper part of last whorl and the entire shell is covered with fine dots of brown. The surface of the shell is covered with fine spiral ridges. 2½″.

3. **Voluta ponsonbyi,** Smith. Coast of Natal, Africa, 40 fath. A striking shell of a shade of red, the upper part of each whorl ornamented with points. There are five or six spiral whorls of white markings. I have had in years past several fine specimens that were taken from the stomach of large fishes by fishermen. About 3″ and very rare.

4. **Voluta punctata,** Sw. Port Jackson, Australia. A rather smooth shell with faint row of knobs on top of each whorl. Ground color russet-red, ornamented with dots of brown and a row of flashes of same color on last whorl. Upper whorls have short perpendicular lines of brown. Always rare. 2½″.

5. **Voluta flavicans,** Gmel. Tasmania. A dark grayish shell, the early whorls marked with faint reddish. Rare. 3″.

6. **Voluta lapponica,** L. Ceylon. A typical shell is completely covered with interrupted spiral lines, which appear like dots and dashes of brown on a lighter surface. Not rare. 3¼″.

7. **Voluta rutila,** Brod. New Guinea and Australia. A shell that is completely covered with irregular markings of red and yellow. One faint row of knobs on upper part of last whorl. 3½″.

8. **Voluta schlateri,** Cox. Tasmania. A typical Voluta shaped shell, that is pure white and 3½″ to 4″. Rare.

9. **Voluta kingi,** Hutt. Kings Id. So. Australia. A smaller shell than Schlateri, of a flesh color with usually three folds in the umbilical region instead of four as in former specie. Rare. 3″.

10. **Voluta sophiae,** Gray. Australia. A grayish shell with two double rows of dots or splashes on last whorl. Top edge of each whorl has sharp prongs, and dark radiating lines. 3″. Rare.

The preparation of mollusks for anatomical purposes has been frequently described. For ordinary work nothing is better than clean 98% denatured alcohol diluted with a proper proportion of water. If the specimens are large they should first be put into a jar kept for that purpose, in which the alcohol is comparatively weak, having, say 50% of water added to it. After the immersion of specimens in this jar for several days the fluids will have been extracted by the alcohol and a specimen can then be removed, washed clean of mucus and dirt, which will almost always be found about the aperture of a spiral shell, and placed in its own proper jar of 90% alcohol diluted in about the proportion of 30% with pure water. Specimens to be preserved for the cabinet require the removal of the soft parts if they are still present, the cleaning off of parasitis or incrusting growths, and, in the case of bivalves, securing the valves in a convenient position for the cabinet.

✦

Dredging

This topic is one that could be dealt on rather extensively and I am not going to take up the required space. There are splendid books on the subject, covering many forms of dredges both large and small. The Biological Supply Houses carry small dredges in stock at reasonable prices or one can "make his own." Most any good blacksmith can weld the iron parts, and then all you need is the rope which can be secured at any hardware store and the nets from Supply Houses or firms handling seines for fishermen.

✦

Harpa crenata, Swain is the only Harpa in Western Hemisphere and rarely found on the beaches. It has been taken at Acapulco, Tres Marias, La Paz, San Juan del Sur. I expect at 5 to 20 fathoms there are plenty of them but if they nestle in coral it is hard to bring them up with a dredge. A diver would have much better luck.

PLATE 27

1. **Voluta guntheri adcocki,** Tate. Middleton, South Australia. The shell much resembles a small **undulata,** having similar undulating lines, with the addition of two brown spiral lines on last whorl. 2″.

2. **Voluta gracilis,** Swain, New Zealand. A small shell, almost an exact miniature of some of the larger forms found in Australian waters. There are five whorls, the four upper prominently ridged and last smooth. Finely marked with brown zigzag stripes, which do not show up well in the cut. 2″.

3. **Voluta costata,** Sow. Moluccas. A handsome shining shell ornamented with parallel ridges. The color is white, with circular bars of red. 2″.

4. **Voluta africana,** Rve. Durban, Africa. A small smooth shell with row of rounded knobs at top of each whorl. Flesh color with faint markings of reddish-brown. Very dark blotch at top of aperture. The shell seems to be rare in really fine condition. 1¾″.

5. **Voluta deliciosa,** Mont. New Caledonia. A smooth shining small shell with thickened lip, beveled along the edge. Ornamented with fine circular lines interrupted with tiny spaces. These lines are more distinct on the bevel edge lip. 1¾″.

6. **Voluta rupestris,** Gmel. China. A rather elongated shell of flesh color, with zigzag lines of brown. The entire shell has very fine spiral lines. About eight folds in umbilical region. 3½ to 4″.

7. **Voluta megaspira,** Sow. Japan. Somewhat resembles **rupestris,** but usually larger, has prominent longitudal ridges on each whorl. Ground color light brown, with irregular splashes of darker color, arranged in form of three bands. Umbilical ridges almost obsolete. 4½″.

8. **Voluta concinna,** Brod. Japan. It resembles **megaspira** but the longitudinal ridges are more sharp, the ground color is flesh, and the brownish lines are perpendicular. There are a number of forms similar to this one in the island empire. About 5″.

9. **Voluta thatcheri,** McCoy. This specimen is from Bromton Reef, Australia but is has been reported from New Caledonia. A slender elongated shell of white, with prominent blotches of red, some much darker than others. Very rare. 3″.

10. **Voluta cathcartiae,** A Ad. Locally unknown, and I have been unable to trace same in the literature at hand. The shell has every appearance of being a rather small, but fully adult **"scapha."** It has the same shiny light yellowish surface, and prominent zigzag markings of reddish-brown. 2½″.

11. **Voluta jamrachi,** Gray. Northwest Australia. A shell that is somewhat similar to **turneri** but the reddish-brown lines are usually broader and farther apart, and quite zigzag near the base. Usually smaller and more rare. 2″.

12. **Voluta praetexta,** Rve. West Australia. A small smooth shell completely covered with a pattern of minute tent-like markings of a pale brown on a white background. Occasionally blotches of brown. A rare little 2″ shell.

13. **Voluta turneri,** Gray. Australia. A smooth glistening white shell with wavy lines of reddish-brown and some patches of same color. Row of dark small blotches at suture. Very handsome 2″ shell.

What is usually termed the Littoral zone is from the existing shore line to 100 fathoms. This is as far as ordinary light usually penetrates and as far as marine vegetation can exist. It is also called the Laminarian zone or region of brown kelp. Beyond the 100 fathoms it is practically certain that no light reaches the bottom and that no sea weeds grow. Outside of this the borders of the continents slope gradually to the bottom of the ocean which is found usually at a depth of 2,500 fathoms or 15,000 ft. But there are various "deeps" in the Pacific and some in the Atlantic which reach nearly six miles.

✦

Tornitinidae are a very large family of small to minute shells mostly white and include Retusa, Pyrunculus and Volvula. I had a great many species all in small vials with label inside. They are like very small Bullas and found most everywhere.

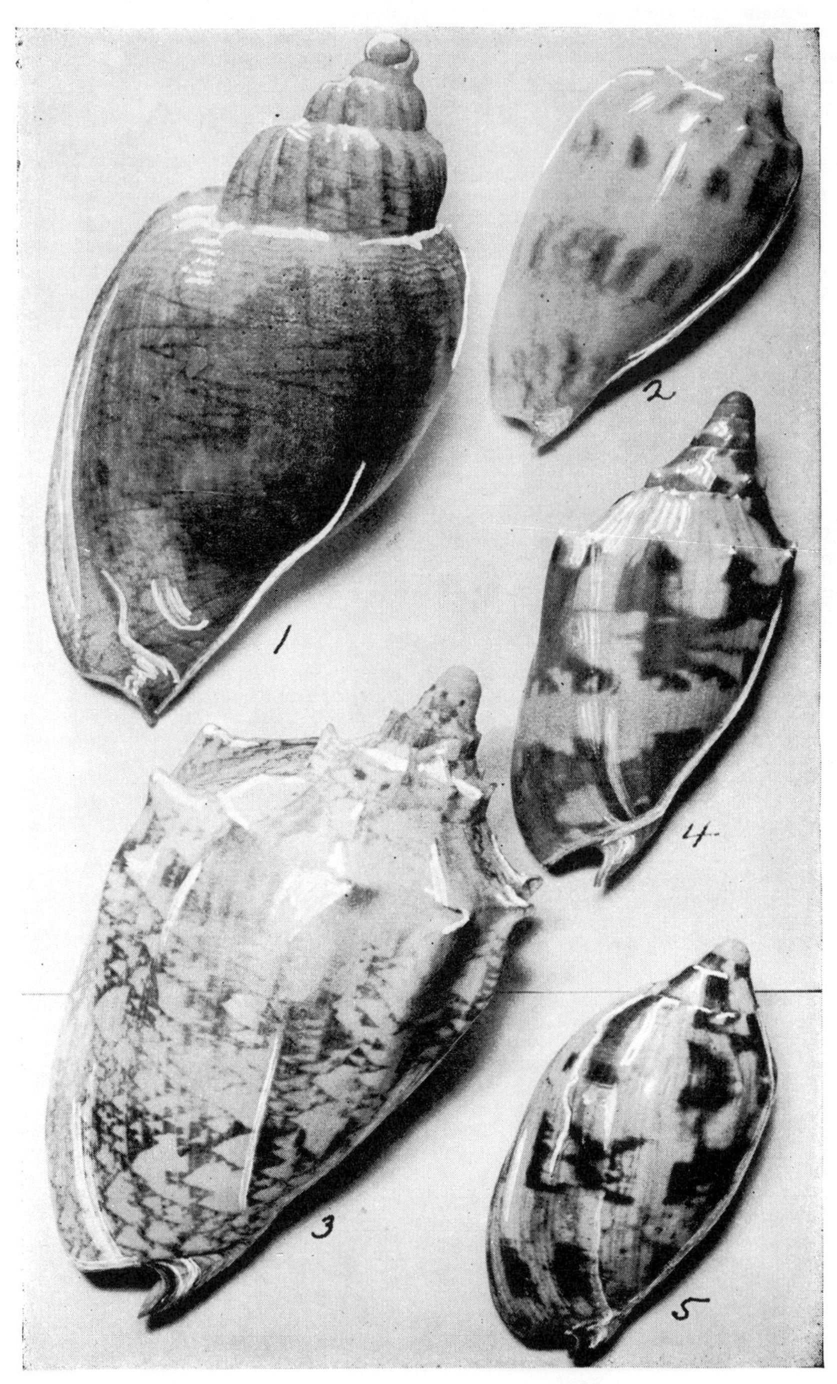

PLATE 28

1. **Voluta Roadknightie,** McCoy. Victoria, Australia. Our friends "down under" consider this species to be one of their rarest shells. For years it was only known from a dead specimen, but a live one is occasionally now dredged from deep water. The last large whorl is rather smooth with prominent sharp pointed markings of brown on a yellowish-red background. The three upper whorls have prominent perpendicular ridges and are adorned with similar markings. 7 to 8".

2. **Voluta marmorata,** Swain. New South Wales. The shell is rather thin for its size, smooth with three rows of irregular brownish markings on a pale yellow background. A faint row of points on upper part of each whorl. 4½".

3. **Voluta rossiniana,** Bern. Isle of Pines, New Caledonia. A rather strong heavy shell with prominent pointed knobs on top of each whorl. The ground color is yellowish, with zigzag markings of russet-brown. Interior yellowish. Averages about 7".

4. **Voluta aulica,** Col. Moluccas and Polynesia. A shell that much resembles deshayesi but is much larger, points are usually more sharp, spire and upper whorls more elongated and the reddish markings more picturesque. A grand shell of striking color. 5".

5. **Voluta ceraunia,** Crosse. New Georgia Id., Solomons. It is very similar in form to many of the smooth shells of the vespertilio complex. Interior is yellowish. Main body color is gray with prominent squarish blotches of brown. Markings extended to apical whorl. 4".

Deep-sea mollusks, of course, did not originate in the depths. They are the descendants of those venturesome or unfortunate individuals who, by circumstances carried beyond their depth, managed to adapt themselves to their new surroundings, survive, and propagate. Many species must have been eliminated to begin with. Others more plastic or more numerous in individuals, survived the shock and have gradually spread over great areas of the oceanic floor. In accordance with these not unreasonable assumptions, we should expect to find among the newer comers at least some characters which were assumed under the stress of the struggle for existence in the shallows, and which, through specific inertia, have not become wholly obsolete in the new environment. We should also expect to find a certain proportion of Archibenthal species to any given area, identical with or closely related to the analogous Litoral region forms of the adjacent shore.

✦

The Naticas and Nassas live on the flesh and juices of bivalves which they seek beneath the sand, drilling in their shells a small circular hole which you so often find on the dead bivalves along the beach. The Eupleuras and Urosalpinx usually called drills are very destructive to young oysters and other small bivalves. The bivalves on which the above genera feed live just beneath the surface of the sand and can usually be detected by the small holes by which water obtains access to their siphons. If the advancing waves or tides uncover them, as it usually does, then the drills get very busy. The Oliva and Olivellas burrow in the sand at low water mark and sometimes burrow quite deep.

✦

Acapulco, West Mexico, has always been a favorable collecting ground but the best work can be done some two miles from town as the beaches nearer are usually bare of shells except after storms and how are you going to pick a time when you are sure to have the storm.

✦

The Strombiformidae consist of genera Stylfer, Eulima, Strombiformis, Apicalia, Mucronalia and several other small genera. There are a very large number of species scattered over the ocean world. The Eulimas are glistening white with sharp point. Some species have the upper whorls distorted and all of them are so very smooth it is hard to pick them up. One of my collectors in Philippines sent me a whole vial of dainty specimens he picked off the bottom of starfish which were acting as host. I had a very large number of species of this great family all in small vials. Many species have been found in very deep water.

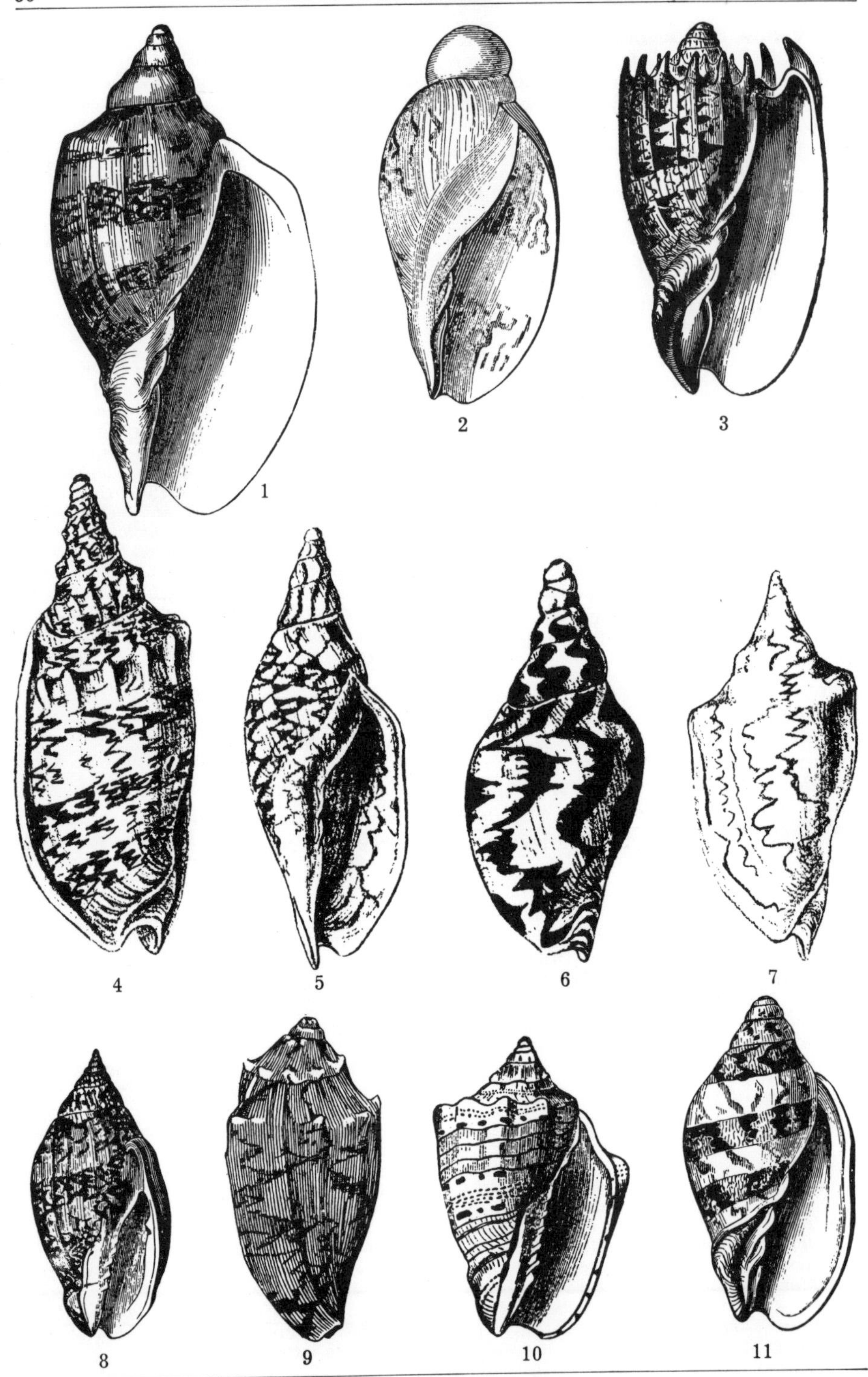

PLATE 29

1. **Voluta magnifica,** Chem. Australia. One of the large shells of this genus ranging 5 to 6″. It has dark chestnut wavy markings over a creamy background. Edge of outer whorl is thin. Markings arranged in four distinct bands.

2. **Voluta mamilla,** Gray. Tasmania. The largest shell of the genus averaging about 9″ with wide open aperture like the Melo shells. Has a very large apical knob or embryonic shell. Chestnut markings on light buff. Shell rather thin for its size. One of my collectors dredges them in 5 to 20 fathoms, says most of the shells brought up are dead and have been inhabited by hermit crabs. A live shell is rare and costly.

3. **Voluta imperialis,** Lam. Philippines. Although found over a wide range, good specimens do not seem to be very common. It is a large shell of distinct form and well marked with brownish tent-like splashes. 6 to 8″ but 2 to 3″ specimens are very attractive.

4. **Voluta pacifica,** Sow. Philippines. A widely distributed species of which there are naturally good varieties. In the Australian region they have dredged similar forms which have new names. The shell is well marked with brownish splashes. 4 to 5″.

5. **Voluta fusiformis,** Swain. Australia and Tasmania. A very fine 6 to 7″ species, adorned with wavy, chestnut markings on a pale background. It is a smooth shell as are most of the species of this genus, very few being found with incrustations like most other marine shells.

6. **Voluta fulgetrum,** Swain. Australia and Tasmania. A noble species showing dark, wavy, chestnut markings on light yellowish-white. The apex is dark and knobby. The embryonic shells of this genus are always of great interest to serious collectors. 5″.

7. **Voluta angulata,** Swain. Patagonia. Fairly common in shallow water in this very cold region. It is almost devoid of color, simply faint traces of russet. Those I received from Uruguay were covered with a thin coating of nacre as if they had been dipped in shellac. 5″.

8. **Voluta harpa,** Barnes. Acapulco, Mexico. A small dark form, one of the smallest of the genus seldom over 1½″. It has longitudinal ridges with three distinct bands of brownish color.

9. **Voluta vespertilio,** L. Bat Volute. East Indies generally. This species varies from brown to reddish with various patterns of mottled design. The shells may be entirely smooth or adorned with sharp spines, small knobs and all the variations between. Several distinct forms have been given varietal names. Most specimens run 2 to 3″ but specimens have been found up to 5″. The most common species of the genus, which is composed largely of rare shells.

10. **Voluta musica,** Lam. Trinidad and other points on the East Americas. The Music Volute is a great favorite with collectors in that the pattern of coloring much resembles the bars of written music. There is a wide variation of this pattern from different localities. Usually 1½″, but specimens have been found to 3″. About six color varieties have been named.

11. **Voluta piperita,** Sow. New Georgia, Solomon Islands. A very handsome 3″ shell adorned with reddish markings forming bands. Seems to be closely allied to the species called **ruckeri.**

As everyone is aware, the sea is the most prolific region for molluscan life, far exceeding in the number of its species the land and fresh water regions combined. A similar disproportion exists between the respective numbers of families and genera. The earliest known mollusks were coeval with the earliest fossiliferous rocks and were marine forms. Air-breathers are not known to have existed before the Carboniferous period, but when the much more ancient Cambrian forms were living, the molluscan type was already old and exhibited development in several of its principal lines. There can be no doubt that the sea has continuously existed since the earliest development of life on the globe, and most naturalists believe that in it the first organic life took rise. Marine mollusks, regarded as a whole, have therefore formed a continuous series, and in the depths of the sea are to be sought those recesses where change of conditions from age to age has remained at its minimum. There linger forms which are of incalculable antiquity, some of which differ little, regarded as generic types from some of those which existed in Paleozoic times while representatives of genera developed in Cenozoic time are numerous.

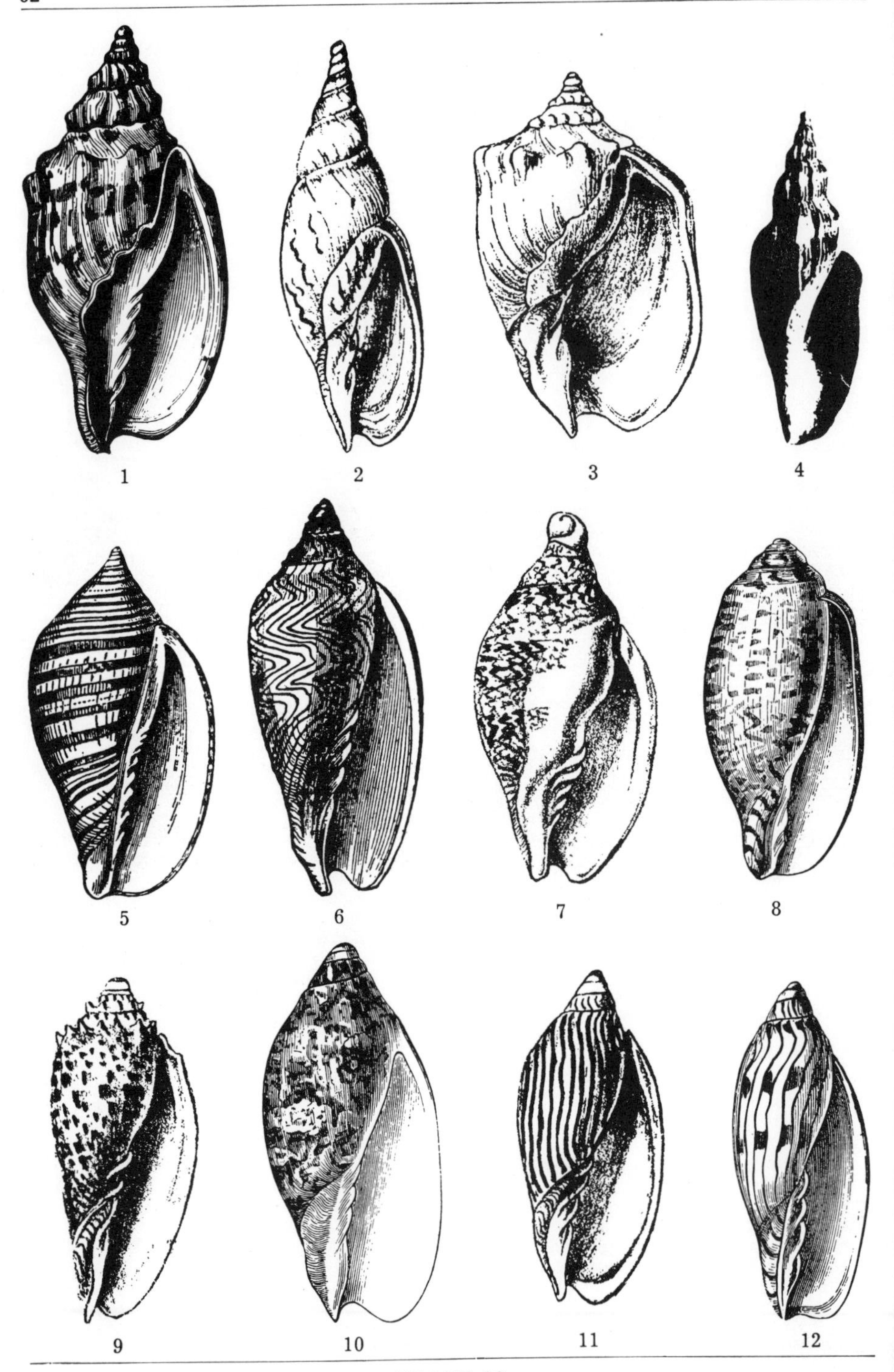

PLATE 30

1. **Voluta festiva,** Lam. S. E. Africa, Natal Coast. One of the rarest of the genus. It is a rosy-white clouded with orange-red. I have never been able to secure a specimen of this shell, but there are many other Voluta few collectors have ever seen.

2. **Voluta ancilla,** Sow. Patagonia. A rather thin shell, devoid of color as is usual with shells living in such cold water. Most specimens that come to me show the effect of the turbulent seas, even to grinding the shell so thin, holes are found in perfectly fresh live specimens. 6″.

3. **Voluta braziliana,** Sol. Brazil southward. Specimens sent me recently from Uruguay were covered with a reddish periostracum. All have wide apertures and are not a very attractive shell. I also had sent me the egg-sacs laid by this mollusc. They resemble gopher eggs of Florida about 2″ diameter, soft, translucent, and those that were developed showed about a dozen embryos in the one sac. Preserved in formaldehyde they are a very interesting curio for the shell den. 3 to 4″.

4. **Voluta prevostiana,** Crosse. Japan. A slender form, thin and attains 7 or 8″. Large specimens show only faint traces of color, usually reddish-brown. The Japs are great shell collectors but they do not find many of this shell.

5. **Voluta vexillum,** Chem. Ceylon. A very striking shell with its regular bands and reddish markings on a white background. About 2½″, is one of the finest of the region and not at all rare.

6. **Voluta undulata,** Lam. So. Australia. A fairly common shell much admired for its zigzag markings of wavy lines. My collector finds them burrowing in sand in shallow water. Usually 3″.

7. **Voluta papillosa,** Swain. South Australia, deep water. A fine robust shell, usually lightly mottled with chestnut. Only occasionally seen in collections. I suspect it lives in rather deep water like a number of forms of this genus. 5″.

8. **Voluta bullata,** Swain. South Africa. A rather small shell and of curious form for this genus. It has a wide aperture and shows only faint markings of brown on a buff background. Live collected shells are rather scarce in collections. 2½″.

9. **Voluta cymbiola,** Chem. Moluccas. It is a very rare shell and only occasionally seen. About 2½″. Diffused blotches on a whitish background. **Voluta sophiae,** Gray from West Australia is somewhat similar and equally rare.

10. **Voluta ruckeri,** Crosse. New Georgia, Solomon Islands. A richly colored shell with its reddish markings which always attracts attention. It has deep blotches on a creamy-white background. My collector in these islands writes me the natives only occasionally find a nice specimen. 3 to 4″.

11. **Voluta elliotti,** Sow. Northwest Australia. A beautiful shell with chestnut, wavy lines on a creamy background. It has a fine smooth polished surface and doubtless comes from deep water. There is another species similarly marked called **Turneri, Gray** from the Australian region, but it is very much rarer than this shell. 2½ to 3″.

12. **Voluta gatliffi,** Sow. Port Keats, No. Australia, named for a prominent collector in that continent. A very fine and rare shell of which I have not seen over six specimens in fifty years. Few collectors have ever been able to procure one and yet is liable to turn up on the market at any time. It has chestnut markings arranged in oblong squares on a creamy or buff background. Smooth 3½″.

Santa Elena Bay on the Costa Rica western border is a rich field for the naturalist. You will likely find such prizes as the big Strombus galea and you may be lucky enough to see the Murex regius feeding on such bivalves as Dosinia and Venus. In collecting these large shells you will find a great many of them worm eaten. Leave such specimens right in the sea as they are not worth bringing home as specimens. The big Fasciolaria granosa which attains 6″ or more will likely be found from here to Panama feeding on bivalves. It is surprising how many fine shells live almost entirely on other forms of sea life.

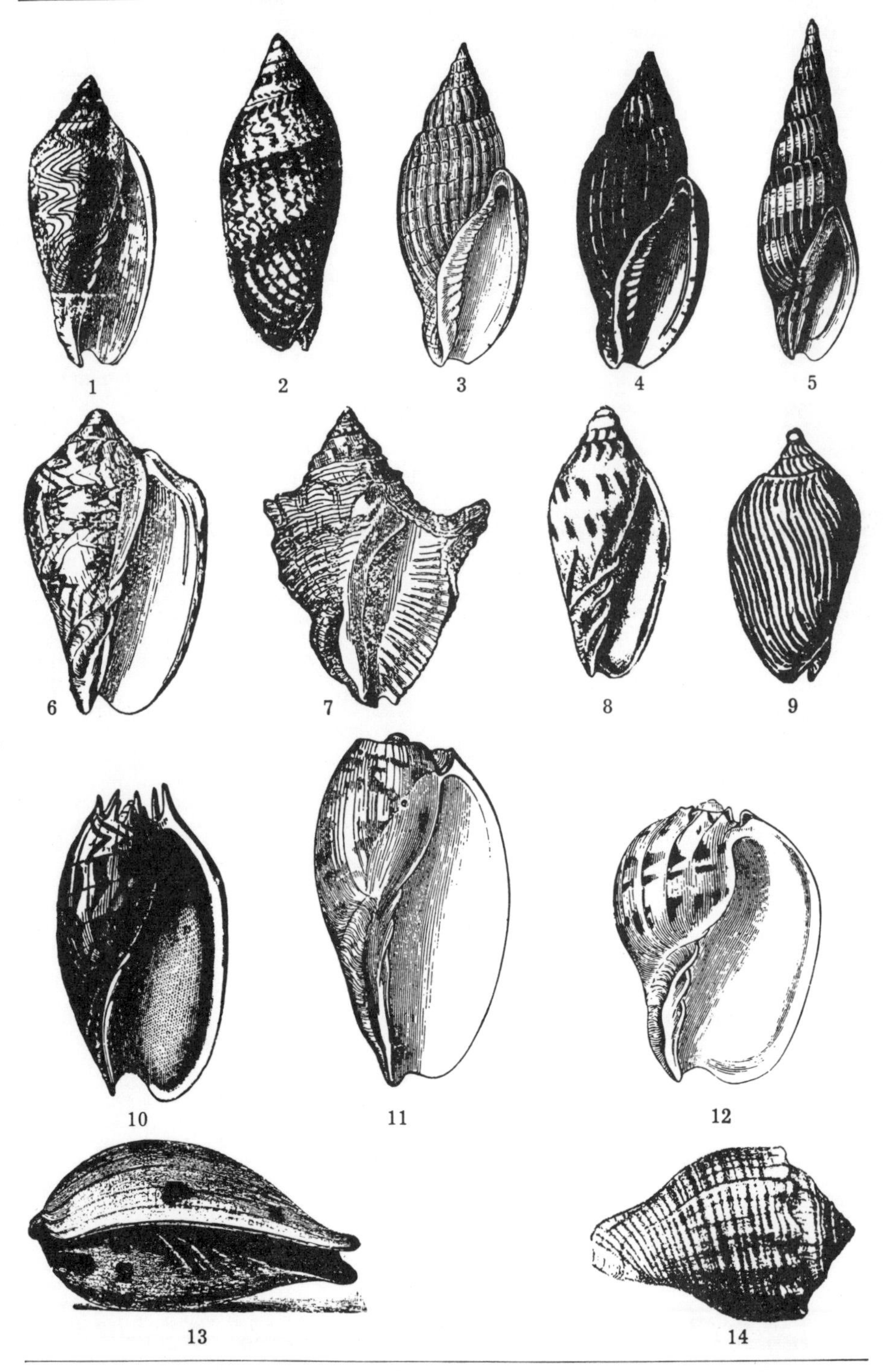

PLATE 31

1. **Voluta undulata angasi,** Sow. Port Lincoln, Australia. This species is similar to type but is adorned with a reddish background. In other respects it is about the same. 3″.

2. **Voluta maculata,** Swain (Caroli, Iredale) Queensland, Australia. This is a rich smooth shell mottled with light chestnut, which is more dense in two distinct bands in the middle of the last whorl. It has a very brilliant natural polish. 2½ to 3½″.

3. **Voluta mitraeformis,** Lam. South Australia. A small but pretty shell of light and dark brownish markings, with ridges the length of the shell. It comes from deep water and not very common. 2″.

4. **Voluta delessertiana,** Petit. Madagascar, north coast. A small 2″ shell with deep reddish ridges and faint white markings and lines. It belongs to the section **Lyria** all the species of which are similar in form.

5. **Voluta lyraeformis,** Brod. East Africa. A small 2″ slender shell with fine ridges and bands of white and brown. Rather rare and not often seen.

6. **Voluta scapha,** Gmel. Singapore on reefs. One of the most common shells of the genus to be shipped into this country commercially. It was always seen in curio stores and now is seen generally in real old homes. It has a natural polish, adorned with various patterns of brown markings and ranged from 2″ to 5″.

7. **Voluta hebraea,** L. Brazil to West Africa. A very fine and variable shell, somewhat allied to **musica.** The brownish patterns are very attractive. There is a finer illustration on plate 24. As it is not rare where found, one should have several specimens with as different pattern as possible.

8. **Voluta caroli,** Ire. Queensland, Australia. A shining, fulvous shell, with transverse bands of chestnut. Rare 2″.

9. **Voluta zebra,** Lam. Zebra Volute. East Coast. A small 2″ form well marked with regular chestnut stripes on a creamy background. There is a variety quite similar called **lineata, Leach.** From Tasmania. See plate 25.

10. **Melo diadema,** Lam. Indian Ocean. One of the largest of the 22 species in the genus. The natives call it a Bailer Shell, as they always carry one in their boats to bail out the water. Natives usually have a big one to carry water for domestic use. I have seen specimens that would hold a full pail of water. The Melos are ovoviviparous. Yellowish color with chestnut stripes. 6 to 10″.

11. **Melo aethiopicum,** L. Australia. Another Bailer of a yellowish color with very large aperture. This species is usually 6 to 8″ but may grow larger.

12. **Melo regia,** Schubert. Indian Ocean. One of the rarest of the Melo's and not often seen in collections. About 4″. Another species quite as pretty and of about same size is **broderipi, Gray.**

13. **Melo indicum,** Gmel. Singapore. In this species the apex of the shell is simply a crown showing the gradual divergence of the genus to the section which for generations has been called **Cymbium.** There are splashes of brown color on a yellowish background but specimens are often unmarked. 6 to 10″.

14. **Melongena paradisiaca,** Mart. Red Sea. A small 2″ shell, smooth and somewhat colored with yellowish shades. Very faintly resembles the Florida form that is occasionally found free from the spines on the crown. There are about 33 species of Melongena in the world and they vary greatly in form.

Mazatlan, West Mexico, has for generations been a favorite visiting place of shell collectors. The town is on a peninsula almost entirely surrounded by water and there are very nice hotels at present. The shore line for five miles from town presents very diversified collecting. To the west of the city the coast is rocky with considerable surf where you will find many Acmaeas, Chitons, Trochidae, Turbinidae, Thais, Murex, etc. To the south and north stretch long sandy beaches, which are covered with fine specimens especially after storms. In the inner bay which is lined with mud flats you will likely find the beautiful Telidora burnetti, Dentalium and Cardiums. Along the mangrove swamps are lots of Arca tuberculosa, Turritella goniostoma and other fine shells.

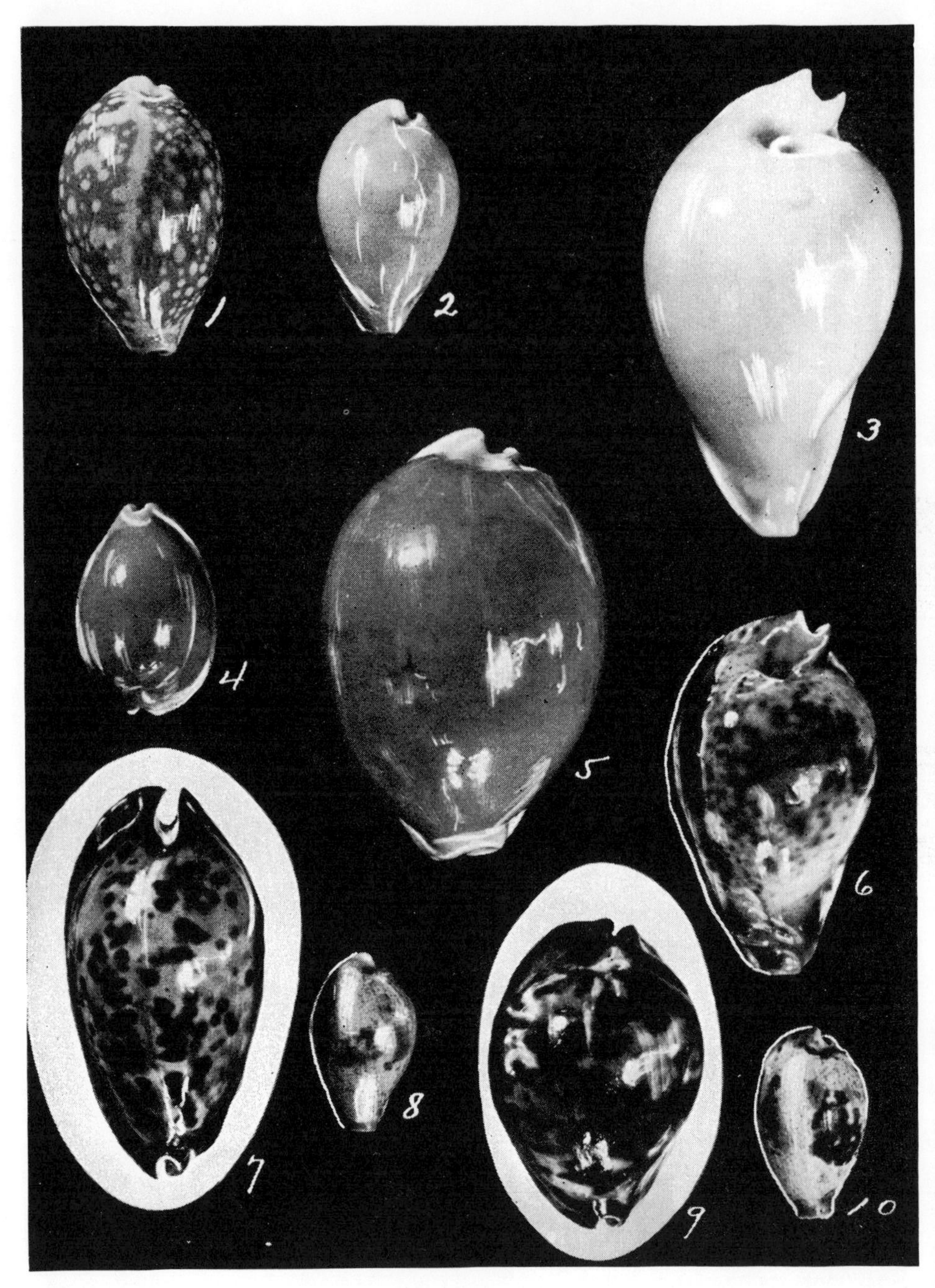

PLATE 32

1. **Cpyraea nivosa,** Brod. Ceylon. The base color is gray-brown with clear cut white eyes throughout. Superficially resembles **vitellus,** but entirely distinct. Base flesh-white. A very rare shell of which only a few specimens are known. 2¼".

2. **Cypraea onyx nympha,** Ducl. Ceylon. The shell is entirely creamy-white slightly darker at end. A very distinct variety of the type which ranges through both eastern oceans. 1¾".

3. **Cypraea howelli,** Iredale. Bass Straight, Tasmania at 75 fathoms. This shell appears to me a clear white species almost identical in form to **umbilicata** of the same territory. 3¾".

4. **Cypraea suldicidentata xanthochryma,** Melv. Hawaii. Similar to type but clear light yellowish throughout, very slightly darker at base. A grand shell as also is the type and only dredged in deep water. 1¾" .

5. **Cypraea aurantiam,** Mart. Fiji Ids. and nearby groups. Used to be called **auroa.** The shell is a rich yellow which it is impossible to show in cut. Base is pure white. Fresh specimens are extremely brilliant. Rather rare. 3¾".

6. **Cypraea hesitata,** Iredale. 50 fathoms off South Australia. I see no difference in this shell from the form umbilicata. It can only be obtained by dredging and as is usual with such forms is rather uncommon. 3".

7. **Cypraea scotti,** Brod. New Holland and other oceanica points. An elongated shell of flesh color ornamented with spotting of rich blackish-brown. Base is dark which extends up on the edges. There are deep sutures at each end. A rather rare shell. 3½".

8. **Cypraea subviridis,** Rve. Australia West. The shell is bluish-white with faint specks of brown and darker patches on top. Base is rounded flesh color. 1⅛".

9. **Cypraea thersites,** Gray. West Australia. The base is flat and dark black which color extends up on the edges. Top is mottled with russet-brown and white. A very rich shiny dark shell of brilliant pattern. Rather uncommon. 2½".

10. **Cypraea pulchella,** Lam. China. Broad teeth across base. Brownish rows of spots on bluish edge and large splashes of reddish-chestnut on top. Scarce. 1½".

The Cypraeidae consists of 200 species and varieties. They are usually called Cowries. Are found in both tropical and sub-tropical seas, often under rocks at low water. There are 100 or more fossil species from crataceous on. From the pliocene beds along the Caloosahatchie river in Florida I have found remarkable fossil forms with real nacre but no such form exists in the Gulf of Mexico today as far as is known.

The very young shell has a sharp lip like a Bulla and often covered with a thin epidermis. When the shell is full grown the mantel-lobes extend up each side, and deposit a shining enamel over the whole shell, and the spire is then entirely concealed. You will usually see a line lengthwise of top of shell showing where the mantel meets.

The shell is built up by various layers of nacre, often a dozen or more and the later layers are of a different color and often pattern than the fully adult shell, showing that when the last layer is deposited the final pattern is complete. These immature shells often are very confusing to collectors.

There is a great difference in size of fully matured shells of the same species, the same as in the human family. The habits of the shell is shy and they crawl around slowly. They will glide along among the coral reefs, with the lateral lobes of their mantle adorned with showy colors, often more beautiful than the adult shell itself.

When you discover a Cypraea in its natural habitat you often do not see the shell at all, simply the mollusk completely enveloping its home; but when disturbed it quickly withdraws into its shell. It has no operculum to help defend itself from its enemies.

The Family includes the Ovula (Now called Amphiperas) Cypraeovula (Only one species from South Africa) Cypraedia (One small species) Trivia (Many very small species of various colors and usually deeply ridged) Pustularia (About 10 varieties some smooth and some ridged) Erato, (All small forms and usually white).

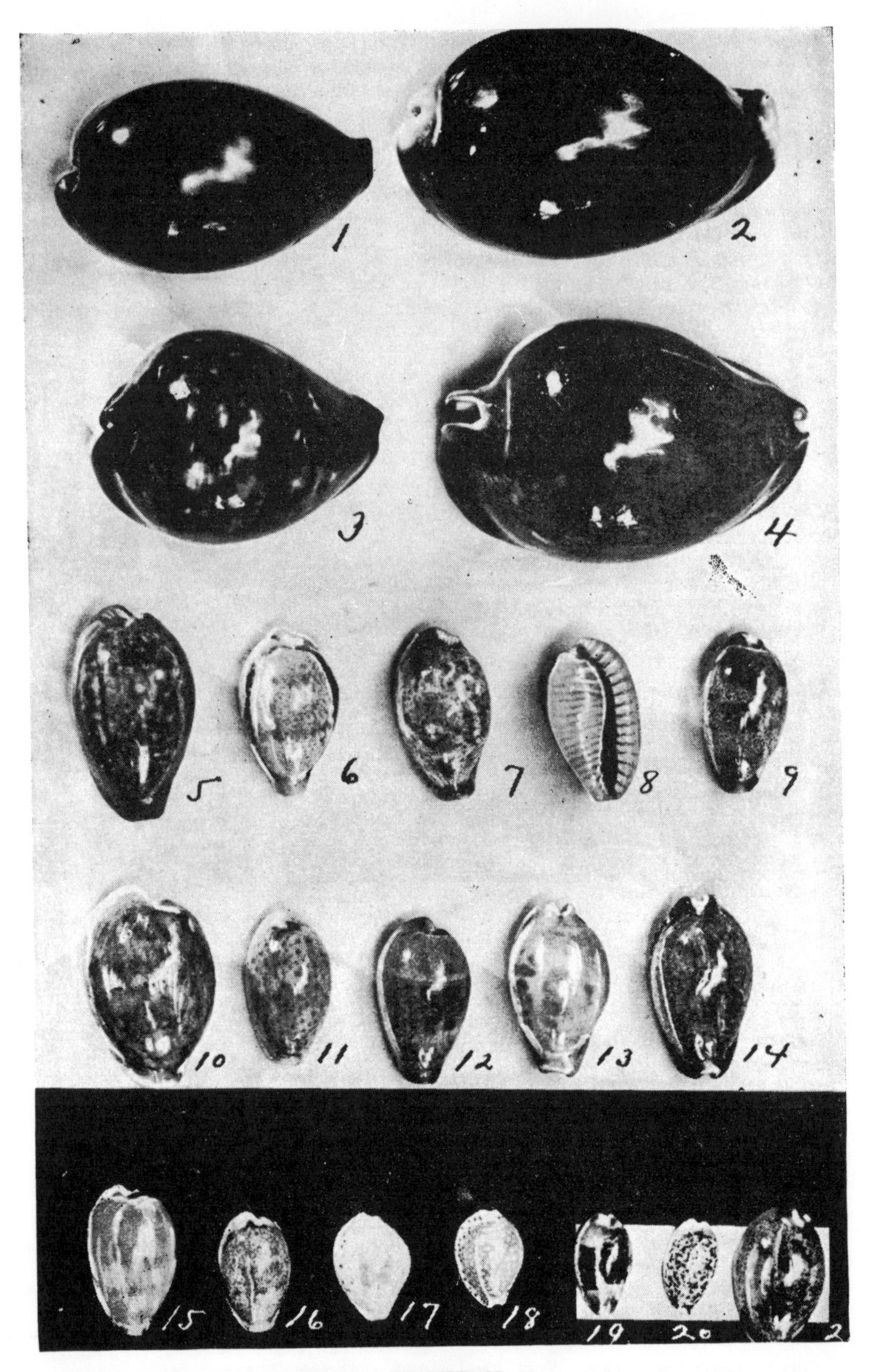

PLATE 33

1. **Cypraea exusta,** Sow. Red Sea. Fresh specimens are a rich shade of glossy brown throughout. A rather rare shell only occasionally seen in extensive collections. 2¼″.

2. **Cypraea pantherina theriaca,** Melv. New Caledonia. This a form of the type in which the brilliant spotting of the shell is completely covered with a dark coating of red enamel. Other unusual coloring in cowries are found in this great island group now under French protectorate. 2″.

3. **Cypraea decipiens,** Smith. North Australia. Base and sides are a rich shining black. A brilliant shell on the order of the larger **thersites** but more thick and chubby. 2″.

4. **Cypraea thersites,** Gray. West Australia. This is simply a form of this shell which is covered with an intense black coating of enamel which completely covers the usual color pattern. It does not show up well in the cut. There are also jet black forms of No. 2 on this plate. 2¾″.

5. **Cypraea pyrum,** L. Algiers. Base is reddish-chestnut. Top is of a lighter color with numerous spots of white and light brown. 1½″.

6. **Cypraea hungerfordi,** Sow. Japan (Kliiensis). Base is creamy-white also edges. Just above the edges is a very dark almost black line and above that faintly mottled with brown and white. 1″.

7. **Cypraea lentiginosus,** Gray. Ceylon. The numerous brownish dots around the edge extend over on part of the bottom. Top is mottled with reddish-brown. Dark blotch at lower end. 1¼″.

8. **Cypraea fuscodentata,** Gray. Cape of Good Hope. The shell is of a drab color throughout, the most prominent feature being the teeth which extend across the base, as shown in cut. Very choice fresh specimens seem to be fairly rare, indicating it may be a deep water shell. 1¼″.

9. **Cypraea cylindrica,** Born. Philippines. The base is flesh color. Top is mottled with bluish-gray, with few dark blotches of darker color. Subcylindrica Sow. from North Australia is similar with two dark spots at each end. 1¼″.

10. **Cypraea fuscorubra,** Shaw. South Africa. A rather thin bulbous shell. Parietal teeth faint, deeper on outer edge. Top curiously mottled with reddish-chestnut. A rather rare shell in American collections and seldom seen. 1½″.

11. **Cypraea boivini,** Kien. Philippines. The shell is bluish-gray covered with numerous small brown dots. Base white. Not common. 1⅛″.

12. **Cypraea walkeri,** Gray. Bohol Id., Philippines. The lower edge is a rich lavender with dots of darker color. Top light brown with band through middle of wide darker blotches. A rather distinct distinct rare shell. 1″.

13. **Cypraea bicallosa,** Gray. St. Vincent Id., West Indies. The base is reddish-chestnut, teeth at each end unusually sharp and deep. Top is mottled with chestnut. A rare shell seldon seen. I have in fifty years only owned two specimens both of which were sold at a high figure for so small a shell. 1¼″.

14. **Cypraea nigropunctata,** Gray. Galapagos Ids. Base light. Teeth with deep edge. Margin mottled with dark spots almost black, top is mottled with reddish-brown. A rather rare shell in fine condition and hard to secure as there are no shell collectors on this island. 1¼″.

15. **Cypraea angustata piperita,** Sow. New South Wales. The edge of base has a row of fine brownish dots. Top light chestnut throughout showing three faint bands. Very choice shells of the **Angustata** complex seem to be rather scarce in USA collections. 1″.

16. **Cypraea bregeriana,** Crosse. New Caledonia. Base brown with prominent teeth. Side flesh color. Top mottled with light brown. Rather rare. 1″.

17. **Cypraea rashleighiana,** Melv. Hawaii. A small broad oval shell with white base and few chestnut dots on top. Rather scarce and likely from deep water.

18. **Cypraea gaskoini,** Rve. Paumotus Ids. Base white, edges dotted with reddish-brown. Top covered with small white dots on a yellowish background.

19. **Cypraea erythraensis,** Bk. Oceanica. Teeth extend over much of base. Top bluish-white with rich heavy reddish-brown patches.

20. **Cypraea crossei,** Cox. British Solomons. Teeth sharp and broad for size of shell, sides flesh-white, top mottled with reddish-chestnut. A slender rather elongated fine rare shall I had sent in recent from above locality.

21. **Cypraea pallida,** Gray. Kurachi, India. Base is flesh colored with brown top, mottled with chestnut.

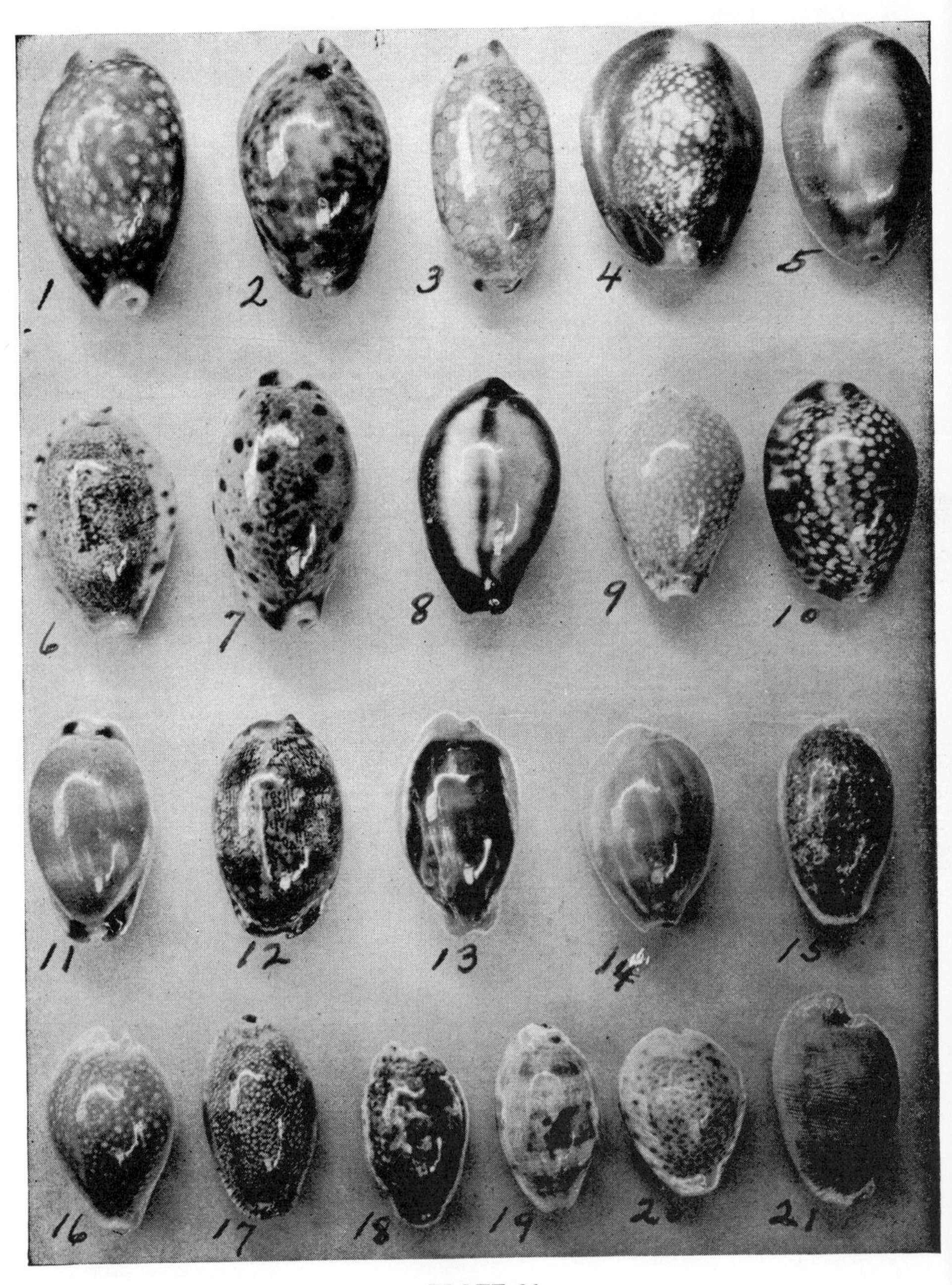

PLATE 34

1. **Cypraea vitellus,** L. Calf Cowry, Philippines. It Is of a light shade of brown with white spots. 1½ to 2½".

2. **Cypraea stercoraria,** L. African Cowry, West Africa. Ground color of a pale bluish covered with irregular brownish spots. Small black patch on one end. 1½ to 2½".

3. **Cypraea scurra,** L. Mouse Cowry, Pacific generally. Back is mottled with white spots on a light brownish background, base covered with dark spots. 1¼ to 2".

4. **Cypraea caput-serpentis,** L. Snake head Cowry, Indian Ocean and Pacific. Of a light brown color, back has numerous white blotches, base uncolored. 1½".

5. **Cypraea ventriculus,** Lam. Pacific. Base white, sides deep brown, top is white and russet color, 1¼ to 2".

6. **Cypraea caurica cairnsiana,** Melv. Ceylon. Base ridged and uncolored, top mottled with shades of brown, the flaring edges have dark spots. 1¼".

7. **Cypraea lynx,** L. Lynx Cowry, Pacific and Indian Oceans. Base white, balance of shell richly spotted, like the animal it is named after. Some of these very common cowries have found their way most all over the world. 1¼ to 2".

8. **Cypraea onyx,** L. Onyx Cowry, Philippines. Base and sides a rich black. Top bluish-white. A very rich looking shell. There is a variety called **adusta** that is entirely brown and another variety **nympha** of rich cream color. They all average 1½".

9. **Cypraea lamarckii,** Gmel. Madagascar. Base white, sides have reddish spots, and top is entirely covered with gray spots. 1½".

10. **Cypraea arabica intermedia,** Gray, Philippines. This variety can always be distinguished from the reticulata by its smaller size and white base. 1¼".

11. **Cypraea lurida,** L. Italian Cowry, Naples, Italy. A light brownish shell, with two jet black spots on each end. 1½".

12. **Cypraea arabica,** L. Arabian Cowry, Philippines. Base and sides are often brilliantly spotted and back is lined with brown on bluish background. This shell has three distinct varieties all of which are fairly common shells. 1½ to 2".

13. **Cypraea spadicea,** Sow. The Chestnut Cowry, California. Base and sides a rich bluish-white and top is a glistening shade of brown. 1½".

14. **Cypraea carneola,** L. Yellow Cowry, Pacific. Base and sides are shades of gray, top is reddish-yellow with bars. 1¼ to 3".

15. **Cypraea errones,** L. Pacific. Base and sides white to yellowish. Top mottled with shades of brown. 1¼".

16. **Cypraea miliaris,** Gmel. Pacafic. Base and sides pure white, top mottled with small round white dots. 1½".

17. **Cypraea erosa,** Lam. Pacific. Base mostly white, sides have a black patch on each side, top is mottled with white dots and occasional blotches of brown. 1¼".

18. **Cypraea picta,** Gray. Cape Verde Islands. Also illustrated and described on plate 35.

19. **Cypraea tabescens,** Dill. Mauritius. Base white and top is marked irregularly with brown. Shell is somewhat pointed at each end. 1¼".

20. **Cypraea turdus,** Lam. Thrush Cowry, Persian Gulf. Base pure white, and top is mottled like the breast of a thrush, hence the name. 1¼".

21. **Cypraeovula capensis,** Gray. Cape Cowry, South Africa. There is only the one species which is completely lined base and top with circular ridges. As all the Cypraea are smooth this variety tends to merge with the Trivias so it is in a genus by itself. 1¼".

The Pyramidellidae consists of Pyramidella (Used to be called Obeliscus) Otopleura, Syrnola, Odostomia, Eulimella and a lot of other genera covering many hundred species from all over the world. The Pyramidella and Otoplura are handsome shells of one inch or less but many of the other genera run down to very small shells usually white. There are many fossil forms. One has to be a real enthusiast to love the tiny shells but there is no more fascinating collection than three or four thousand vials filled with all classes of very small adult shells.

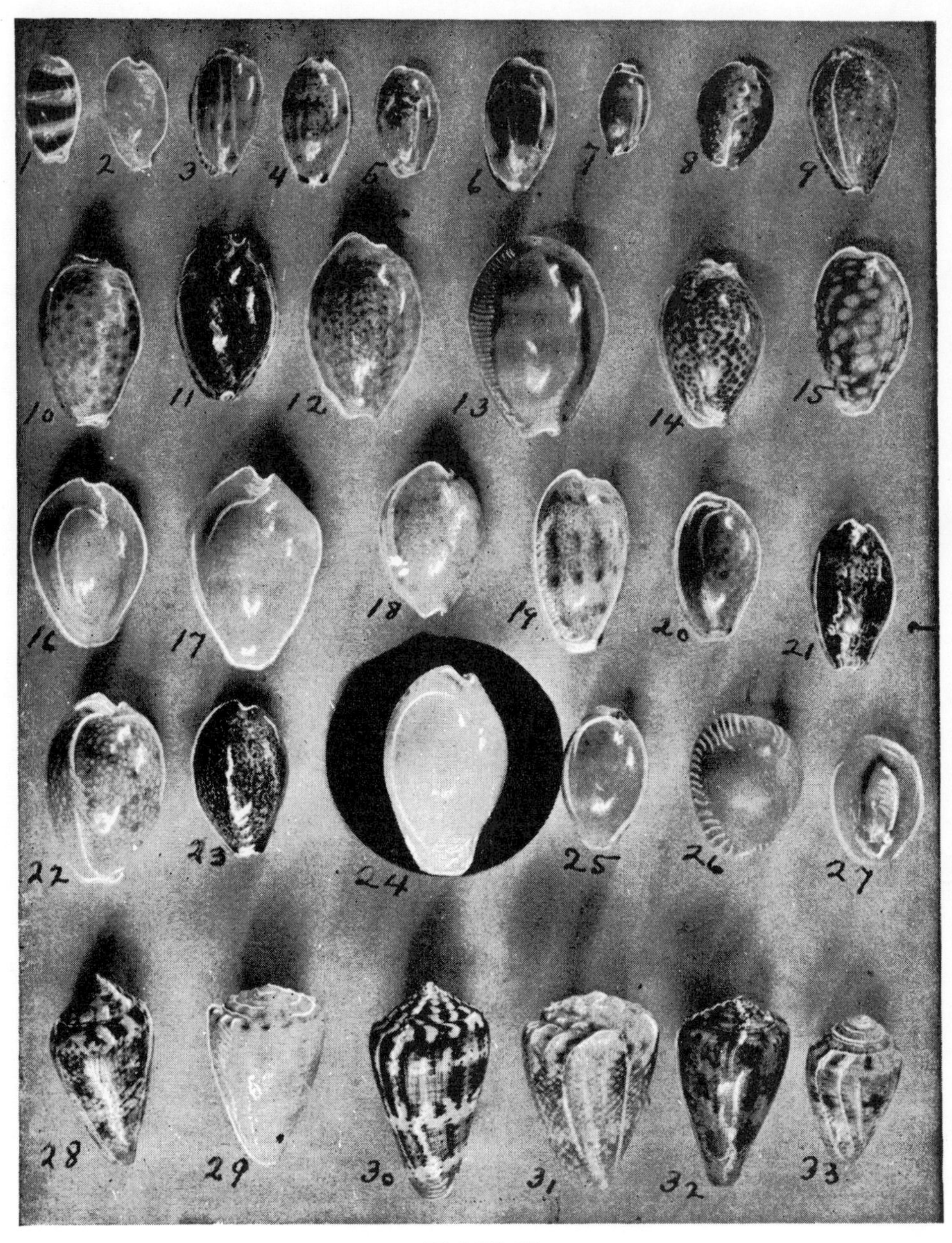
1
2
3
4
5
6
7
8
9
10
11
12
13
14
15
16
17
18
19
20
21
22
23
24
25
26
27
28
29
30
31
32
33

PLATE 35

1. **Cypraea asellus,** L. Philippines. A small white shell with usually three broad brown bands across its back.

2. **Cypraea clandestina,** L. Ceylon. A small white shell not usually colored in any way.

3. **Cypraea angustata comptoni,** Gray. New Holland. A small light brownish shell with spots along the base of the shell and usually three indistinct bands across the back.

4. **Cypraea gangrenosa,** Dill. China. A small gray shell with white dots, few brown blotches and two small dark spots at each end.

5. **Cypraea macula,** A.Ad. (notata Gill) Japan. A small bluish shell with few brown splashes on top, small edge occasionally extended over the bottom.

6. **Cypraea hirundo,** L. Philippines. A bluish-brown shell with splashes of yellowish-white, dark spots on each lower edge with base usually white.

7. **Cypraea neglecta,** Sow. Mauritius. A small bluish-gray shell with few white blotches, dark spot in umbilical region, two at other end, and teeth extending mostly over the base.

8. **Cypraea helvola argella,** Melv. Marquesas. A true little broad shaped **helvola** but the back is mottled with bluish-gray edges with brown and few brown spots over the surface. A real little gem of a variety.

9. **Cypraea ocellata,** L. Ceylon. A small shell with yellowish surface completely covered with white and brown spots, like a real ocelet. Base also spotted.

10. **Cypraea helvola,** L. Ceylon. Base of shell reddish-brown, bluish-gray above mottled with brown spots and much smaller white dots. There have been a number of color forms found, some with an even color and unspotted.

11. **Cypraea arabicula,** Lam. Panama. The black is bluish, with irregular workings of deep brown, around the lower edge a richer shade of brown covered with black spots. The base is rather flat and the teeth much finer than in No. 23 on this plate, with which it is often found associated on the beaches. About 1″.

12. **Cypraea cruenta,** Gmel. Ceylon. The top is yellowish with reddish-brown minute blotches. Below all around it is white with reddish spots which extend over the base. 1″ or larger.

13. **Cypraea arenosa,** Gray. Anna Ids. Base white, sides very light brown, top yellowish with four indistinct brownish bars. 1¼″.

14. **Cypraea albuginosa,** Gray. Gulf of California. Base white, sides bluish, top is yellow, completely covered with small brown dots. 1″.

15. **Cypraea cribraria,** L. Mauritius. Base white. Sides and top yellow covered with small round white spots.

16. **Cypraea annulus,** L. Ring-top Cowry, Philippines. A white shell shading to gray with an orange stripe around the top. Called the Ring-top Cowry by sailors and collectors alike, and has been an article of commerce for generations, where it is used in the manufacture of novelties.

17. **Cypraea moneta,** L. Ceyon. The Money Cowry. A yellow shell which varies from smooth to knobby and occasionally specimens are found with a ring top like the preceding species. Has been an article of commerce for many generations where it is used in Africa as money by the native peoples and in this country in the manufacture of novelties. But most native people of the world now have modern copper or nickel coins supplied by the countries who govern them.

18. **Cypraea cruenta coloba,** Melv. Ceylon. Exactly like the type No. 12 but is more flat and less oblong. Has been recognized as a variety for many years, whether it is or not.

19. **Cypraea tabescens latior,** Melv. Mauritius. White below, the top is mottled with light brown, usually lighter color than type. 1″.

20. **Cypraea flaveola,** L. Lifu Ids. The base is pure white, upper parts gray with numerous small spots of brown.

21. **Cypraea picta,** Gray. Cape Verde Islands and Gambia. The top is mottled with white and brown, shading to a rosy color below which has a row of spots thru same. Base slightly shaded.

22. **Cypraea listeri,** Gray. Philippines. Upper part grayish-yellow with numerous white dots and few larger spots of pink. Sides spotted with pink which extend over on the base.

23. **Cypraea puncticulata,** Gray. Panama. Top marked with irregular brown on a bluish base, sides rosy-red surrounded with numerous black spots.

24. **Cypraea eburnea,** Barnes. White Cowry, Philippines. The only pure white shell of the genus which attains 1½″.

25. **Cypraea cinerea,** Gmel. West Indies. The top is light brown with a couple indistinct bands, sides lighter colored with some brown dots on same or plain diffused color. 1″.

26. **Trivia ovulata,** Lam. Australia. A very large variety of a genus in which nearly all the shells are small. Above, it is pink, below white, with ridges over the base and extending over the sides. Top smooth. About 1″.

27. **Cypraea obvelata,** Lam. Tahiti. A white shell in which the top is bluish, with orange ring around same, and depressed almost even with the sides. A rather peculiar shell for a genus which always runs fairly true to type as regards shape.

28. **Conus articulatus,** Sow. Mauritius. A shell mottled with white and shades of brown, top conical and also mottled. 1 to 2″.

29. **Conus miliaris,** Hwass. Viti Ids. A white shell finely marked with light russet and lined with darker lines. It also comes in other shades of color and is usually about 1″.

30. **Conus vittatus,** Lam. Pamama. A small shell richly marked with white, brown and black. The upper part shows the lines of each whorl mottled with black and white in regular stripes 1 to 1½″.

31. **Conus cinctus,** Swain. Mauritius. The entire shell is covered with white and reddish-russet markings. It also varies greatly in these markings. 1″ or a little larger.

32. **Conus purpurascens,** Brod. Panama. The entire shell is covered with reddish-brown and blue shades of color, in the form of blotches. Top the same 1½ to 2½″.

33. **Conus alveolus,** Sow. Moluccas. A small bluish-gray shell with brown marks in regular bandings. Top the same. About 1″.

PLATE 36

1. **Cypraea camelopardalis,** Perry. Red Sea. Uniformly gray throughout with very faint spotting. Base and edges white. Teeth reddish. 2″.

2. **Cypraea reevei,** Gray. New Holland. A faint shade of brown, teeth white, both ends of aperture pink. Rather globose shell and fairly thin. 1½″.

3. **Cypraea caput-serpentis,** L. Peru. This is an undescribed very dark brilliant variety, of which I have seen no similar examples from any other part of the world. See plate 34.

4. **Cypraea sulcidentata,** Gray. Hawaii. A shell of deep water always. Fresh dredged specimens are rosy above, lighter around the base and the teeth are deeply beveled which will always distinguish the shell, as it has no near relatives in that respect. 1¾″.

5. **Cypraea cervinetta,** Kiener. Panama on Pacific side only. Very closely allied to similar specimens from the Florida region. The teeth are more deeply brown color. White spots on a brownish background. 2 to 3″.

6. **Cypraea erosa phagedaina,** Melv. Mauritius. This form is completely covered with minute white dots on a drab background. Base light with reddish streaks on teeth. 1 to 1¾″.

7. **Cypraea sowerbyi,** Kien (annettae, Dall) Gulf of California. The entire shell is deeply splashed with blotches of reddish-brown. Base lighter and not spotted. Edge has small spots. 1¼″.

8. **Cypraea caurica,** L. Indian and Pacific Oceans generally. Very common everywhere. Back is light brown, edges strongly developed and base teeth ridges. There are varieties. ¼″.

9. **Cypraea onyx adusta,** Chem. Ceylon and likely Philippines and elsewhere. A variety that is uniformly brown, usually with two lighter bands across top. Base reddish-chestnut. 1¼″.

10. **Cypraea caurica oblongata,** Melv. New Caledonia. Simply a very large elongated variety from this locality and likely found generally over Pacific in various localities. Mr. Melville described many of the unusual varieties of this genus. 1½″.

11. **Cypraea moneta icterina,** Lam. Indian and Pacific oceans generally. The form is usually more slender than the type and uniformly yellow. 1″.

12. **Cypraea moneta ethnographica,** Roch. Pacific generally. A nobby variety, some specimens being very unusually so. You will find very small specimens that seem to be fully adult.

13. **Cypraea cinera clara,** Gask. Differ from the type in usually more rounded and not so long. Flesh color with three bands.

14. **Cypraea zonata,** Chem. Corsico, West Africa. The back is splashed with brown on a lighter background. Usually an indistinct wide band across top. Spotted edges, base light flesh. 1″.

15. **Cypraea errones sophiae,** Brazier. Australia. Very similar to the parent shell, but usually larger with very prominent wide yellowish-white edges and flesh colored base. 1½″.

16. **Cypraea similis,** Gray. So. Africa. This whitish shell is rather rare as most specimens I have seen were collected in a dead state. The shell is smooth but base is finely ridged, teeth strong. 1¼″.

17. **Cypraea physis,** Broch (achatidea, Gray) Deep water off Sicily. A rather fat rounded shell, with chestnut markings on top, sides and base flesh-white. Teeth very minute or entirely lacking. Rare 1″.

18. **Cypraea moneta mercatoria,** Roch. Pacific generally. A rather broad yellow form, with the ringed top, similar to annulus.

A visit to LaPaz, Lower California, is always a great treat. You reach it usually by steamer across the Gulf of Mazatlan, West Mexico. The town has long been famous for the fisheries of black pearls, produced by Margaritiphora mazatlanica. The town faces the west on a beautiful landlocked bay. In every direction from town there is a wide expanse of sandy beaches and mud flats with occasional sand bars only exposed at very low tides. Lots of Pinna rugosa are found here. You will also find Architectonica granulatum, Dentalium splendidulum, Lyria cumingi and Cardium aspersum. You will likely find many choice Murex princeps and Conus princeps.

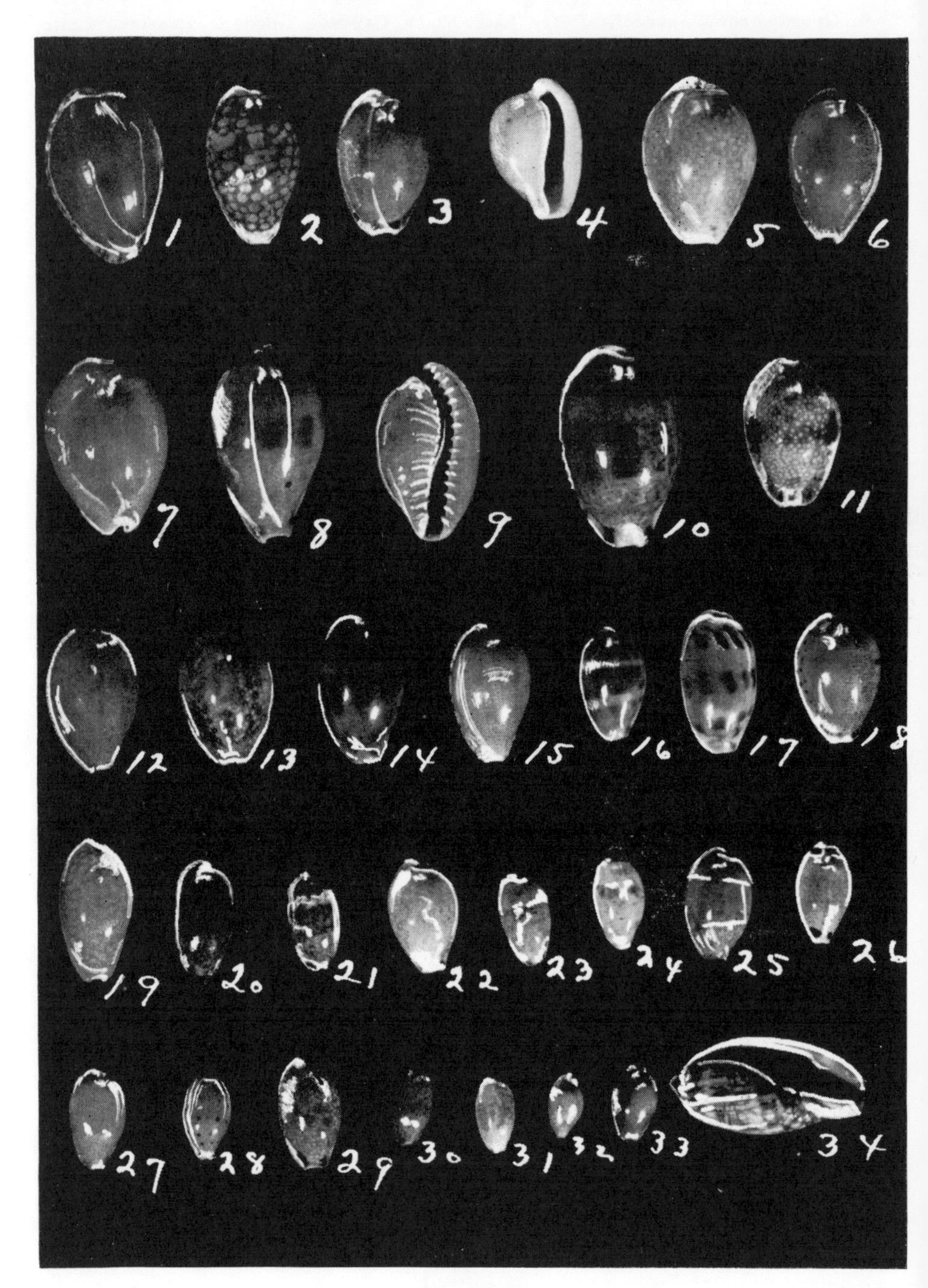
1
2
3
4
5
6
7
8
9
10
11
12
13
14
15
16
17
18
19
20
21
22
23
24
25
26
27
28
29
30
31
32
33
34

PLATE 37

1. **Cypraea angustata,** Gmel. New Holland. Uniformly light brown above with numerous spots around edge, base white.

2. **Cypraea esentropia,** Ducl. Philippines. Completely covered with white spots on a yellowish background. Edge and base white.

3. **Cypraea spurca,** L. Cape Verde Ids. and West Indies. A yellowish shell with fine dots, row of brown spots around edge, base white.

4. **Cypraea edentula,** Sow. S. E. Africa. Rather thin and bulbous, top flat prominent rounded lips, with no teeth. This latter feature will always distinguish same.

5. **Cypraea cernica,** Sow. Mauritius. Rather deeper yellow than **spurca** with which it is often confounded. Numerous fine white spots, base rounded and thick, white and very strong teeth.

6. **Cypraea helvola hawaiiensis,** Melv. Hawaii. A brilliantly smooth polished shell of uniform light brown, base lighter and wide teeth. A very good variety.

7. **Cypraea carneola propinqua,** Garret. East Indies. Always smaller than type, strong, the light brown top ringed with purple and white. Teeth purple as usual. The type is found up to 2" or more but this variety usually around 1".

8. **Cypraea pyriformis,** Gray. Ceylon. Rather closely allied to **pulchella** being pyriform, with two faint bands across top, umbilicated, teeth reddish and thin. Very rare.

9. **Cypraea fuscodentata,** Gray. West Africa. This shell is illustrated on another plate.

10. **Cypraea xanthodon,** Gray. Queensland. Back is reddish-chestnut, sides light with brown spots. Base light. Not common.

11. **Cypraea erosa nebrites,** Melv. Andaman Ids. and likely elsewhere. A typical small erosa with very dark reddish-black blotches, brilliant dots and base covered with dots.

12. **Cypraea declivis,** Sow. Tasmania. Uniformly light brown with faint markings, edge sparsely spotted, base light.

13. **Cypraea poraria,** L. Tahiti and Central Pacific. Pink with numerous ocellated brown spots, edge pink. Quite distinct.

14. **Cypraea sanguinolenta,** Gmel. Gambia. Light brown with wide band of same, edged deeply spotted with brown, base flesh-colored. Scarce.

15. **Cypraea nebulosa,** Kien. Gambia. Flesh color, with faint spots around edge and white base.

16. **Cypraea walkeri rossiteri,** Dautz. New Caledonia. A fine colored shell of light brown, two white bands edged with spots. Base light. Rare.

17. **Cypraea interrupta,** Gmel. Ceylon. Light bluish with 3 rows of brown blotches across back. Umbilicated, base white, fine teeth.

18. **Cypraea angustata bicolor,** Gask. Victoria, Australia. Brilliant shiny brown back with numerous spots around edge, base flesh color. Usually smaller than type.

19. **Cypraea peasei.** Sow. Mauritius. Entirely covered with white spots on a yellow background. Edges white, as is base. Teeth most prominent on lip.

20. **Cypraea felina,** Gmel. Indian Ocean generally. A narrow darkly mottled shell with very promient spotted edges which extend over part of the base.

21. **Cypraea felina ursellus,** Gmel. Viti Ids. Similar to type but usually smaller and lighter colored, teeth rather more prominent.

22. **Cypraea zigzag,** L. Mozambique. A rounded small white shell with zigzag markings over the top, prominently spotted around the edge with yellow base.

23. **Cypraea fimbriata,** Gmel. Mauritius. A small light colored shell with faint markings, prominent spots at each end, base white.

24. **Cypraea microdon,** Gray. Philippines. A light colored shell with faint markings over back and not spotted around edge. White base.

25. **Cypraea lutea,** Gron. New Caledonia. North Australia. Light brown above with two tiny white bands, edge and flesh-colored base spotted with dots of brown.

26. **Cypraea quadrimaculata,** Gray. Mauritius. Back gray, with two prominent black spots at each end, base white and ridged by extended teeth.

27. **Cypraea oweni,** Sow. Philippines. Back flesh color, edged with reddish spots. Base white.

28. **Cypraea punctata,** L. Philippines. A small white shell with numerous small brown dots over surface. Base white.

29. **Cypraea flaveola labrolineata,** Gask. Japan. Back is covered with white dots on a gray background, ends with smudge of brown, base white.

30. **Cypraea irrorata,** Sol. Navigator Ids. A very small finely dotted shell, reddish dots along edges, base uncolored.

31. **Cypraea goodalli,** Gray. Lord Hood Ids. Central Pacific. Very small shell with diffused brown markings, base white.

32. **Cypraedia adamsoni,** Gray. Philippines. A very small finely ridged shell. The only species in the genus. Somewhat resembles a **trivia** but ends are more elongated, the aperture is deeply curved with fine teeth. Rather scarce.

33. **Cypraea semiplota,** Migh. Hawaii. A small light brownish shell with white edges and base. Much resembles some of the forms of **Pustularia staphylaea** from same locality.

34. A young **Cypraea tigris** in the bulla stage. Most of the larger cowries are of this form when very young. A collection of the young in different stages of the great Cypraea genus are very interesting.

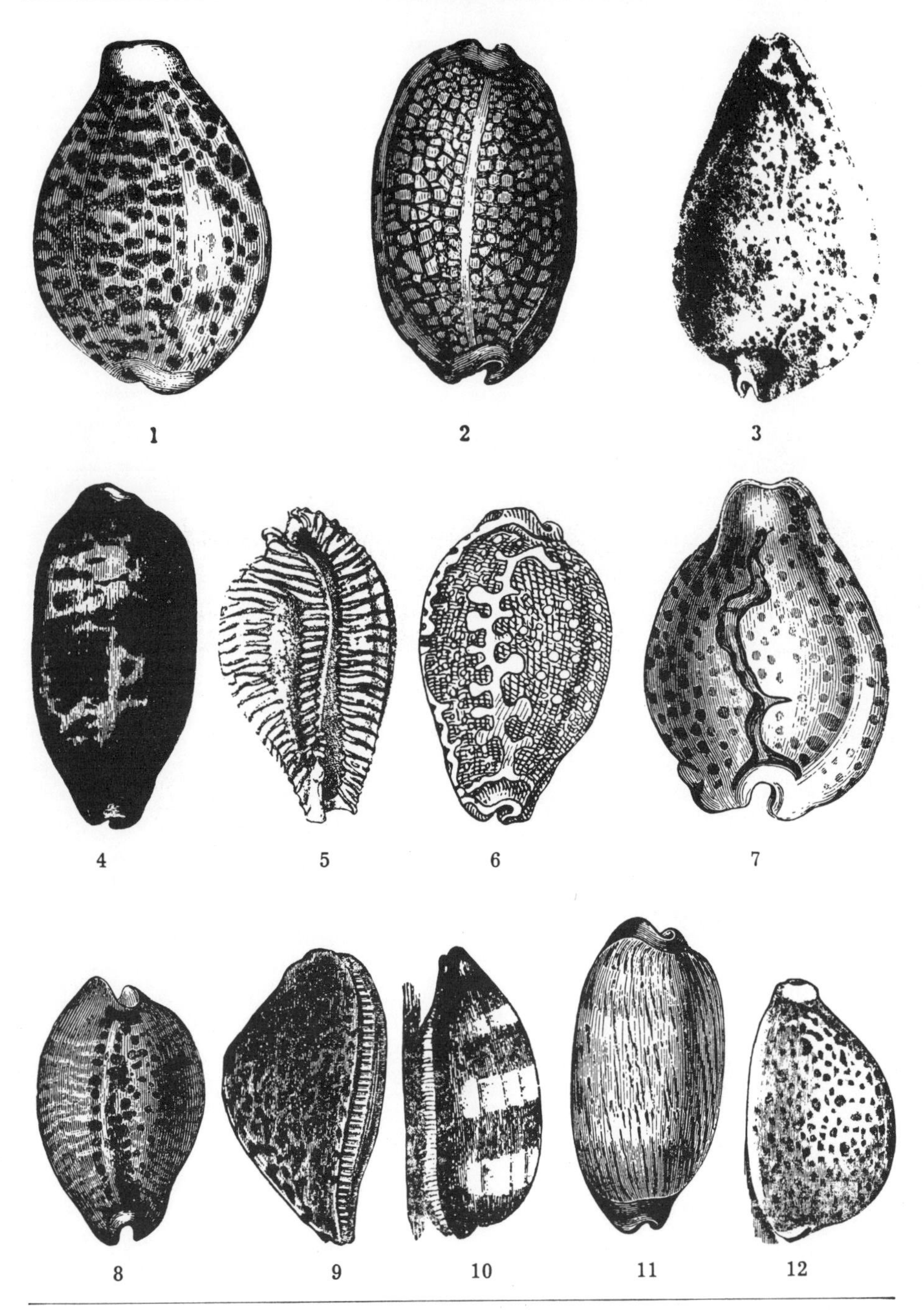

PLATE 38

1. **Cypraea tigris,** L. The Tiger Cowry. All over the Pacific. A large fine shell which varies greatly in spotted pattern and also in color. There are forms with a yellow, whitish and reddish background but usually rare. You will also see in old collections specimens with Lords Prayer on back or Souvenir of some place. You will find specimens in almost every store in the world where shells are sold. Like all other Cypraea some forms have no spots and these are usually immature. 3 to 3½".

2. **Cypraea arabica reticulata,** Mart. The Reticulated Cowry. Fairly common in Hawaii and Pacific generally. The white markings of the back are typical, the flaring base with black spots and dark splash of color in middle of base. There is a variety **histrio** that is somewhat similar but the base is more narrow and pure white with no blotch. The spots on top are more distinct. 2 to 2½".

3. **Cypraea umbilicata,** Sow. South Australia and Tasmania. Usually dredged in 10 fathoms or more. A fine large shell of yellowish pattern, with a slight hollow in umbilical region. Has always been considered rather rare. 3 to 3½".

4. **Cypraea testudinaria,** L. Madagascar, East Africa and other parts of Pacific. I have had numerous specimens from Japan area. Of a peculiar dark pattern, covered with minute white dots that look like dust. For this reason the shell is never as brilliant as other forms. No other Cypraea has a similar appearance. 3 to 5".

5. **Cypraea guttata,** Gray. Admiralty Group, Pacific. It is a well spotted yellowing shell and can always be instantly recognized as the teeth extend clear across the base of the shell. It has been reported from various localities in the Indian and Pacific Oceans. The latest record is one found in the above group of islands during the war. It is still very rare but may turn up fairly common at any time. 2½".

6. **Cypraea mappa,** L. Map Cowry. Philippines and likely all over the Pacific. It is a fairly common shell with a brilliant design on back, the cut showing white patches where the two edges of the mantle of the mollusk meet. There is a variety **panerthyra, Melv.** with a rich pink base and a variety **subsignata, Melv.** with violet-purple base. This type shell always has white base. The above two varieties are rather scarce also a variety that is richly adorned with red. I saw one collection recently where the owner had displayed 80 shells most all of which differed in some respect. You must have a large series of some form of Cypraea to make a real display.

7. **Cypraea leucostoma,** Gray. Persian Gulf. Of a grayish color with faint chestnut markings. The pattern of the top differs from all other shells I have seen. A fairly rare shell which brings a good price.

8. **Cypraea mus,** L. Mouse Cowry, Medititerranean Sea. Is of a light brownish color with rows of dots on top where the two edges of the mantle meet. There is a curious variety from Africa called **bicornis,** with two small humps on back near widest end. 1½ to 2".

9. **Cypraea mauritiana,** L. Mourning Cowry. Mauritius and elsewhere. Of a rich dark color, there is much variety of pattern, some specimens being almost black. The immature shell is likely to be rich brown with no pattern whatever, the teeth will also be immature and only show slightly. 2 to 4".

10. **Cypraea talpa,** L. Mole Cowry. Philippines and elsewhere in both Pacific and Indian Oceans. A brilliant shell when first collected as are most all of this genus and they fade if constantly exposed to light. Of a rich varying shade of brown with distinct bands. Some of the largest and finest shells I have seen come from Madagascar. 2 to 3".

11. **Cypraea isabella,** L. Pacific generally. Of a grayish color with reddish tips. There is a variety **controversa** from New Caledonia and a variety **Limpida,** Melv. from Hawaii. Also a variety **Mexicana,** Stearns from West Mexico. There is little difference in the three mainly shades of color, doubtless due to different temperatures of water and feeding conditions. 1 to 2".

12. **Cypraea pantherina,** Sol. The Panther Cowry. Indian Ocean, Red Sea and elsewhere. A very brilliant spotted shell of which the cut only gives a faint idea. There is a variety **albonitens, Melv.** from Persian Gulf that is of a reddish color, a variety **obtusa,** Perry from Mauritius that is somewhat similar, a variety **therica** which is illustrated on another plate and variety **Syringa, Melv.** Red Sea that is almost all white. A large series of this shell from many localities is always desirable. 2 to 2½".

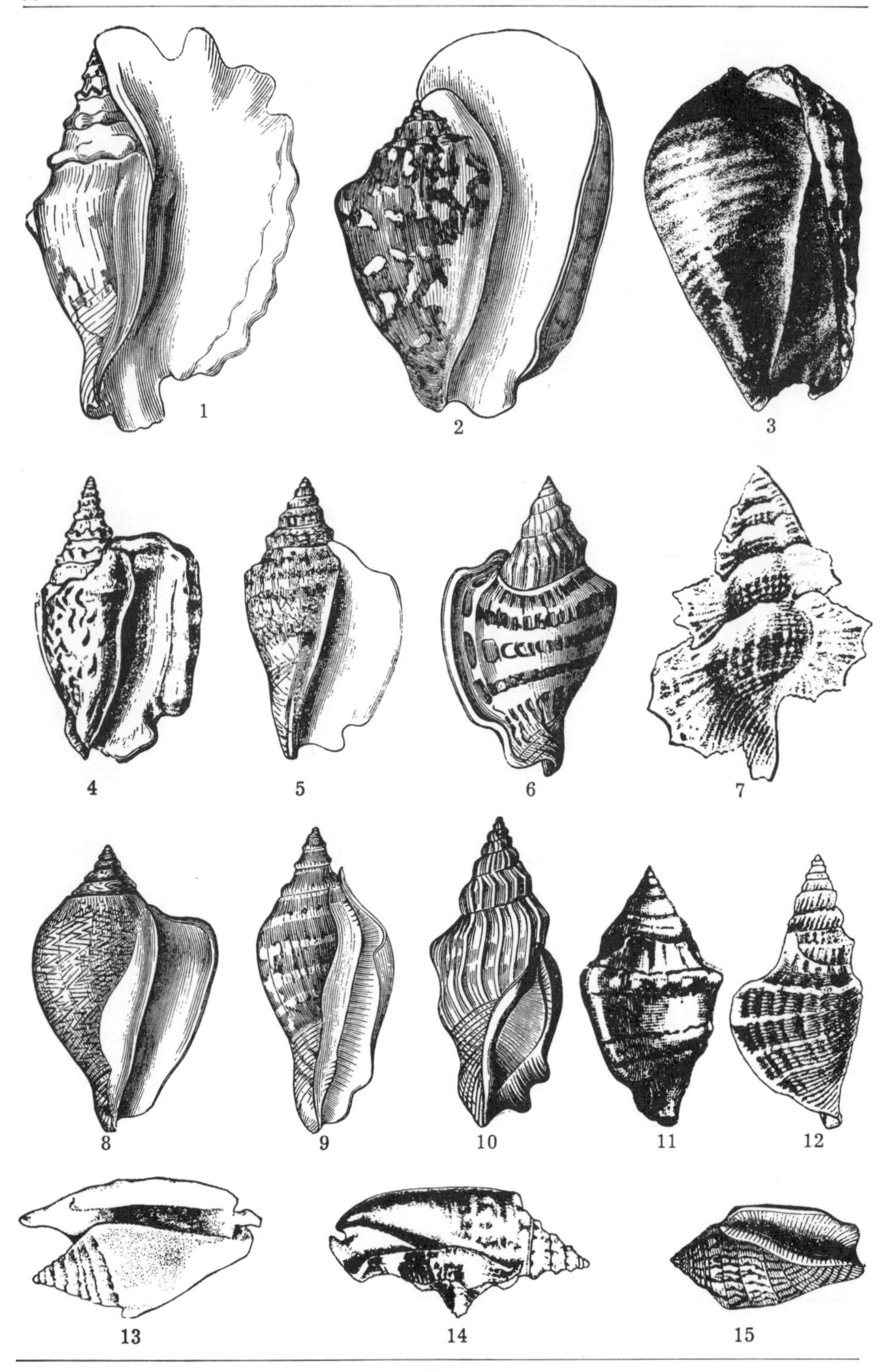

PLATE 39

1. **Strombus laciniatus,** Chem. Philippines. A very strange formed shell as the aperture which is richly colored with reddish-purple, extends to the top of the spire. A strong solid species which is rare and choice. 3½ to 4″.

2. **Strombus latissimus,** L. Philippines and Pacific generally. A real solid and heavy shell and which attains 7″. The aperture extends outward and inward, and a full inch above the spire. Has chestnut stripes on body whorl. I have also had specimens sent me from Fiji and other Oceanica islands. Rather rare.

3. **Strombus galeatus,** Sow. West Mexico. This is the Big Conch of the Mexico-Pacific region and takes the place of the Big Conch of the Bahamas. Has heavy periostracum which when removed shows a plain white shell on back and just a shade of russet in the aperture. Lives just below the tides and pushes around in the mud everywhere. Fairly common in some localities. 8″.

4. **Strombus thersites,** Gray. Australia. It has tall spire, back ornamented with reddish blotches. Is 5″ with the thick lip. Very distinct from all other species and a rather rare shell.

5 & 6. **Strombus variabilis,** Swain. Australia. It is well named as it is very variable. You can hardly find specimens from two localities that are exactly alike. I illustrate two types and there are many more. 1½″.

7. **Ranella pulchra,** Gray. Winged Ranella. China Seas. Of a slight buff color, the wings are exactly opposite sides of the shell. It is very distinct from other species and is not rare. 2″.

8. **Strombus canarium,** L. New Caledonia. A fat chunked species with thick white lip and faint markings. A few similar types are found all over the Pacific. ½ to 2″ and fairly common.

9. **Strombus succinctus,** L. Philippines. The back of the shell is covered with yellow coating, ornamented with chevron white markings. Lip is sharp and aperture very distinct from other species. ½″.

10. **Strombus minimus,** L. Viti Islands. A very variable and handsome small shell which extends over a wide territory. You will find many types quite similar around the Pacific islands. All of the small Strombus make fine cabinet specimens, and look best in series of 4 to 6 specimens of a kind. 1½″.

11. **Strombus fasciata,** Born. Indian Ocean. Has a row of nodules at top of body whorl which is ornamented with five heavy black lines. Aperture is usually brilliant orange. 1½″.

12. **Strombus Campbelli,** Gray. Port Darwin, Australia. A fine small 2″ species which is quite distinct in color pattern, being ornamented with chestnut markings. There are over a hundred species of Strombus in the world which range from the giant S. goliath from Pernambuco, Brazil which is very rare, to the small one inch forms. This chap is about 1½″.

13. **Strombus auris-dianae,** L. Philippines. The lip rolls over showing up the rich reddish aperture to fine advantage. The shell is quite smooth and glossy but there are varieties which are not so. Much more attractive than the cut would indicate. 2½″.

14. **Strombus granulatus,** Gray. Panama. A common form from Lower California southward. If you love to collect shells, secure a good boat and helpers and cruise over a thousand miles of this shore making frequent stops and you will be surprised at the collections it is possible to make. A friend of mine has been doing this for years, keeping careful notes on every species and the pamphlets he has published are very fascinating. Go thou and do likewise if you need a real rest.

15. **Strombus floridus,** Lam. Philippines. I never knew why this name was applied to a shell not found in Florida. It is a small species which must be very common over a wide territory as it is sent to this country in quantity and used in manufacturing novelties. Very variable in color. 1″.

The Strombidae usually follow the Cypraea in the nomenclature of shells. The shells have an expanded lip, which is notched near the canal. There are likely 100 species or more extending from West Indies to New Zealand. They are found on reefs in shallow water and down to 10 fathoms. Only a few species are found fossil starting with the cretaceous.

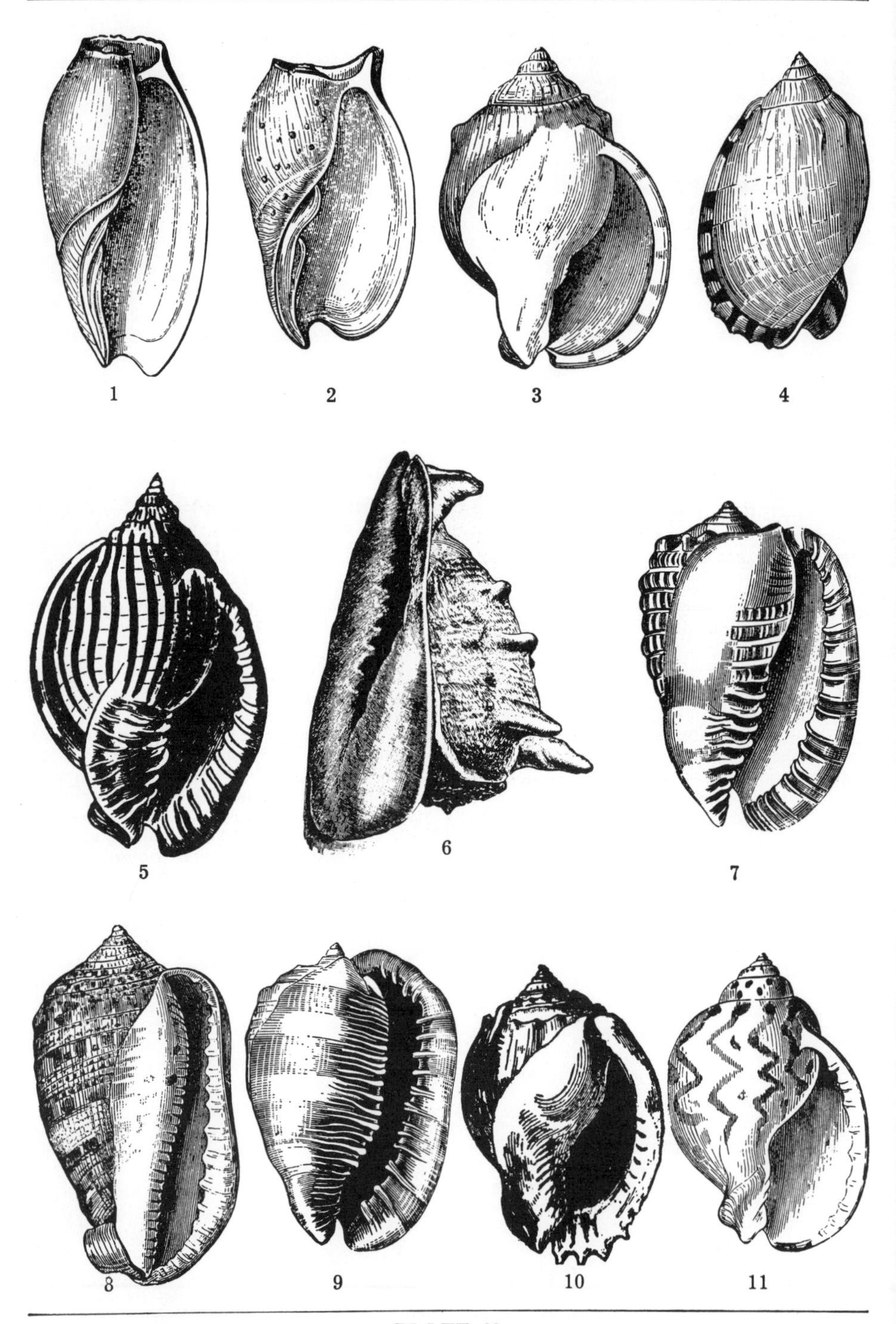

PLATE 40

1. **Cymbium proboscidale,** Lam. West Africa. Most of this section of the Melo family are from this region. This species is usually 6 to 8″ but in the American Museum in New York there is a specimen easily 14″. I suspect the largest known. All are of a horn color.

2. **Cymbium porcinum,** Lam. West Africa. Most specimens seen in collections are 3 to 4″ but it really attains a much larger size. It has the peculiar concave apex and is of a grayish color.

3. **Cassis pyrum,** Lam. Mauritius. A small round species of 2″ with smooth surface inclined to reddish-brown color. Closely allied to **Achatina** which comes from the Cape Verde Islands.

4. **Cassis vibex,** L. Mediterranean Sea but it has a very wide range. There is a good named, variety from the Philippines and shells of a similar form and color are found in all oceans. It is elongated, smooth 2 to 3″, not very heavy, but has a brilliant polish and striking coloring. The lip is ornamented with regular black stripes. The variety is similarly marked but more stubby and smaller.

5. **Cassis strigata,** Gmel. Philippines. The Striped Cassis attains 3 to 4″. The dark parallel stripes are more prominent on some shells than others but always well defined. It is a strong robust species.

6. **Cassis cornuta,** L. Mauritius. The Yellow Helmet is one of the Three largest Cassis of the world, the two others being West India shells. For generations this shell was shipped into our country and sold commercially in shell stores but of recent years few have been sent. It is a big attractive yellow shell 10″ or more in size and you often see them in very old homes, brought here in the 18th and 19th centuries in sailing ships as part of their ballast.

7. **Cassis tenuis,** Say. Galapagos Islands. A thin 3½″ shell very similar in markings to the Cameo shell. The reflexed lip is strongly marked with splashes of black, arranged in pairs. It lives on mud flats and deep water, fairly rare.

8. **Cassis coarctata,** Gray. West Mexico. At first sight it resembles the common Cameo shell but it is more elongated and thinner. The color pattern is similar. I suspect it could not be used for carving Cameos as is the Indian Ocean form. 4″.

9. **Cassis rufa,** L. Zanzibar and East Africa. The Bullmouth or Cameo Shell has been an article of commerce for a long time. Immense quantities are collected, shipped to points in Italy where they are carved in exquisite designs. Most of the carvings are exact copies of famous paintings in their local museums. The large 6″ shells are often completely carved, an electric light inserted and used for mantle lamps when they command a fairly high price. The shell is of reddish color and matures from 3 to 6″.

10. **Cassis glauca,** L. Philippines. Of a grayish color with strong reflexed lip and small points near the base of the aperture. Usually 3 to 4″, smooth, it makes a fine cabinet specimen.

11. **Cassis turgida,** Rve. Mindanao, P. I. The zigzag markings of this 2½″ shell usually differentiates it from most other species. It is closely allied to vibex, also shown on this plate.

The Family Cassidae are found from the West Indies to Australia and about 40 fossil species are known beginning with the Eocene. The shells are usually thick and solid, with the last whorl usually very large. The operculum is oval and narrow.

Some of the shells like rufa, madagascarensis tuberosa, etc., are much used for cameo cutting and before last war were commonly seen on our market in the finer art stores. The substance of the shells are often made up of differently colored layers and of different hardness.

The word Cameo derived from the Arab word, signifying bas-relief was for centuries restricted to hard stones such as onyx, engraved in relief but in the last century, has been extended to include gems cut on shell, lava and other substances.

I well remember some years ago visiting a very small shop in London that handled cameos exclusively. I purchased several dozen at the time for my friends who were shell collectors. There are still many thousand persons in Italy and elsewhere, who spend their lives cutting cameos mainly from shell.

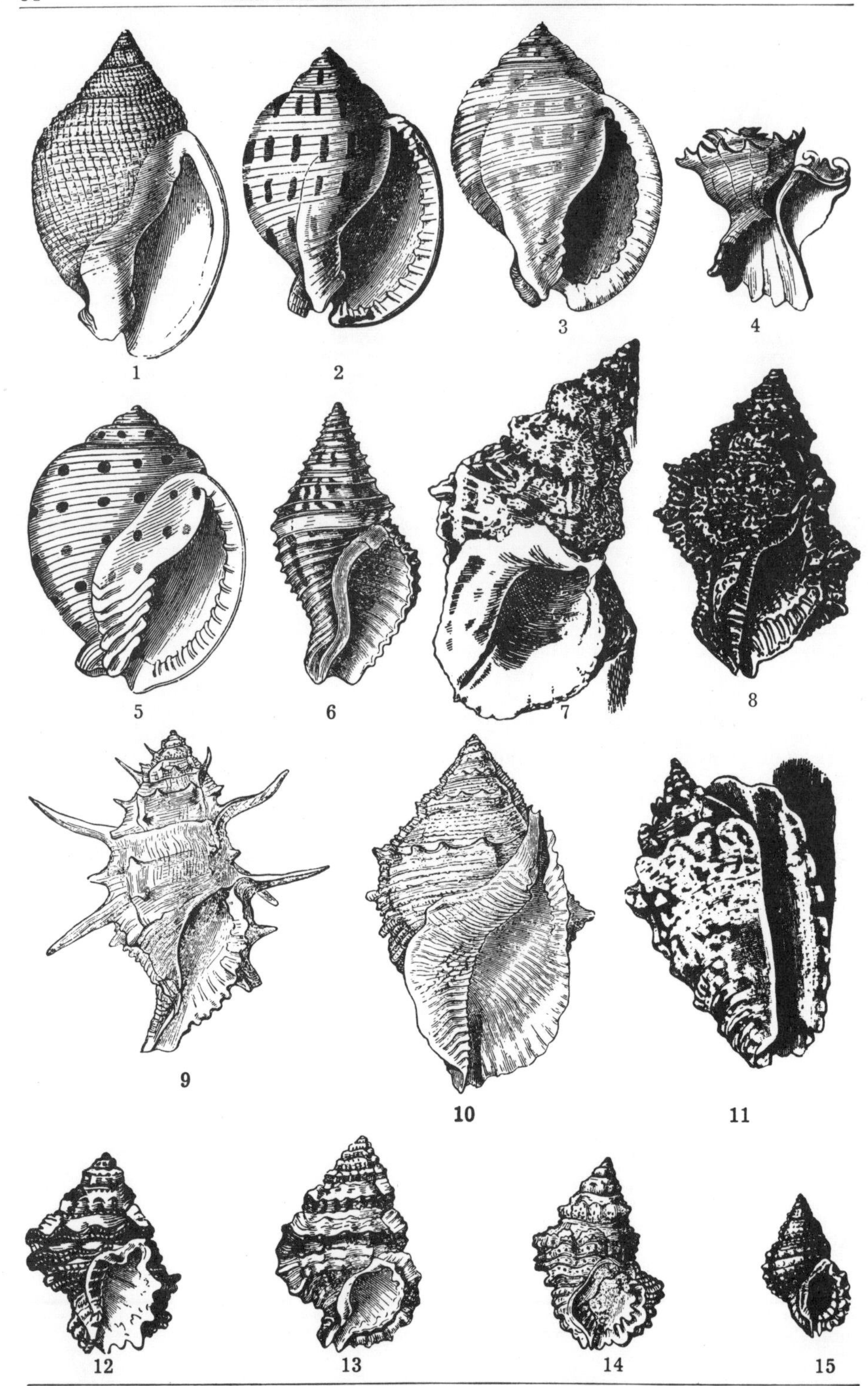

PLATE 41

1. **Cassis semigranosa,** Wood. Victoria, Australia. More elongated than most of the many species of this genus and completely covered with fine reticulations that add to its attractiveness. 2″.

2. **Cassis canaliculata,** Brug. Philippines. Shells of this type with regular blotches and usually 2 to 3″ seem to be found in all oceans and usually fairly common. One could easily form an entire draw of them. This species has more elongated spire than some of the others.

3. **Cassis saburon,** Ads. Mediterranean Sea. A round neat shell covered with faint blotches in regular patterns. Fairly common in the whole region. 2″.

4. **Latiaxis mawae,** Gray. China Sea. Pure white 1 by 1½″. It has a flat apex, last whorl partly disconnected. Umbilical region open to end of whorl. Last whorl has curved frills. The other species from this region have elevated spires, pure white and real little Pagoda like form, often ornamented with curved spines. They are a fascinating group and often quite rare as is this species.

5. **Cassis bisulcata,** Schub. Philippines. A small round 1½ to 2″ shell with granular surface and dots of chestnut in regular pattern over the surface.

6. **Fasciolaria fusiformis,** Val. New Holland, Oceanica. A small 1½″ shell with several ridges and row of small spiny points around middle of whorl. Of a light brownish color. 1½″.

7. **Ranella lampas,** Lam. Pacific everywhere. The Frog shell as it is often called, attains 8″ and is usually white when well cleaned. Small shells have a reddish aperture. It is the largest species of the genus and has been sold commercially in shell stores for many years.

8. **Ranella albivaricosa,** Rve. Philippines. A fine white variety with chestnut markings, the surface completely covered with small and large points. Typical of about 85 species of the genus found all over the world. They are scattered here and there over all oceans. 3″.

9. **Ranella spinosa,** Lam. China Seas. Ranges from 2 to 3″ ornamented with spines and nodules of various length. It is the only species so ornamented and not very common anywhere.

10. **Ranella crumena,** Lam. Ceylon. The aperture of this shell is the distinguishing feature, as it is usually orange color and well developed. The body is covered with chestnut blotches. Lives under rocks at low tide and is a rather attractive species. 1½ to 3″.

11. **Strombus lentiginosa,** L. Silver Lip. Philippines and Pacific generally. The shell is mostly white and covered with small knobs. Surface always irregular. It has been sold in a commercial way for generations. 3″.

12. **Ranella bufonia** Gmel. Philippines. A handsome small form of 1½ to 2″. Highly ornamented with knobs, and well marked with reddish-brown band. Aperture has a dash of red. A very difficult shell to clean satisfactorily.

13. **Ranella gyrina,** Lam. Australia. It has a wide russet band in middle of each whorl and is a rather bright colored shell that is quite variable. 1 to 1½″.

14. **Ranella rhodostoma,** Sow. Cape Verde Islands. A neat small species that will attract attention in any cabinet, on account of its crumpled, knobby appearance and fairly bright color. It has a dark aperture. 1½″.

15. **Ranella granifera,** Lam. Philippines. The rows of small knobs are quite typical of a number of other similar shells in form. It is fairly common over a wide territory. 1½″.

The Ranellas are commonly called Frog Shells and are found world wide in tropical seas. About 25 fossil species from Eocene. The shells are ovate or oblong, with two rows of continuous varices, one on each side. Aperture is oval, canal short, outer lip crenated.

The tenticles of the mollusk are usually somewhat closer together than in the Triton (Cymatums) and the head is longer and narrower than in the Murex. The foot is large. Operculum ovate, horny. The shells mainly inhabit warm seas. Those of the typical group with winged varices usually live in deep water, while the nodose species are found in more shallow water around coral reefs and rocks.

PLATE 42

1. **Turbo marmoratus,** L. China and many other places in Pacific — Called Green or Pearl Snail. The natural color is green but this is often ground off down to the pearl which takes a very brilliant polish. Curio dealers have the prominent ridges ground down to the pearl when you have a green shell with silver pearl stripes. Has been a commercial shell for generations and still fished extensively for its pearl. 3 to 8″.

2. **Turbo torquatus,** Gmel. New Holland, Oceanica. A very distinct species which I have never found very common and seldom see in cabinets. Much the same form as the very common **sarmaticus** but has the row of ridges on top of the body whorl, open umbilicus and light aperture. 2 to 3″.

3. This is a typical shelly operculum which is found in all Turbos but in some species like petholatus, this operculum has a brilliant green polish. These operculums on the big Turbos will be 3 to 4″ and 1″ thick. Sailors call them Cats eyes.

4. **Turbo petholatus,** L. Philippines and Pacific generally. This shell has a very high natural polish. It is very richly ornamented with different shades of black, green and brown. Some shells will be all one shade. I had a very distinct variety all gray patterns from Sula Sea and I suspect from many other island groups different patterns could be obtained. Those sent me from the British Solomons were quite distinct and very rich in color. Has bright green operculum. 2 to 3″.

5. **Turbo argyrostoma,** L. Pacific Generally. The Silvermouth is named for the brilliant silver aperture. The back is ridged and varies from gray to greenish, with stripes of brown and white. When polished it is all pearl. 2½″.

6. **Turbo saxosus,** Wood. Panama. It is covered with ridges which are cut into tiny segments like ruffles. There is a thick periostracum which if removed shows the pearl. Lives under stones between tides. Of a grayish color. 1½ to 2″.

7. **Turbo imperialis,** Gmel. Imperial Turbo, Indian Ocean. A round greenish shell, smooth surface and fine pattern. Very attractive. 4 to 6″.

8. **Turbo cornutus,** Gmel. Japan Seas. The Spiney Turbo is a large shell with numerous horns. The aperture as usual has a heavy operculum. Color is green. If polished it is a brilliant pearl but is not as thick a shell as **marmoratus** and is seldom fished for commercial purposes. 4 to 5″.

9. **Turbo ticaonicus,** Rve. Philippines. A handsome shell and very variable. Usual color is green, the whorls being lined with minute pebbly surface. Many specimens are spotted with white. 2″.

10. **Turbo natalensis,** Rve. South Africa. The surface is covered with mottled ridges and rich russet color, which makes it a rather attractive small shell. 1½″.

11. **Turbo lamellosus,** Brod. Australia. A very depressed shell completely covered with wrinkles. Very little color. A shell which is very rarely sent me by Australian collectors and I suspect is not very common. 2″.

12. **Turbo stramineus,** Mart. South Australia. One of the odd forms of the genus. Rather flat with ridges and pebbled surface and light aperture. Quite distinct shell. 2 to 3″.

I had over 300 volumes on Mollusca in my library but there was very little information on where the shells were found. Of the 18 species of Conus found in Western America from Panama to Lower California here is data I have been able to collect. CONUS ARCHON, *Brod. dredged at 20 fath.* ARCUATUS, *Sow. dredged.* BRUNNEUS, *Mawe. in rock ledges at extreme tides.* COMPTUS, *Gld. spawning on sandy mud flats.* DALLI, *Stearns. usually dredged but an occasional live shell on beach. It is a "tent" cone and very attractive.* EMARGINATUS, *Rve. dredged at 20 fath.* MAHOGANYI, *Rve. living on mud flats at low tide.* NUX, *Brod. nesting in crevices of rocks at low tide.* PRINCEPS, *L. on coral reefs at low tide.* PURPURASCENS, *Brod. on coral rocks at low tide.* REGULARIS, *Sow. on mud flats at low tide.* TORNATUS, *Brod. dredged at 20 fath.* XIMENES, *Gray. dredged at 20 fath.* PYRIFORMIS, *Rve. living in crevices of rocks.* VIRGATUS, *Rve. on mud flats.* LINEOLATUS, *Val. on mud flats but more common in deeper water.*

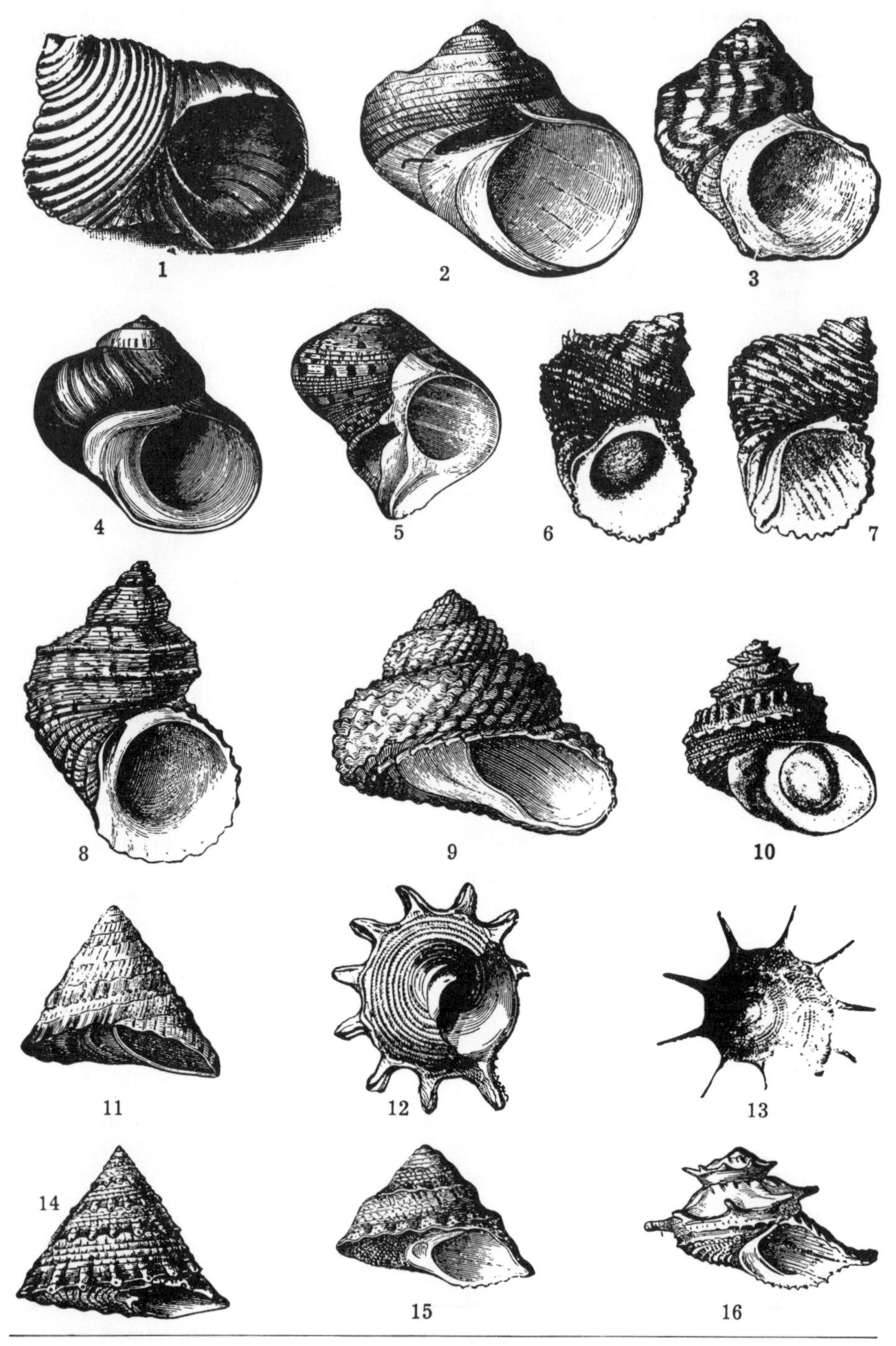

PLATE 43

1. **Turbo chrysostoma,** L. Goldmouth. Philippines. The aperture is a rich golden color which gives it its name. The surface is covered with regular ridges which show greenish color. Ground down to the pearl it used to be sold as a pearl shell. 2 to 2½".

2. **Turbo sarmaticus,** Gmel. Turks Cap. Algoa Bay, Africa. A shell that must be very common as it used to be sold polished commercially in vast quantities. It is of a black color and when polished some of the black was usually left on, to make a striking contrast with the pearl surface. 3".

3. **Turbo fluctuosus,** Wood. Lower California. The species is well marked with wavy diagonal stripes and lives among stones between tides. Is 2½".

4. **Turbo smaragdus,** Gmel. New Zealand. It is of greenish color with a rich deep green shelly operculum. Aperture is pearl as usual. 2 to 3".

5. **Turbo lugubris,** Rve. New Holland. A finely mottled greenish shell of peculiar shape. Smooth surface. 2".

6. **Turbo intercostalis,** Mke. Hawaii. It is a finely marked shell of about 2". Ridged surface and brownish color. There are a number of quite similar forms in the Pacific. 2 to 3".

7. **Turbo setosus,** Gmel. Moluccas. A very variable shell always covered with ridges and small nodules but a wide variety of color patterns. Varies from black diagonal stripes to yellow background with brown stripes and some specimens entirely brown. 2".

8. **Turbo radiatus,** Gmel. New Caledonia. A richly colored shell with small tubercles and adorned with broad longitudinal stripes of reddish-brown. The aperture is golden pearl. 1½ to 2".

9. **Astraea sulcatum,** Mart. New Zealand. A strong solid shell 2½ to 3" covered with ridges and has strong periostracum. Of a grayish color it is of fine iridescent pearl when the periostracum is removed.

10. **Astraea modestum,** Rve. Japan Seas. A richly colored pink shell with two rows of spines on each whorl. It has a large orange patch on the smooth base at edge of lip. 2 to 2½".

11. **Astraea inermis,** Gmel. West Indies. Last whorl is carinated with only a suggestion of spines. It is close to other species of the region but always distinct. Lives among coral rocks. 2".

12. **Astraea calcar,** L. Philippines. A small shell devoid of color with row of spines on the carinated edge of last whorl. 1½".

13. **Astraea triumphans,** Phil. Japan Seas. It is quite unusual to see a reddish shell with a row of sharp spines as long as these on the carination of the last whorl. It has a dainty operculum that fits the curious aperture. Not rare. 2½".

14. **Astraea buschi,** Phil. Gulf of California. It is of a greenish color when periostracum is removed. Serrated edge on base of last whorl. Has ear shaped operculum of white. 1½".

15. **Astraea fimbriatum,** Lam. Victoria, Australia. Very much resembles some of the forms from the West Indies. Is of a light greenish color with pearl beneath. 2".

16. **Astraea stellare,** Gmel. New Caledonia. A very unique shell with spines on edge of whorls and a rich yellow aperture. Quite distinct from other species. 1½".

The Pyrenidae are almost exclusively composed of the genus Pirene or as I have always called them Columbella. Why change the common name of a genus that has been used most two centuries. There are 300 to 400 species and every one is a handsome small shell. They have been found fossil in the Tertiary. Their form is so diversified they have been divided into 14 subgenera. They are found all over the world and there are naturally many varieties of different colors, etc. Some European scientists have described many of the varieties that are common in their vicinity but leaving out all of these varieties you will have plenty of work if you secure even half of the known forms in the world.

The Turbinidae cover a large number of species including the typical Turbo which range up to 10" down to the gorgeous colored Phasianellas, the tiny Leptothyras of brilliant colors, Astraliums, etc.

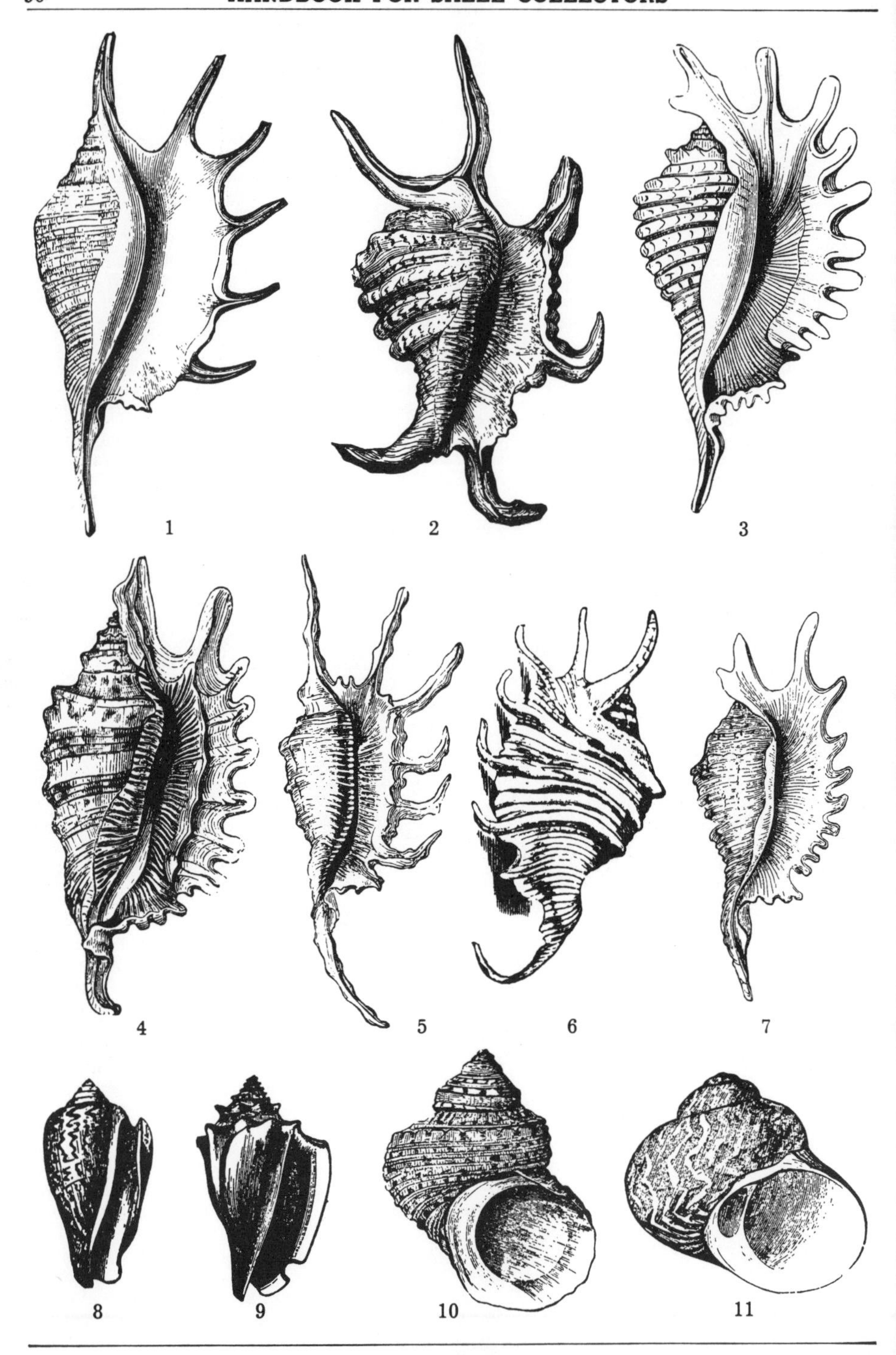

PLATE 44

1. **Pterocera lambis,** L. Spider Shell. Pacific generally. One of the most common shells of the genus and has been sold commercially for many years. Somewhat variable in shape but the general pattern is the same. Usually marked with blotches of dark color on back. Aperture white. 5 to 6″.

2. **Pterocera rugosa,** Lam. Scorpion Shell. East Indies. It is a very common shell and usually shape of cut. Has been and is sold commercially. The six hooks in the form indicated usually identify the shell, but there is another shell very similar with reddish aperture that attains twice the size of this shell. 5 to 6″.

3. **Pterocera violacea,** Swain. Philippines. This is a lovely white shell with rich violet aperture and must be fairly rare as none of my P. I. collectors have ever sent me a specimen. The numerous fingers along the edge differ from all other known forms. 3½″.

4. **Pterocera elongata,** Swain. Mauritius. The stubby spines, flat lip and rich aperture with two spikes at top, will always easily identify this shell wherever found. I suspect it is fairly common but never seen on our market as yet in any quantity. 3½″.

5. **Pterocera scorpio,** L. Philippines. A highly colored Scorpion shell which is sold commercially but not always on the market. The aperture is a rich violet color and the 6 or 7 arms make it a very attractive shell. 4″.

6. **Pterocera aurantia,** Lam. Philippines. The yellow Scorpion is a slender shell, more so than any of the other forms of the genus. Unique in having real hooked fingers. The aperture is a brilliant yellow, fairly common.

7. **Pterocera millepeda,** L. Philippines. I had a large number of these shells sent me from Sulu Sea, P. I., and while the fingers were stubby as shown in cut they were usually slightly curved upwards. The back is lined with brown and aperture same. There are usually 10 or more spikes. 3½″.

8. **Strombus luhuanus,** L. Japan Sea and Pacific generally. A very pretty 2″ shell with a brilliant red aperture bordered with black on the body opposite the lip. Quite common. 2″.

9. **Strombus gracilior,** Sow. West Mexico. Of similar shape and habits to the very common **pugilis** of Florida waters. This shell is almost invariably a light yellow with fuzzy periostracum. 2 to 3″.

10. **Turbo tesselatus,** Kien. Lower California. The shell is of a grayish-brown color with dots and waves of darker color over entire surface. The base of the shell has a greenish cast. Fairly common. 2 to 3″.

11. **Turbo undulatus,** Chem. Tasmania. The entire surface is covered with zigzag markings over a fine green color, it seems to be a hard shell to clean so as to show all of its very attractive features. This is true of many of the shells of this great genus. 2 to 3″.

The Genus Pterocera contains only about 12 species most of which are fairly common in some parts of the Pacific and Indian Oceans. I have illustrated seven forms. P. bryonia Lam. is the largest and grows to 13 inches with 5 or 6 heavy prongs. I have found it more common about 10 degrees south of the Equator in Pacific. P. chiragra, L. is common to both Pacific and Indian Oceans. I have had very fine ones from Philippines. It is almost exactly like RUGOSA *but grows twice as large. There is a form P. Multipes Chem. from Red Sea of which I have only had one. The other forms I have not seen. When young the outer lip of the shell is simple, resembling the young Strombus. The claws are gradually formed with the growth of the shell and are at first open canals, which afterwards become closed and solid. Fossil forms have been found in Jurassic and Cretaceous. The operculum is similar to that of the Strombus.*

The Terebridae include only the genus Terebra. It is a large one over 200 species and 25 or more fossil, commencing with the Eocene. They are divided into 9 subgenera and range from small forms of 1″ or less to the mammoth Terebra maculata, which attains 7 to 8″. Most of them are slender with many whorls, some unusually so. A few that run 4 to 5″ are beautifully mottled with spots. The aperture is small and notched in front. The operculum is annular. Every collection should have a fair assortment as the many whorls put them in a class by themselves only approached by the Turritella.

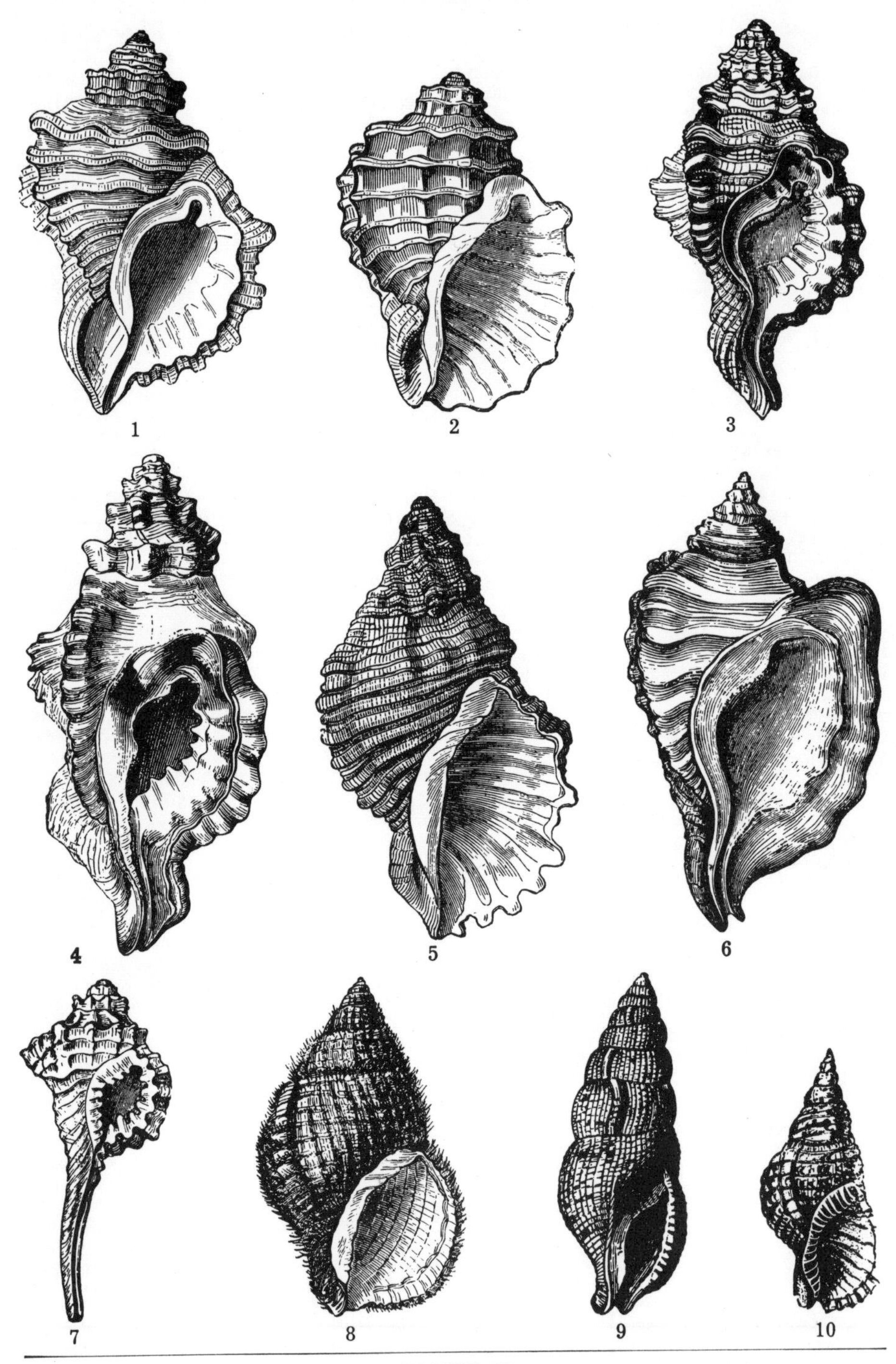

PLATE 45

1. **Cymatium cutaceum,** L. Mediterranean Sea. It is of a horn color throughout with prominent nodules in middle of each whorl. Aperture is pure white with five or six nodules on edge. I have had similar specimens from Australian region. 3″.

2. **Cymatium doliarum,** L. Cape of Good Hope. Of similar color to preceding species but differs some in form and usually of smaller size. I had numerous specimens sent me from Natal. 2 to 3″.

3. **Cymatium grandimaculatum,** Rve. Philippines. A noble solid shell of a russet color and somewhat resembles the next species. Of a general knobby structure. 4″.

4. **Cymatium lotorium,** L. Pacific and Indian Oceans. One of the attractive species of the genus. Of a russet color with dark bands on the aperture and a general knobby appearance. 4″.

5. **Cymatium spengleri,** Chem. Australia. Of a light gray color, the aperture is white and the whorls are covered with circular ridges and nodules. Very desirable shell. 4″.

6. **Cymatium tigrinus,** Brod. West Coast of Central America. A large russet-brown species of angular form, and peculiar shaped aperture. It lives under rocks and is quite rare, only occasionally seen in collections. 4 to 5″.

7. **Cymatium exilis,** Rve. Philippines. It has been found living in sandy mud at 10 fathoms. Of a brownish-white color. Not very common. 2½″.

8. **Cymatium scabrum,** King. Chili. The shell is covered with a deep brown hairy periostracum. When this is removed the whorls are seen to be covered with reticulations which run both ways. Aperture white. 2 to 3″.

9. **Colubraria tortuosus,** Rve. Burias Id., P. I. One of the several fine species of this genus which belong to the Triton complex. The shells are covered with ridges and the upper whorls are often distorted. After this genus, in regular order come **Craspedotriton, Caducifer, Maculotriton,** from various places in both Pacific and Indian Oceans. Many of them are real Baby Tritons as they range down to half inch when full grown.

10. **Cymatium tritonis,** L. South Seas. This is the real Trumpet Shell of the Tropics and is used by millions of natives, as a trumpet to call clans together for pleasure or war. They make a hole in one of the upper whorls and blow it the same as a cornet. The shell has a natural fine polish and richly ornamented with brownish colors. It attains 15″ but 5 to 8″ specimens are fine for cabinet. The name Triton for this genus has been changed by systematic writers to Cymatium, but I like this old name TRITON which has been used for two centuries and should never be changed to another.

The Cymatidae includes all the shells formerly called Triton, some of which attain a foot or more in length, but also include smaller forms under the following generic names: Distorsio, Eugyrina, Crossata, Columbraria, Craspedotriton, Caducifer, Malulotriton, etc. The latter genera, some of them, include small "Tritons" down to half an inch to one inch.

The varices (which denote former apertures) form a continuous row. The operculum is annular. The fossil forms first appear in the Eocene strata. There are over 100 species of recent forms and likely 100 fossil. The many species are distinctly tropical, none being found in cooler water. The many species present a great range in variation both in size and color. There are cancellated forms only found in very deep water. The West American species are usually covered with a rough coated epidermis and are found in sandy mud in from 6 to 30 fathoms. Some few species have a world-wide distribution, which may be due to their free swimming larvae.

Much has been written about the Tritons. In some islands they use the Trumpet shell for a tea kettle. The operculum forms the lid, the canal answers the purpose of a spout and the shell is suspended by a wooden hook over the fire. One collector found the T. nodiferous capable of reproducing an amputated tenticle. The people of Sicily and the Algerians eat this mollusk and consider it a delicacy. At Nice the fishermen and country people make a hole in the apex of the spire and use the shell as a trumpet. This is also a common practice to this day through many of the South Sea Islands.

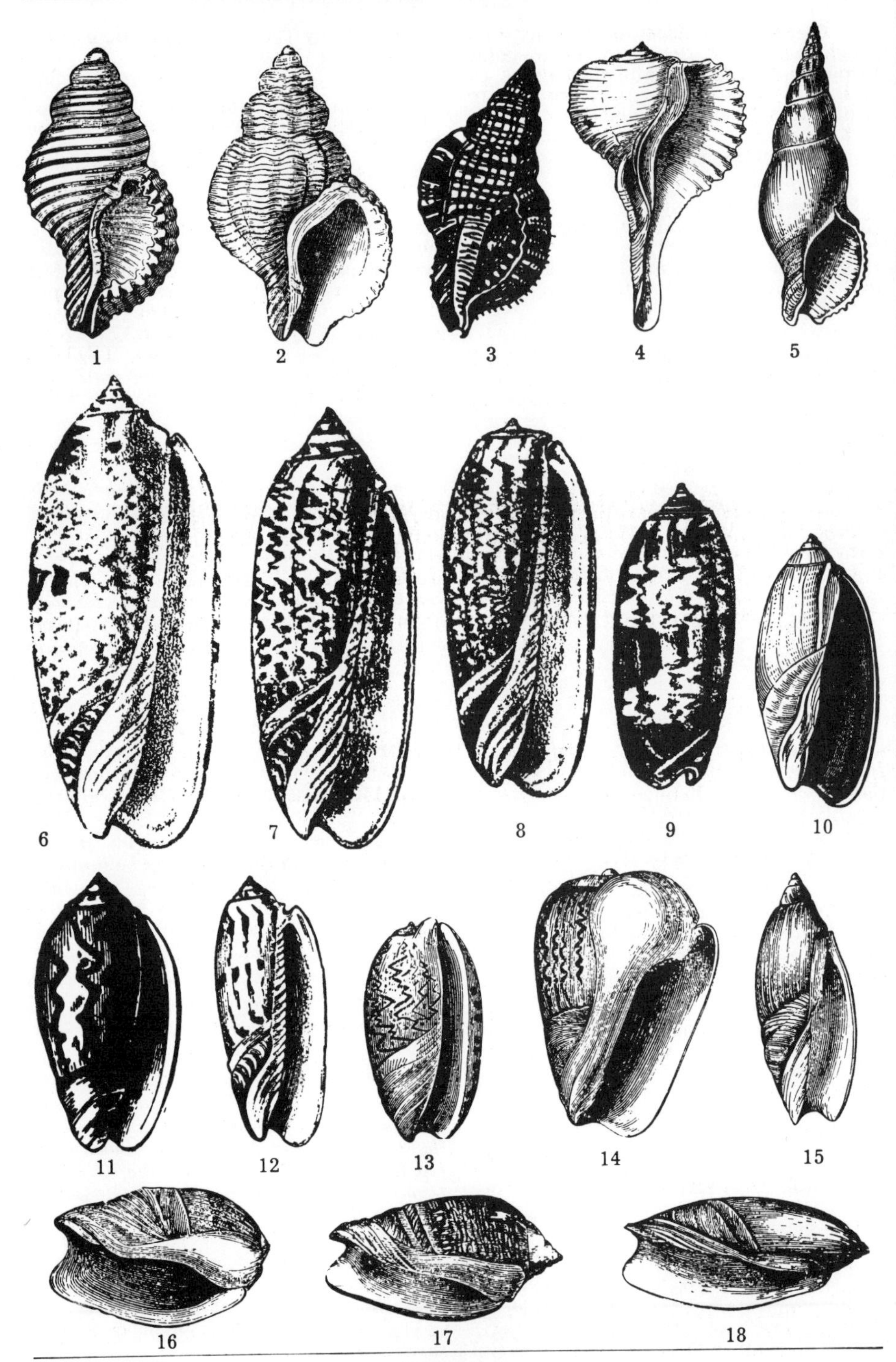

PLATE 46

1. **Cymatium clandestinum,** Chem. Australia. A small russet colored shell with circular rows of ridges close together completely covering the specimens. 1½".

2. **Cymatium rude,** Brod. Peru. Of a grayish-white color, the shells live in mud and sand 6 to 10 fathoms and while the species somewhat resembles scabrum, it lacks the periostracum of that species. 2".

3. **Cymatium pilearis,** L. Philippines. It is typical of many species in this great genus, which comprises some 175 forms over the ocean world. A real collection will show shells from 1 to 15" or more and in all of them you can easily trace the various periods of growth, as the shell always leaves a little of the lip exposed. This species has a fine hairy periostracum which is quickly removed by the waves and sand of the sea, if the shell is picked up on a shore line. 2 to 3".

4. **Rapa papyracea,** Lam. Moluccas. A pure white shell of the peculiar shape shown in cut. There are 7 known species mostly from Oceanica and none are very common. 2 to 3" scarce.

5. **Northia northiae,** Gray. Panama. A smooth elegant olive colored shell of which there are five known species in the genus. This one is a beauty and usually 2".

6. **Oliva sericea,** Bolt. Philippines. A fine large brilliant species which usually lives below tide lines hence not found so easily as many other forms. They fish for them with night lines. 2 to 3".

7. **Oliva minacea tremulina,** Lam. Indian and Pacific Oceans. Everywhere in the tropics you collect this and other forms of the minacea complex. You find many color forms, some of which are very difficult to determine satisfactorily. 2 to 3".

8. **Oliva irisans,** Lam. Pacific generally. A brilliant shell of high polish, of which there are numerous varieties. Very variable although can usually be identified. All of the Olivas are very attractive shells with a natural high polish, mostly inexpensive, so that collectors can easily secure six to a dozen or more of a kind. I saw one collection recently where the owner had an entire drawer of some of these very variable species. 2 to 3".

9. **Olive pica,** Lam. Philippines. It is a white shell mostly covered with blotches of chestnut. Very showy. 2½".

10. **Olivancillaria steerae,** Rve. Senegal. This group of shells are all Oliva but of a different form than most of the species of that genus. Most of them are the form of this cut but I have illustrated others. They are usually grayish shading to brown. 1½".

11. **Oliva peruviana,** Lam. Peru. It is a wavy striped species in typical form, but there are color forms such as **castanea** is brownish-black, **coniformis** is grayish and shouldered, **fulgurata** is wavy and **livida** is gray. I had a series of a hundred specimens showing many fascinating color forms. 2".

12. **Oliva minacea,** Bolt. Pacific generally. In old collections you find it labeled erythrostoma the variety with the orange aperture. I never felt there was any sensible reason for changing names that had been used for centuries to those that were unknown. But you know the scientists always gave a good reason for all of their ideas no matter how revolutionary. 2 to 3".

13. **Oliva bulbosa,** Bolt. Old name was inflata, Pacific generally. A strong robust shell of which there are numerous color varieties. Extremely common and has often been imported for commercial uses. 1½".

14. **Olivancillaria brasiliana,** Lam. Brazil. With flat top, curious thick nacre aperture, this is one of the extreme forms of the genus but there are others even more so. A very desirable shell. 1½".

15. **Olivancillaria acuminata,** Lam. South Australia. The largest species of the genus of a gray-brownish color. 2 to 3".

16. **Olivancillaria auricularia,** Rve. Brazil. Rather odd and curiously distorted, it is not at all common. I had very fine specimens sent from Uruguay. Only occasionally seen in cabinets. 1½".

17. **Olivancillaria gibbosa,** Born. Brazil. One of the most common species of the genus and is imported commercially. Typically it is brownish with white splashes but there are at least four good color forms that are very distinct. 1½ to 2".

18. **Olivancillaria testacea,** Lam. Mazatlan. I had very attractive specimens sent from Panama where the shells were of a rich uniform brown color and highly polished. 2".

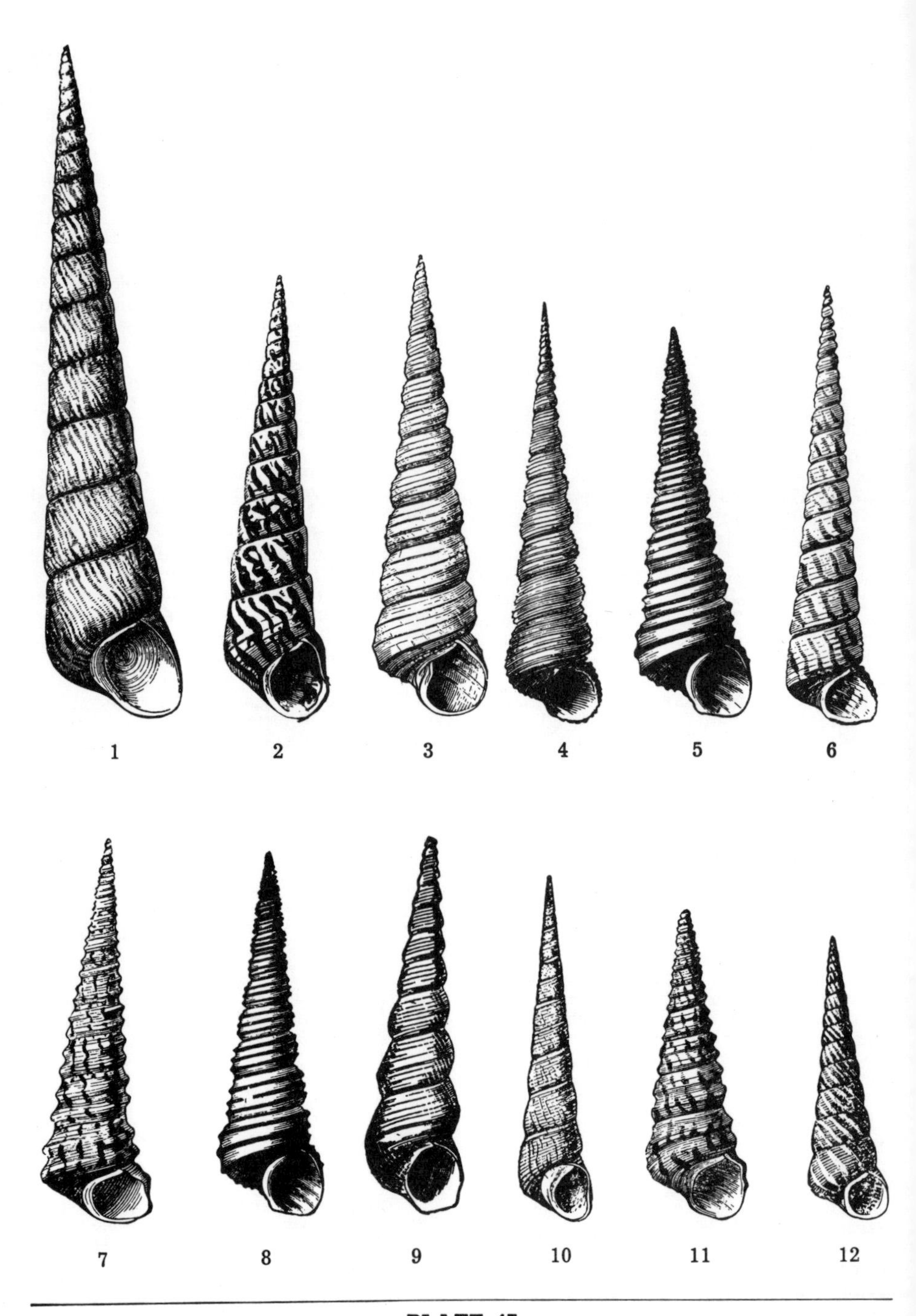

PLATE 47

1. **Turritella broderipiana,** Sow. Peru. One of the finest of the genus, tall and rugged, brown or reddish color, it has been found from Mazatlan southward. 4 to 5".

2. **Turritella goniostoma,** Val. West Mexico. An elegant shell of large proportions. Rather smooth with faint traces of ridges. Completely mottled with shades of brown. Lives buried beneath the surface of soft ooze of Mangrove Swamps often in company with Area tubersulosa. 4 to 5".

3. **Turritella bacillum,** Kien. Ceylon. A light brownish shell with about five lines to each whorl. Some specimens may run larger. 3 to 4".

4. **Turritella terebra,** Lam. Philippines. A fine lined shell of horn color but sometimes showing reddish. It is well-ribbed, with very fine point. 4 to 5".

5. **Turritella cingulata,** Sow. Peru. A shell of many fine ridges and coated with a shade of brown. It is typical of others of the genus. 3½".

6. **Turritella columnaris,** Kien. Ceylon. The whorls are finely mottled with stripes of brown. A very trim looking shell. 4".

7. **Turritella maculata,** Rve. China. A deeply ridged shell beautifully marked with shades of brown. One of the finest of its size. 3".

8. **Turritella attenuata,** Rve. Penang. Finely ornamented with deep circular ridges. One of the forty varieties known in the world of this genus. 4".

9. **Turritella duplicata,** Lam. Indian Ocean. A very heavy well developed shell, that is of horn color. The ridges are beveled, and about 1" thick at base. Total length 4 to 6" ranging one of the largest of the genus.

10. **Turritella vittata,** Hutt. New Zealand. A species with fine striae across the whorls and faint circular ridges. 2 to 3".

11. **Turritella bicingulata**, Lam. Cape Verde Islands. Ornamented with fine ridges and dark spots on a lighter background. 3".

12. **Turritella nodulosa,** King. Central America. The usual ridges are crossed with stripes and brown mottled markings. 3".

The Turritellidae consist almost entirely of the one genus. The shells are always spiral, usually not umbilicated, spire very long with numerous whorls with revolving striae or carinations. Operculum corneous, multispiral.

There are around 100 species of world wide distribution. About 200 species in the triassic.

✦

The Family Cavoliniidae consists of the Genus Cavolina (Hyalaea) Diacria Balantium, Cleodora, Creseis Cuveria. These are all Pelagiagic small forms of shells that float around on the surface of the warm currents of the various oceans. Of the Cavolina there are about 20 or more species from Atlantic to Indian Ocean and about half as many fossil from Miocene. The small shells are globular and translucent. The Diacria are somewhat similar. I never had but two species. The Balantium are triangular and depressed and the type B. recurvum is one of the handsomest of the Pteropods, swims steadily instead of flitting about in a lively manner, like the Cavolina. The Cleodora are narrow and conical. The Creseis are slender, pointed and straight or curved. Seen on the surface towards the decline of day they shine like myriads of shining spicules. Their progress through the water is very irregular. The Cuvieria are cylindrical and transparent. These genera do not belong to the Gastropoda but are in a Class called Pteropods.

✦

The genus Fusus covers around 75 species and there are over 300 fossil forms from Cretaceous to Oolite. They are all fusiform, spire long many whorled, and the canal long and straight. I have had specimen of Fusus colus of about one foot but most of the larger forms which are usually white or faintly marked are 3½ to 5 inches and are often covered with a fuzzy coating of yellow. Operculum is ovate. There are no other marine shells quite like them and they make a fine display in the cabinet.

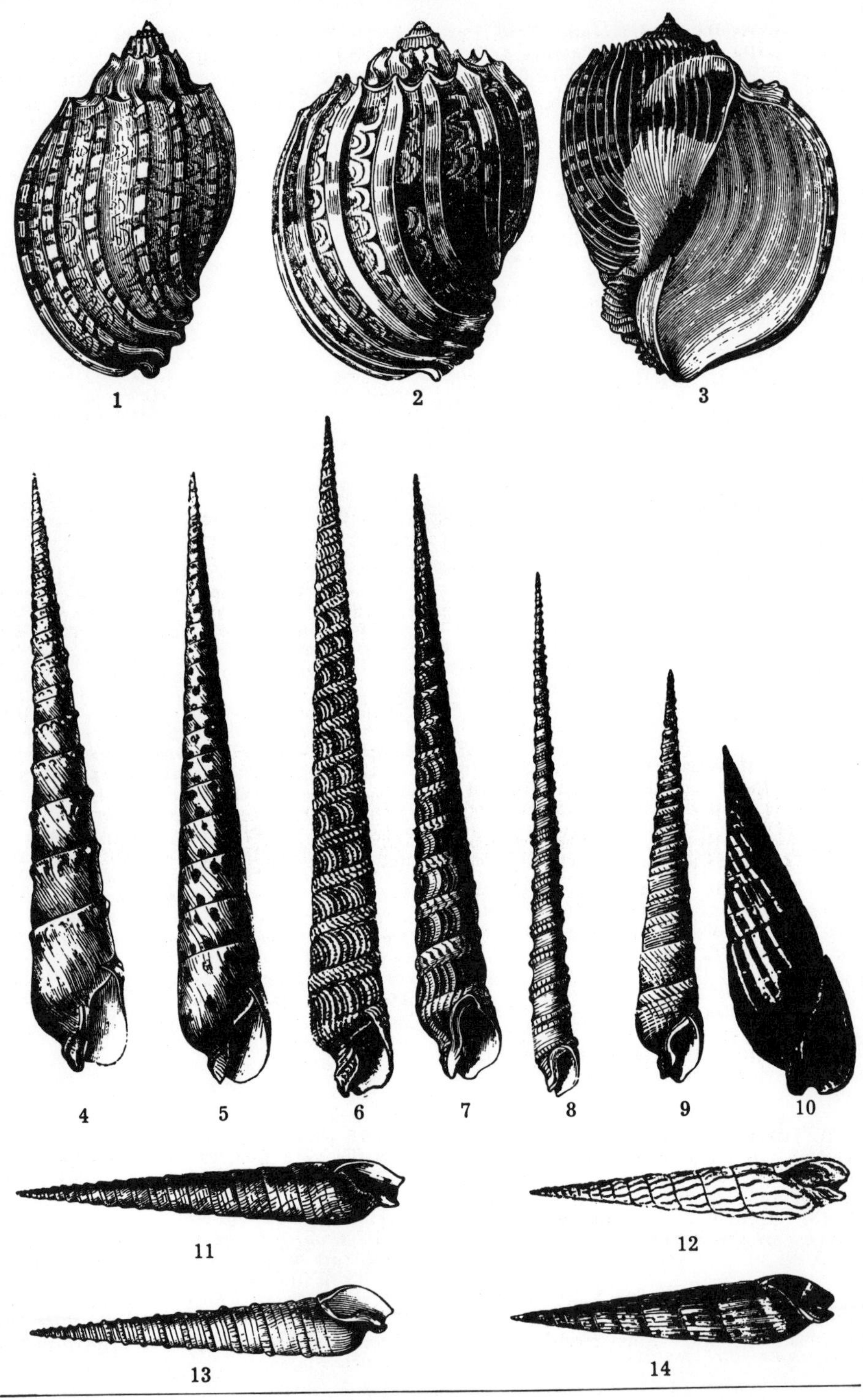

PLATE 48

1. **Harpa articularis,** Lam. Philippines. The Harps are all unusually beautiful shells which are much admired. This species usually has 12 ribs which are ornamented with dark black markings. 2 to 3".

2. **Harpa conoidalis,** Lam. Mauritius. This species attains the largest size of any of the dozen or so known forms in the genus. The 10 or 12 ribs are marked with 4 or 5 bands of darker color. Aperture white with darker shadings and two dark splashes on the body whorl. But there are many shells which exhibit very little color and mostly gray. 3 to 5".

3. **Harpa imperialis (costata),** Lam. Mauritius. Has about 20 ribs set close together, which is the main distinguishing feature. On the last whorl there are faint chestnut markings on a highly natural polished surface. Always has been rare. 3½".

4. **Terebra crenulata,** L. Andaman Ids. and Pacific generally. It is of a grayish-yellow color, ornamented with knobs along the top of each whorl, and faint brown dots. Fairly common. 4 to 5".

5. **Terebra oculata,** Lam. Mauritius. A very attractive shell of a light russet color adorned with white spots in regular rows. Rare. 4 to 5".

6 & 7. **Terebra pretiosa,** Rve. Philippines. A fine slender elongated species with curved ridges on each whorl. It is quite variable adorned with shades of brown and russet. Both cuts are the same species. 4 to 5".

8. **Terebra triseriata,** Gray. Japan Seas. One of the most slender of all the species of the genus with many whorls. Of light color its main beauty is its elegant form. Not often seen in collections. 3½".

9. **Terebra pulchella,** Desh. Philippines. It is of a yellowish-brown color and has faint striations. 2½".

10. **Terebra maculata,** Lam. Polynesia. The Marlinspike, as it has been termed in commerce, has been brought into this country in quantity, as its is the largest and heaviest of the genus. The whorls are striped with regular brown blotches. There are 330 species in the genus but this one always attracts attention by its huge size of 6 to 8".

11. **Terebra cingulifera,** Lam. New Caledonia. A tall slender species with fine corrugations and reddish color. Very attractive. 4".

12. **Terebra cingulata,** Sow. Peru. There are fine regular dark markings on each whorl, and they have a glistening natural polish. Not common. 3".

13. **Terebra monilis,** Quoy. Philippines. A tall slender species with ridges on each whorl, and faint color which adds to its attractiveness. Usually 3".

14. **Terebra strigillata,** L. Hawaii. It has regular rows of dots in the top of each whorl. A smooth shiny species, that is a real gem. 2½".

The Harpidae consist of only a dozen or so species and systematists place them between the Oliva and Marginella. They are commonly called Harp shells throughout the world. All of the recent species are from the eastern hemisphere except one form, H. crenata which is from Western Central America and usually scarce. Fossil forms are found in Eocene.

The mollusk that forms this grand shell is variegated with beautiful colors. It crawls with vivacity. The front of the foot is crescent-shaped and divided by deep lateral fissures from the posterior part. Unable to withdraw completely within the shell, it is said, when irritated, to spontaneously detach a portion of the foot.

The Vermetidae are just about the strangest lot of shells in the world and consist of the true Vermetus and the allied Tenagodus. The shells are tubular and always attached. They are sometimes regularly spiral when young and always irregular in the adult growth. A collection of these shells from all parts of the world attracts wide attention as so very few of them are ordinarily in collections. I had a very great variety of them. The species from our own west coast of Florida now called V. spirata is one of the finest of the elongated forms as they attain 6 to 7 inches. The Siliquaria consist of around 25 species and about that number fossil in Tertiary. The typical species as well as several others occur imbedded in sponges. The shell is tubular, spiral at first, afterwards irregular and the shell has a continuous longitudinal slit, not found in any other form of mollusca. The Siliquaria are usually much harder to secure than the Vermetus.

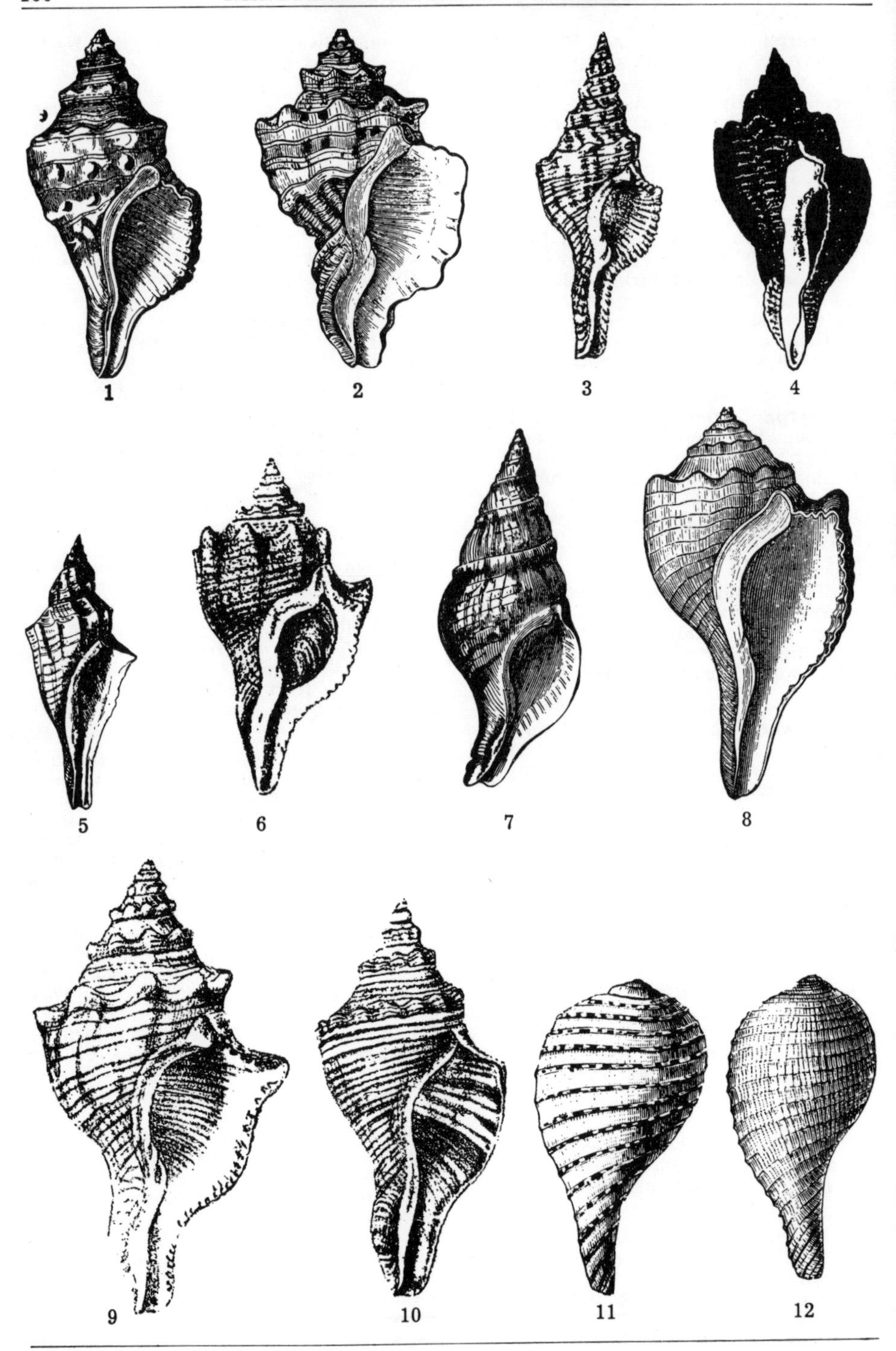

PLATE 49

1. **Fasciolaria aurantiaca,** Lam. Brazil. A fine shell of 3 to 4″ and well marked with red and yellow, more so than other species of this genus.

2. **Fasciolaria aurantiaca,** Lam. Another variety of this species found along the coast of Brazil. Colors are reddish-orange and white. 3 to 5″.

3. **Fasciolaria filamentosa,** Lam. Philippines. A fine elongated slender form widely distributed. It is usually brownish, but reddish forms are found and traces of other colors. All of the shells of this genus seem to propagate easily and are fairly common. 4″.

4. **Melongena pugilina,** Born. Philippines. A brownish shell of uniform color 2 to 3″ and fairly common. Its form suggests some of the smaller Fasciolarias or Siphonalias but is really an intermediate in the complex. The generic name of Melongena has been changed to Galeodes but as the name is not generally known or used, it may be eventually discarded. I like the old designation best and as do plenty of others.

5. **Melongena (Hemifusus) ternatana,** Gmel. Philippines. This species is now included with the Melongenas. There are other species which bridge the gap between this form and the true Melongenas. Usually has a fuzzy yellow periostracum. 6 to 7″.

6. **Fasciolaria granosa,** Brod. Panama. A large 6″ shell that usually has a rough periostracum, which when removed shows a yellowish specimen, often with rosy aperture. Lives on mud flats and often brought in by fishermen. Has strong operculum.

7. **Fasciolaria lignaria,** L. Mediterranean Sea. A small 2½″ shell almost devoid of color and fairly common. Seems to be found along the southern shores of Europe everywhere.

8. **Fasciolaria salmo,** Wood. Japan. Another fine robust shell which is usually seen in collections of about the same size as granosa from Panama. Most of the 35 species cf this genus are strong shells but there are a few small 2 to 2½″ forms. They seem to hold their own everywhere and are well represented in all seas.

9. **Fasciolaria trapezium,** L. Nicaragua. A noble species which attains 6″ or more, heavy, solid with spiral lines, which add to its beauty, when well cleaned. There is quite a little variation in the shells. Similar shells of a darker reddish color from the Philippines are called **audoni Jonas,** and I have had very attractive specimens from the Sulu Sea. 5 to 6″.

10. **Siphonalia dilatata,** Quoy. Australia and New Zealand. I put this shell here to show the close resemblance in form to the Fasciolarias. The Siphonalias are mostly small 1 to 2″ shells but this form ranges up to 6 or 7″ and is the largest species of the region. Only sparsely colored with brown and sometimes uncolored. They closely are related to the whelks or Buccinums.

11. **Ficus decussata,** Wood. Lower California. All of this genus are similar in form to the Florida Fig shell, the main difference being in size and form of reticulations of the surface. There are ten known species ranging from Florida to Australia. Lives on sandy beaches and always quite common.

12. **Ficus reticulata,** Lam. China. A very attractive shell with chestnut wavy lines and finely granulated surface. 3 to 4″.

The Family Fasciolaridae consists of the genera Fusus, Fasciolaria and Latirus. The Fasciolaria contains some of the largest univalve shells known, as F. gigantea of the Gulf of Mexico reaches 26″ in length. There are not many species of recent forms but over 30 from the Cretaceous.

The mollusk is quite similar to that of the Fusus and while the Fusus shells are more slender than the Fasciolaria, there is really no great difference and the Latirus while most all shells of 1 to 2½ inch are of same general appearance, except they are usually covered with very fine ridges.

The operculum of the Fasciolaria neatly fits the shell and should always be saved and fitted in its place which greatly adds to the attractiveness of the specimen.

✦

The Rissoiidae consist of the genus Rissoa, Rissoina and a host of other small forms of sea shells and are allied to the fresh water family of Hydrobiidae. In fact all fresh water forms of the world tie up with marine forms and in a properly arranged, systematic collection they are always included with the marine. Just one more reason why every collector should have a general collection rather than a specialized one. We only pass through this vale once, why not get all out of it we can.

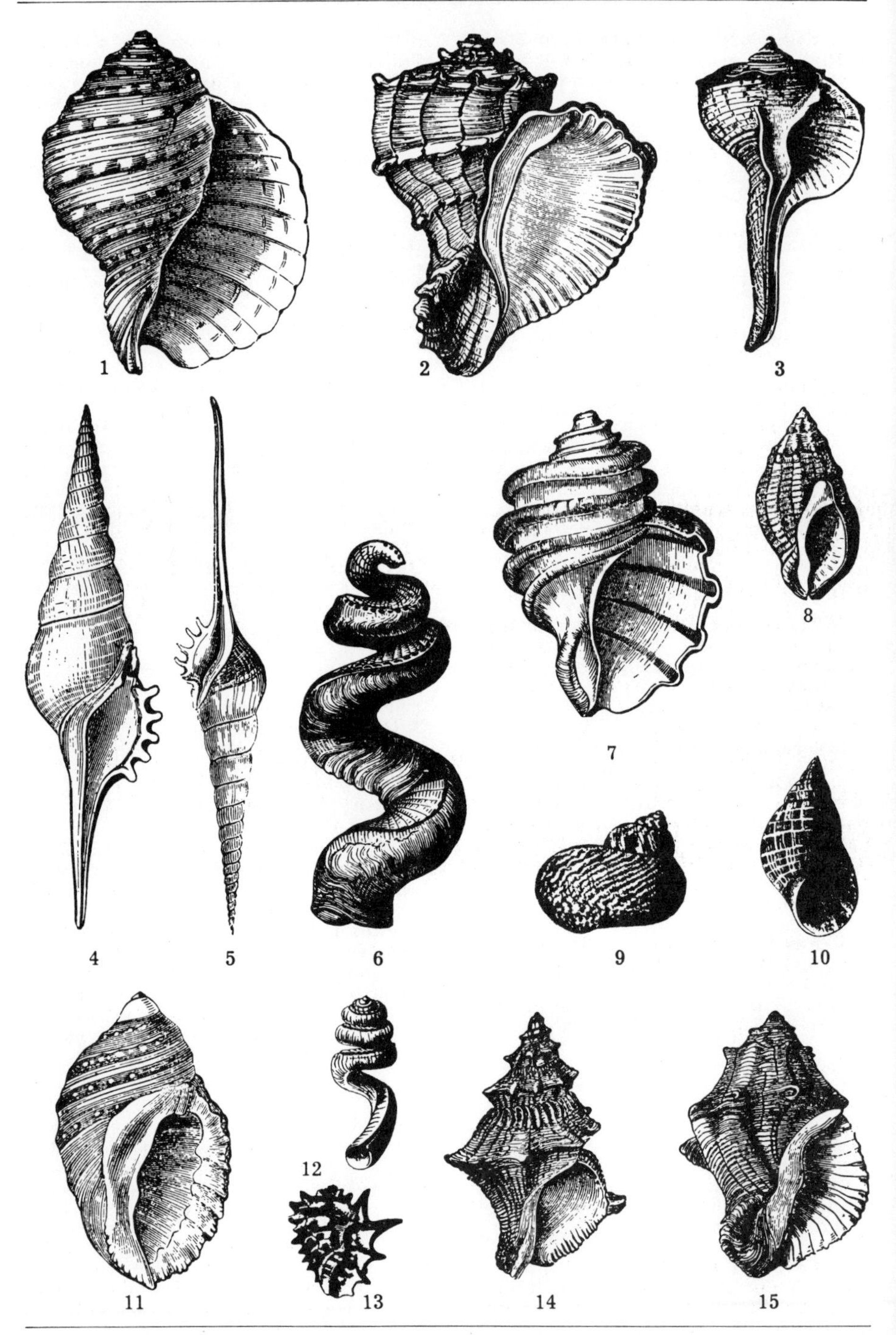

PLATE 50

1. **Dolium (Tonna) maculatum,** Lam. Singapore. A fine round thin shell ornamented with ridges and splashes of light yellow color. About 4″.

2. **Rapana bulbosa,** Sol. Japan. Rather solid shell of a brownish color. They are faintly allied to the Murex and are somewhat similar to them. There are 18 species known but only 2 or 3 are ever seen in the average cabinet of shells. 4″.

3. **Tudicle spirillus,** L. Tranquebar. A drab white shell that is not closely related to any other marine species. It is placed between the Vasum and Fulgars. Only two species are known and they are both uncommon shells. 2½″.

4-5. **Tibia (Rostellaria) fusus,** L. Celebes. The long slender spire and equally long and more slender base, make this one of the real gems of any shell cabinet. It simply must be seen to be appreciated. I illustrate two different forms of the same shell but which used to have different names. Rather rare. 5 to 6″.

6. **Tenegodus (Siliquaria) australis,** Q&G. Australia. This is a remarkable 3″ shell in which the spirals have an opening the entire length. The genus is allied to the Vermetus or Worm shells. There are over 50 known species many of which are of very peculiar form.

7. **Thais cingulata,** L. South Africa. A very unique shell with deep ridges which circle the shell. The ridges are shaded with brown. Much resembles Neptunea decemcostata of our own coast. Rather rare. 2½″.

8. **Thais amygdala,** Kien. Australia. A small species with light colored ridges on a darker background. It is not very common. There are 160 species of this genus. 1¼″.

9. **Phasianella venosa,** Rve. Australia. There are some 80 species in this genus and most all small. Many are almost minute but nearly always of the most brilliant colors imaginable. This form is 1½″ and has a richly mottled surface.

10. **Phasianella australis,** Gmel. The Pheasant Shell. South Australia. An extremely highly colored species, that baffles all attempt at description. It is very fragile in texture but not really thin. Ground color gray, ornamented with patterns of red, white and other colors in intricate design. 2½″.

11. **Thais columellaris,** Lam. Galapagos Ids. and may be found at Panama. A remarkably solid thick shell for its size, white, distinct from all other forms of the genus. It has slight circular ridges of a brown color throughout. Lives on surf beaten rocks in company with **patula** but is a much rarer species. 2½″.

12. **Tenegodus (Sliquaria) anguina,** L. Moluccas. A very singular shell with opening entire length of coils. Does not seem to be very common, as so many cabinets do not have it. Uncolored. 2″.

13. **Drupa digitata,** Lam. Fingered Shell. Philippines. Fairly common with other species of the same genus on the same beaches. Spiney with peculiar white aperture. Other forms have pink, violet and yellow apertures and are much desired by collectors. Sometimes all four will be found in the same bays. About 100 species are known mostly from Eastern Seas. 1″.

14. **Thais kiosquiformis,** Ducl. Panama. A handsome deep brown shell with a row of sharp short spines through the middle of the whorls. Lives on rocks in muddy situations and on mangrove roots. Feeds on oysters. 2″.

15. **Cuma tectum,** Gray. West Mexico. An angular grayish shell with row of stubby tubercles around the middle of the last whorl. Quite distinct from all other forms. There are 21 species of this genus and while all are allied to Thais, they are usually very distinct in form. Live on rocks between tides. 2″.

The Delphinulidae consist of the genus Delphinula of which there are only a dozen or so species and Liotia which are exactly like them but all very small forms around a half inch in diameter or less. The Delphinula are found fossil in the Jurassic and on. They live on reefs at low water which accounts for most of the forms being fairly common as it is easy for collectors to find them but I have had species that range to $10 each likely being hard to find. They inhabit mainly the Eastern tropical seas. The Liotia are numerous and mainly inhabit tropical seas.

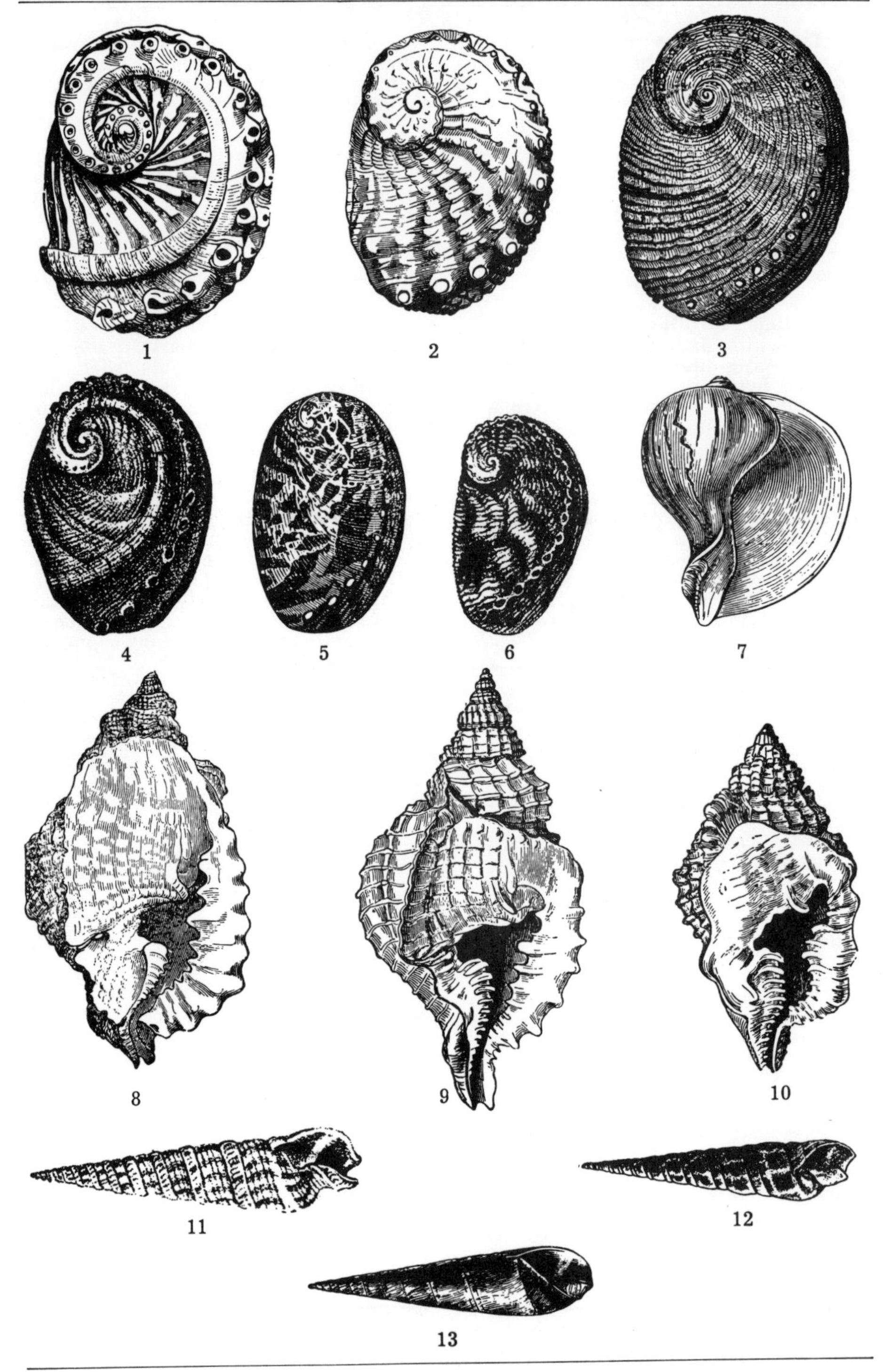

PLATE 51

1. **Haliotis tricostalis,** Lam. Australia. A remarkable shell with ridges and plaits adorned with reddish and other colors. Quite rare. 3½".

2. **Haliotis ovina,** Chem. Polynesia. Back is corrugated and adorned with green and reddish shades. Very attractive when the numerous incrustations are removed. 2½".

3. **Haliotis roei,** Gray. Australia. Finely ridged with mottled surface. Not very common. 3 to 4".

4. **Haliotis improbula,** Iredale. South Australia. A handsome wrinkled shell as per cut. Rather thin. 3".

5. **Haliotis glabra,** Chem. Moluccas. A neat 2" shell, finely lined and colored with brown and white. 2".

6. **Haliotis pustulata,** Rve. New Caledonia. A small shell of brownish color with undulated surface. 1½".

7. **Melapium lineatum,** Lam. South Africa. A small odd shaped shell with small dark lines on light surface. It is allied to the Fulgars. There is one other variety that differs slightly from the type. 1".

8. **Distorsio (Persona) anus,** L. Philippines. The aperture is flattened out and covers the whole surface of the shell. Of light color, it is of very unusual shape. There are eight known species in this genus. 2".

9. **Distorsio (Persona) constrictus,** Brod. Panama. A fine white form with the very peculiar aperture of the genus and finely reticulated. 2".

10. **Distorsio (Persona) ridens,** Rve. Hong Kong. Slightly smaller than the other two species, it is white and otherwise quite similar. 1½".

11. **Terebra variegata,** Gray. Lower California. The surface is finely marked with variegated colors. One of the medium sized forms of 3".

12. **Terebra subulata,** Lam. Philippines. A very handsome spotted shell, long and slender with fine brown markings. 4 to 6".

13. **Terebra lanceata,** Gmel. New Caledonia. The species is smooth and polished with regular dark spots near the top of the whorl. One of the real little beauties of the genus. 2½".

The Family Haliotidae consists of only one genera, Haliotis. There are likely over 100 species of recent forms, some of which attain nearly a foot in diameter. They are all ear-shaped, with a small flat spire, aperture very wide, usually iridescent. The exterior often is striated and dull but there are many exceptions. Some few being beautifully colored on the back and perfectly smooth. The outer angle of the shell is perforated with a series of holes, those nearer the spire usually being closed.

The shells cling to the rocks with great tenacity and many stories have been printed about animals being trapped by putting their foot under the shell and the rising tide finished them off. The genera Stomatella, Gena, Broderipia and Stomatia which immediately precede the Haliotis are about the same shape but all are small, being under 1 inch. They are very neat shells, often of brilliant color and much sought by collectors.

The Trochidae consists of the genus Trochus which contains the largest shells and then there are 30 or more other genera which contain beautiful shells of a smaller size. There are around 300 species of Trochus proper and they range from shallow water to 100 fathoms and more. There are 400 species or more fossil from Devonian on. If you include all the genera allied to the Trochus like Clanculus, Cantharidus, Monondonta, Chlorostoma, Umbonium, Gibbula, Calliostoma, Euchelus to mention only part of them, you would have many hundred other species as some of these genera are very numerous. Nearly all of them are very attractive small shells, some among the most gorgeous known. Many of the shells are of a pearly nature and are made into attractive necklaces and used to adorn objects of art. The most famous of the Trochus is T. niloticus, called Trochi all through the East Indies, where it is an article of commerce. Before the war on Thursday Island, N. E. Australia, there were 60 or more boats engaged in fishing for Trochi and about 85% of the haul came to the USA where the pearl was used in commerce. All shells under 3½ inches were thrown back and only the larger mature shells were shipped. The whole operation was controlled by Australian government. As the boat owners employed Japs as divers and helpers in preference to the native bushmen, I have often wondered what became of them, but I suppose the industry will get going as soon as possible.

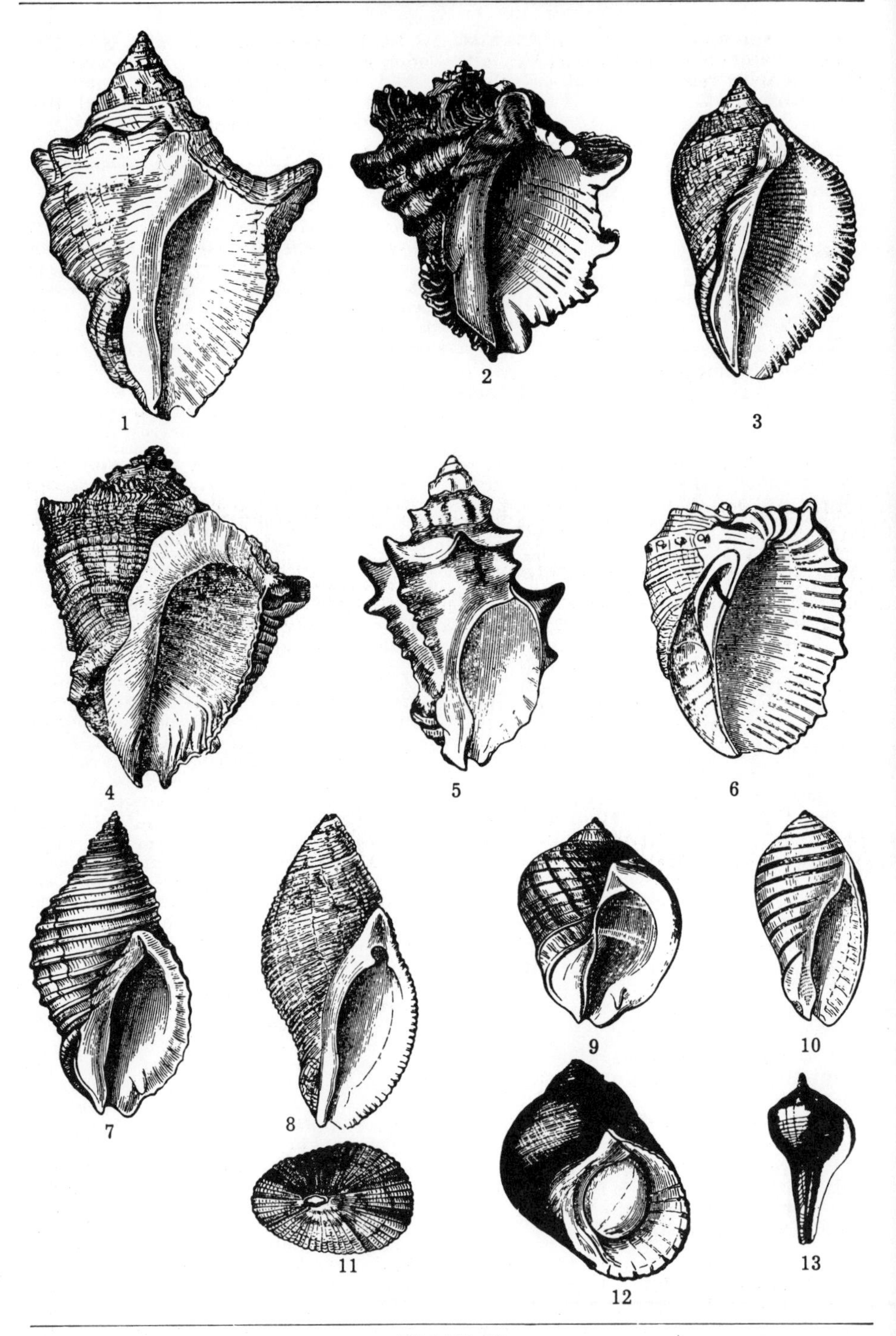

PLATE 52

1. **Thais consul,** Lam. Galapagos Ids. A strong solid shell showing very little color and prominent knobs. There are many unusual types of marine shells in these isolated islands. 2½".

2. **Acanthina muricata,** Brod. Mazatlan to Panama. A strong shell with deep ridges, scalloped edges around same, large aperture and wrinkled periostracum. Lives on rocks between tides and is not common. 3".

3. **Thais persica,** Lam. Philippines. One of the most attractive shells of the group. The white aperture is shaded with deep color, the back is fairly smooth and adorned with dots and splashes of blackish-brown with white dots. 2 to 3".

4. **Rapana bezoar,** L. Japan. A rather solid shell of horn color, rough back and attains 4 to 5".

5. **Thais bituberculari s,** Lam. Singapore. A very solid shell of almost black color covered with three rows of pointed tubercles. Common on East African shores and much of Pacific. 1½".

6. **Thais planospira,** Lam. Galapagos Ids. Rather unique shell of a reddish-russet color. The aperture is deep and wide adorned with reddish stripes and longer than the shell. There is a black spot on left side of aperture. A rare shell living on rocks at extreme tides. 2".

7. **Acanthina acuminata,** Mont. Cape Horn, South America. A finely reticulated shell, almost devoid of color, as it lives in very cold water. Most of the species of this genus are from South America. 1½".

8. **Iopas sertum,** Brug. Pacific generally. A rather common fine shell of which there is the one species and one variety of the genus. It is of a reddish-brown color with dark apical whorls. 1½".

9. **Acanthina crassilabrum,** Lam. Chile. A short fat stubby, heavy, shell that is a light shade of brown. I suspect it is from deep water as most all shore specimens are white and dead when found. 1½".

10. **Vexilla vexillum,** L. Hawaii. A neat little striped shell of which there are only three forms known. All from Pacific territory. About 1".

11. **Fissurella fascicularis,** Lam. Bahamas. The back is finely reticulated and interior white. It is fairly common in the little coral tide pools that surround some of the islands. 1".

12. **Chlorostoma nigerrinum,** Gmel. Coquimbo, Peru. A very fine round black shell of neat proportion. All the specimens I have had seem to hold the operculum in place very tightly. The shell is occasionally included among the Turbo. 1¼".

13. **Turbonella pyrum,** L. Chank Shell. Ceylon. This species is the sacred shell of the Hindus, on which you will find much information in the old Natural Histories. 3 to 5".

The Thais (Old name Purpura) are part of the great Family of Muricidae and they differ from the Murex mainly by having no varices but nodules and the operculum with a lateral nucleus. It is almost impossible to entirely differentiate them from the Murex as there are so many forms which are closely allied, specially among the small species. Some of the smooth varieties of Murex are very close in form to the Thais. There are a very large number of species known and they are found in all parts of the world. When they are thoroughly cleaned many are of considerable beauty but most of the species tend to be dark in color, some fully black.

The ancients also obtained various dyes from the forms found so abundantly along the sea shore. The shells prey on all sorts of mussels, limpets and barnacles being specially fond of oysters. Occasionally one is found sinistral but that also happens in many marine genera but they are always rare, except in the Fulgar perversa of the Gulf of Mexico which is always sinistral, and attains one foot in size, being as far as I know unique in that respect among marine forms.

Very much has been written in English scientific journals on the life history of their most common form Thais lapillus which attains a much larger size than is found on the American shores of New England and elsewhere.

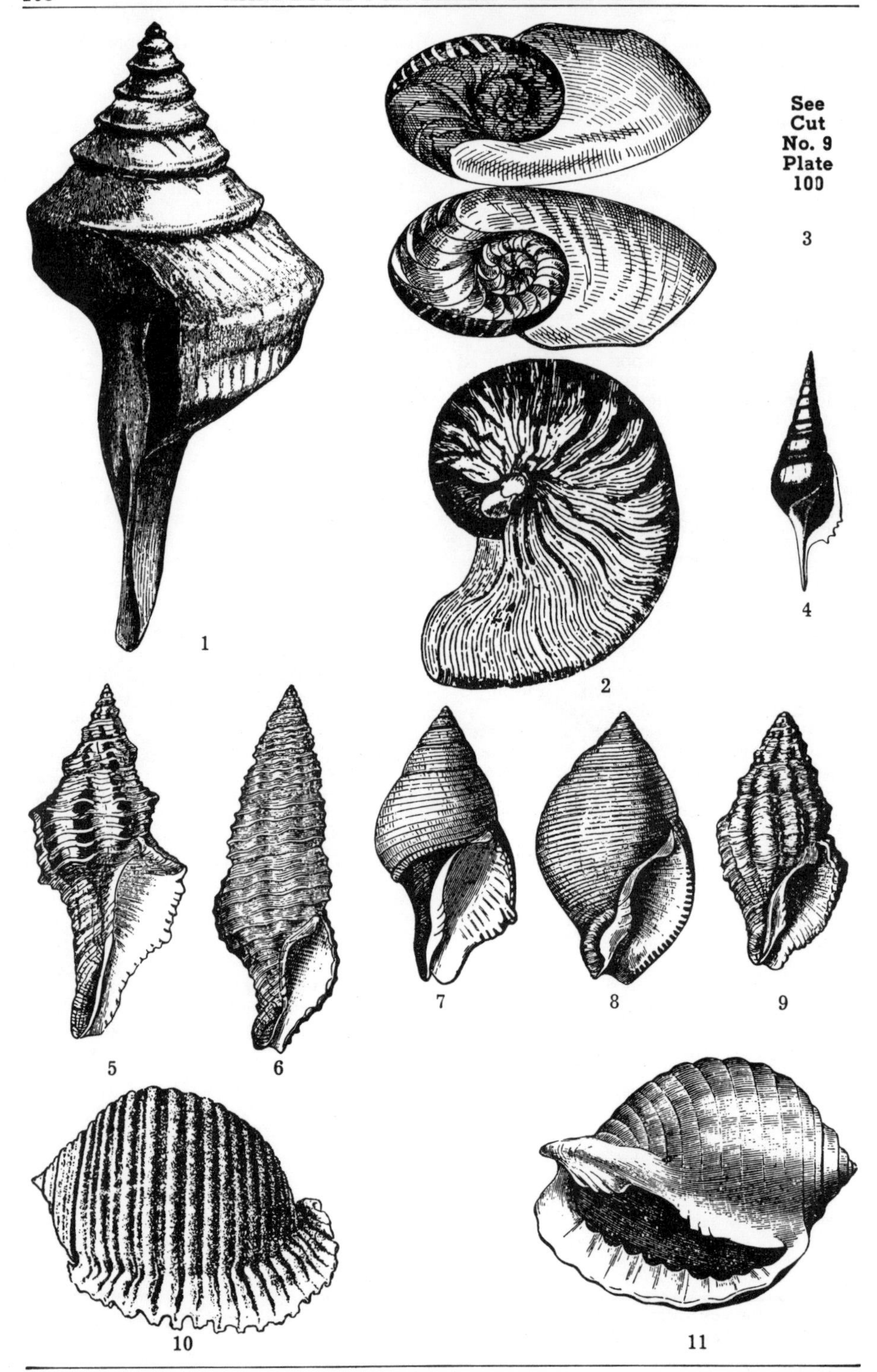

PLATE 53

1. **Megalatractus proboscidifera,** Lam. Australian reefs. One of the two largest marine shells in the world. Of a uniform yellow color it attains 20″ or more but fine 15 or 16″ shells are usually the best color and upper whorls are more apt to be perfect. It is a comparatively light and thin shell for its immense size. The other unusual marine univalve shell is the Fasciolaria gigantea of Florida. The pair shows the highest development of size in univalve shells. It is placed with the Melongenas.

2. **Nautilus pompileus,** L. East Indies. When this shell is polished it is commonly called the Pearly Nautilus. In the New Hebrides and other island groups of the South Seas the fishing for Nautilus is a regular business of the natives. They fashion a barrel of bamboo, with a curved inward opening at each end, place a rock inside to make it sink, put in the bait and drop it to the bottom in 30 to 50 ft. of water. Sometimes there will be a long string of these traps connected with ropes and buoys. The Nautilus mollusk crawls along the bottom, goes into the barrel after the bait and seldom knows how to get out. Twenty million years ago there were several hundred species but now there are only two living forms. Oliver Wendell Holmes wrote a poem on the Nautilus which is often read in schools. You will find much information on this species in the August 1935 number of the National Geographic Magazine. The upper cut shows a shell cut in half with the air chambers. Size 6 to 9″.

3. **Columbrarium pagoda,** Less. Japan. A very odd shell with spire like a Pagoda and long slender basal appendage. There are three forms known. It seems to be allied to the Turris. Is of brownish color. 3″.

4. **Tibia (Rostellaria) curvirostris,** Lam. Philippines to Red Sea. This is a strong robust highly polished shell with elongated spire and stubby spiral base. A brilliant yellowish species which has always been much admired by shell collectors but so often not on the market in sufficient quantity. 6″.

5. **Latirus polygonus,** Gmel. Mauritius. A finely marked striking species and one of the largest of the genus. It is yellowish-white and the ridges are splashed with rich dark brown. 2¼″.

6. **Latirus craticulatus,** L. Mauritius. This species has spiral lines of red color and prominent ridges. I have had it from Philippines. 2″.

7. **Latirus leucozonalis,** Lam. West Indies. A comparatively smooth species of a brownish color. Most of the shells of this genus are ridged in all directions. 1¼″.

8. **Latirus smaragdulus,** Lam. Philippines. A dark blackish shell which is often completely covered with bryozoans of all sorts. Usually must be well cleaned to know what you have four.d. 1¼″.

9. **Latirus nassatulus,** Lam. Mauritius. A short stubby species with prominent ridges of a whitish color. The aperture is deep pink. 1½″.

10. **Dolium (Tonna) ringens,** Swain. Panama. The Cask Shell. Must be fairly common from Lower California southward as they used to be sold commercially in all sizes from 3 to 8 inch. They are almost perfectly round with flaring aperture. There are 35 species of this genus in the world most of which are comparatively thin shells but this one is a solid fellow. Lives under edges of rock at low tides.

11. **Dolium (Malea) pomum,** L. Philippines. A handsome shell of pure white or flesh color. They are round, with a slightly flaring aperture. Burrows in sand bars. 2 to 3″.

The teeth of all classes of mollusca have been studied for many generations. In the best scientific works they are illustrated. There is a wide difference in form and number. In the genus Umbraculum they reach their greatest number believed to be about 750,000. It seems hard to believe that a small mollusk of only 3 to 4 inches could ever use such a vast number in its daily life.

✦

It has been reliably reported that new species established on a certain shore have migrated as much as 60 miles a year. Littorina littorea appeared at Salem, Mass., in 1872. It reached New Haven, Conn., in 1880, averaging about 23 miles a year. Ocean currents are responsible for many species being widely distributed when in the larvae stage. This form of distribution has been going on for countless ages.

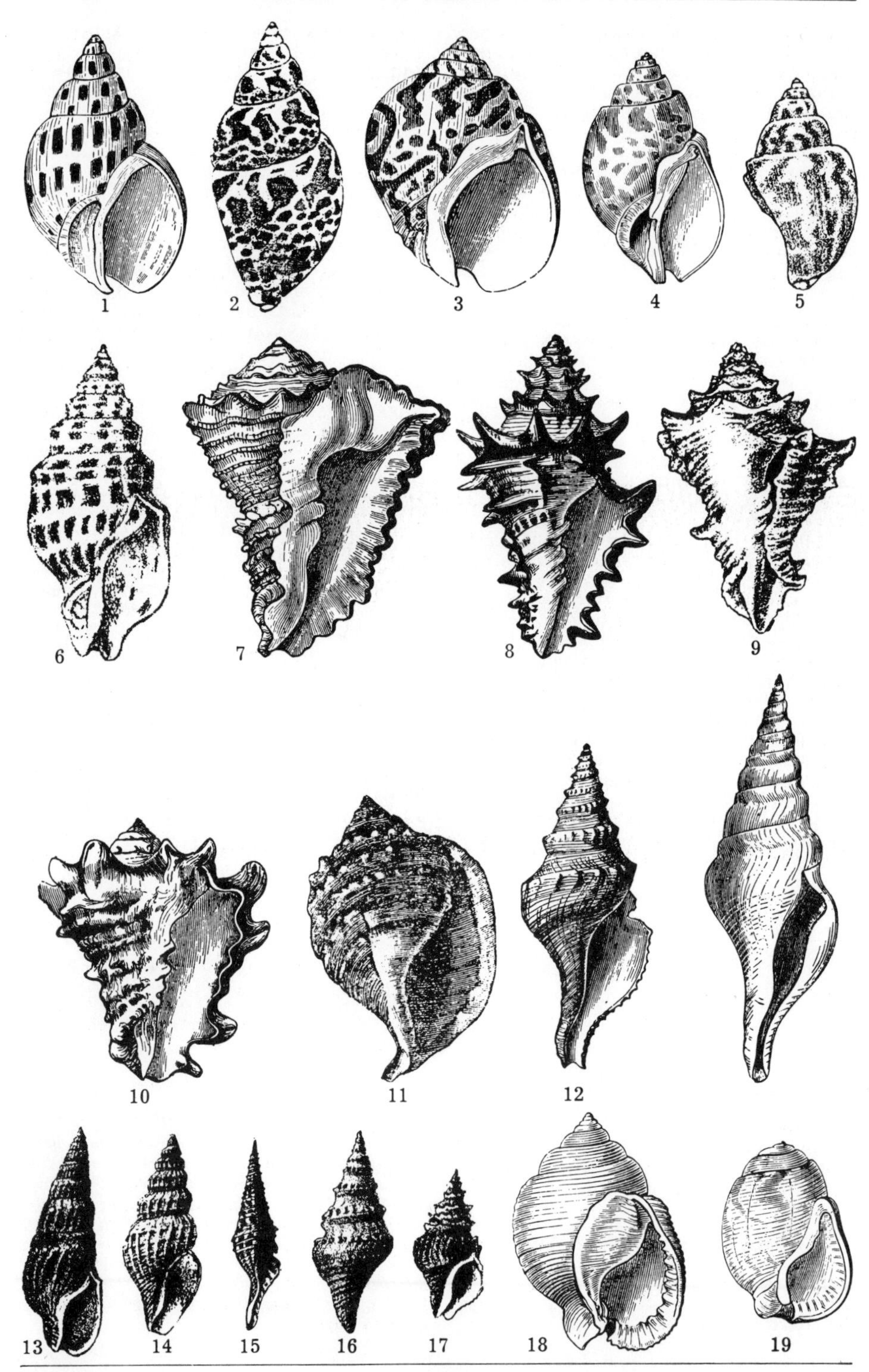

PLATE 54

1. **Eburna areolatus,** Lam. Hong Kong. A round, elevated spotted shell that is not real common. All of this genus are called Ivory shells and I have never heard the reason. They have a periostracum usually brown which must be removed to show the color or pattern. 2″.

2. **Eburna japonicus,** Sow. Japan Seas. This is the best Ivory Shell of the genus, and has been most widely distributed. They are so common they are manufactured into novelties, such as whistles, etc. 2½″.

3. **Eburna valentianus,** Swain. Karachi, India. Similar to the others but has a tendency to be shorter and wider, ornamented with dark irregular blotches. The whorls are curved inward. One of the rarest of the group. 2″.

4. **Eburna spiratus,** Lam. Ceylon. A very distinct species with splashes of faint drab over the surface. The top of each whorl is curved inward, about ¼ inch. Size 2″.

5. **Erburna lutosus,** Lam. New Zealand. A small more slender form from the southern part of the world, and ornamented with irregular splashes of yellowish-white, the last whorl is humped at top. There are 16 species in this genus which are widely distributed. 2″.

6. **Pusionella nifat,** Brug. West Africa. This species is smooth and finely ornamented with splashes of yellowish-brown. It is 2″. There are about 14 species of this genus and this one is the largest. Most of them are from the same region and seldom seen in American collections. There are just no live collectors in the territory where found.

7. **Vasum cassidiformis,** Val. Brazil. A very strange form of shell which much resembles some of the wonderful Pliocene fossil species found in Florida. It has little ridges of knobs over the surface and wide white aperture. 3″.

8. **Vasum ceramicum,** Lam. Moluccas and Pacific generally. This species is widest in middle and tapers to each end. Finely ornamented with spines along the top of whorl. Not common. 3″.

9. **Vasum capitellum,** L. West Indies. All of the 13 known species of this genus are strong robust shells and of very odd form. They are usually covered with marine growths which must be removed to show their real pattern. This species is of light brownish, has a few knobs. 2½″.

10. **Vasum cornigerum,** Lam. East Indies generally. This species seems to be one of the most common of the group and the one most often seen in collections. It is wide at the top tapering to base, covered with knobs of black and prominent rounded tops. 3″.

11. **Cassidaria echinophora,** L. Mediterranean Sea. Of the 14 known forms of the genus this is perhaps the most common one seen in collections. It is of horn color, looks much like a Cassis, with which it is closely affiliated. 2½″.

12. **Turris Javana,** L. China. These two cuts show the true Javana and a variety that used to be called nodifera. They are of a grayish color, fairly common and somewhat variable. 2 to 3″.

13. **Turris fusca,** H&J. Gulf of Omar. This little fellow comes from 150 fathoms. It has the cross ridges found in many other species. Almost black. 1½″.

14. **Turris coffea,** Smith. Cebu, P. I. A little brownish shell with usual ridges. 1½″.

15. **Turris grandis,** Gray. Philippines. A princely shell even if the cut is small. One of the largest of the genus ranging to 5½″. It is ornamented with circular ridges and hundreds of small reddish-brown dots. It is truly a grand Turris.

16. **Turris muricata,** Lam. West Africa. A neat little white shell with rows of sharp spines and perpendicular ridges on the whorls. 1½″.

17. **Turris bijubata,** Rve. New Caledonia. A sharp pointed little chap of a dark color and typical of many of the small forms of this genus. 1¼″.

18. **Desmoulea retusa,** Lam. Liberia. It is a small round lined shell with white aperture about 1″. There are 14 species in the world, all of which are more or less rare and seldom seen in collections.

19. **Desmoulea abbreviata,** Gmel. Natal, So. Africa. Differs slightly in form from preceding species but the spiral lines and aperture immediately place it in this genus. This is the form most commonly seen in collections and my collector in Natal found it freely on the beaches of that Colony.

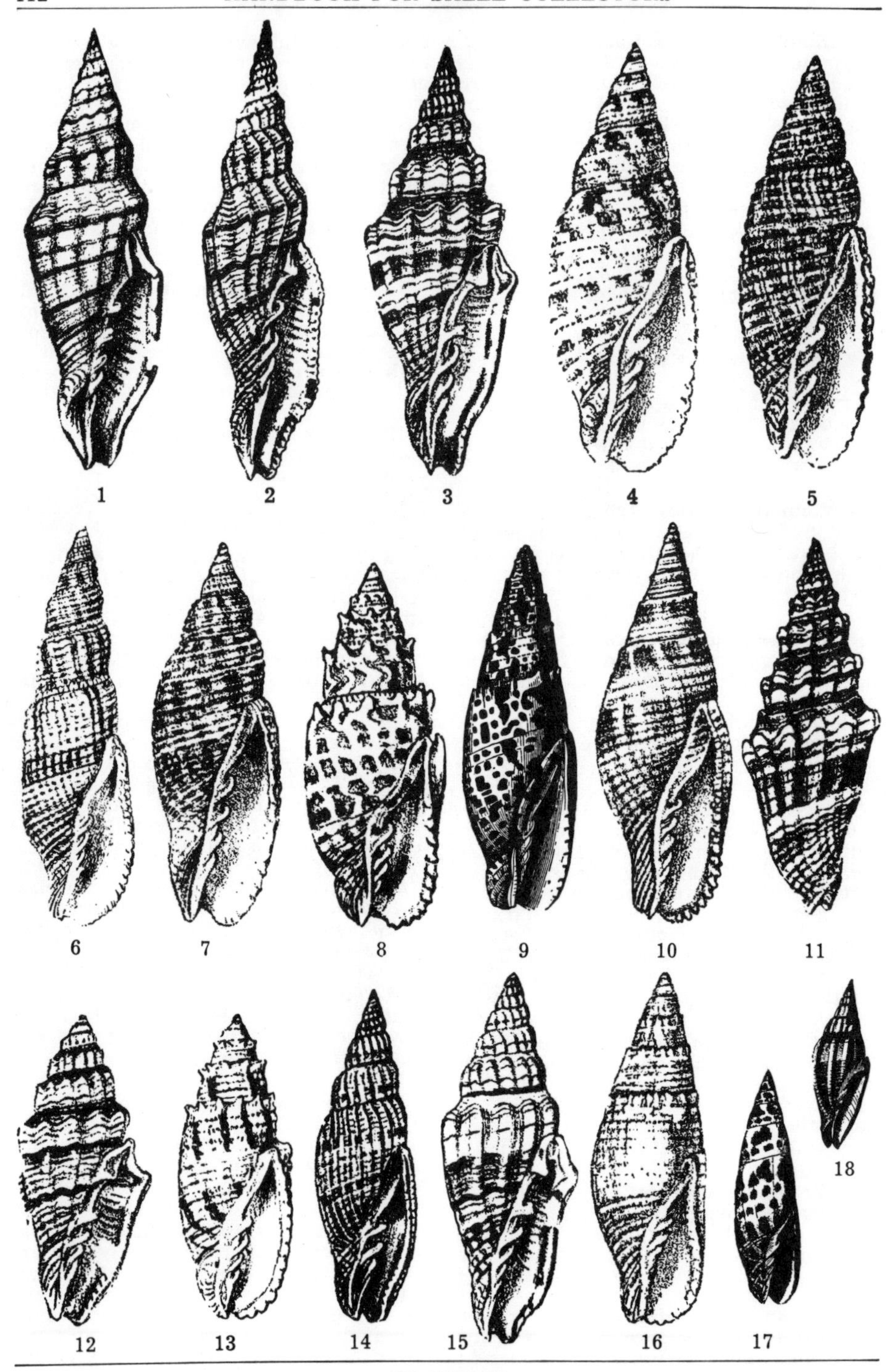

PLATE 55

1. **Mitra melongena,** Lam. Andaman Ids. An elegant narrow species adorned with numerous bands of white and shades of brown. Longitudinal ridges throughout. 2½″.

2. **Mitra regina,** Sow. Andaman Ids. One of the highest colored forms known, richly ornamented with shades of white, yellow and orange in the form of bands. 2½″.

3. **Mitra costellaris,** Lam. Andaman Ids. A slender well marked brownish shell that is quite distinct from other species. 2″.

4. **Mitra serpentina,** Lam. Philippines. Lamarck when he named this species over 150 years ago evidently believed it had a color pattern of a serpent which is quite true. It is mottled with shades of russet. 3″.

5. **Mitra scabriuscula,** L. Philippines. This species is completely covered with circular ridges of russet color and white spaces between. Very attractive. 2″.

6. **Mitra tessellata,** Mart. Philippines. A long slender shell with circular ridges and splashes of russet coloring throughout. Rather uncommon. 3 to 4″.

7. **Mitra sphoerulata,** Mart. New Caledonia. It is completely covered with circular ridges which are ornamented with dark dots. Very distinct and unusual. 2½″.

8. **Mitra pontificalis,** Lam. Philippines. It has a row of nodules at the top of each whorl and the body is richly ornamented with splashes of red on a white background. 1½ to 2½″.

9. **Mitra papalis,** Lam. Ceylon and Pacific generally. A noble species attaining 4″ or more. The Papal Mitra is one of the finest large forms. Richly ornamented with russet markings of various shades.

10. **Mitra variegata,** Rve. Red Sea. Of a general yellow color it is covered with corrugations. 2½″.

11. **Mitra intermedia,** Kien. Philippines. An elongated shell with bands of brown and white and spiral lines at base. 2″.

12. **Mitra corrugata,** Lam. Philippines. A fine deeply ridges shell of dark color. All of the Mitras of the world are very handsome shells regardless of size. 1½″.

13. **Mitra digitalis,** Chem. Mauritius. A splendid shell ornamented with different shades of russet markings and splashes of white. ½″.

14. **Mitra stigmitaria,** Lam. Philippines. A small shell completely covered with a shading of white to gray, and three brilliant red bands. Very showy in its brilliant dress. 1½″.

15. **Mitra vulpecula,** L. Moluccas. A shell of rich coloring and very great variation. It is ornamented with many shades of brown, yellow, orange, russet and white. Frequently four shades or more on same shell. Most of the 625 species in this genus are small, ranging from minute to 1″. It would take a 300 pp. book to properly monograph them all. Many species look like the small Columbellas and others much resemble other genera. 2″.

16. **Mitra adusta,** Lam. Philippines. A yellowish shell finely reticulated that is fairly common in the coral reefs of that territory. 2″.

17. **Mitra episcopalis,** Lam. Philippines. A brilliant shell ornamented with reddish-orange color on a pure white background. Has been sold commercially and more widely distributed than any other species.

18. **Mitra cinctella,** Lam. Moluccas. A very distinct shell, with ridges and indistinct spiral lines, ornamented with russet bands. 1½″.

The Mitridae are composed almost exclusively of the genus Mitra which is a very large one. The Mitroidea and Diabaphus are included but I never had only one species of each.

There are several hundred species of Mitra known, many of which are very small, some closely resemble Columbellas. Most of them are tropical and sub-tropical. Over a hundred fossil forms are known commencing with the Cretaceous. Many of the most beautiful forms are found in the Philippines and range down to the Celebes. Many forms have been dredged to 50 fathoms or more. Some species are strictly reef forms where they hide in holes and under sea weed, stones and rocks. A few burrow in the sand or mud at various depths. Several forms hide under stones during the day and only come out at night to feed.

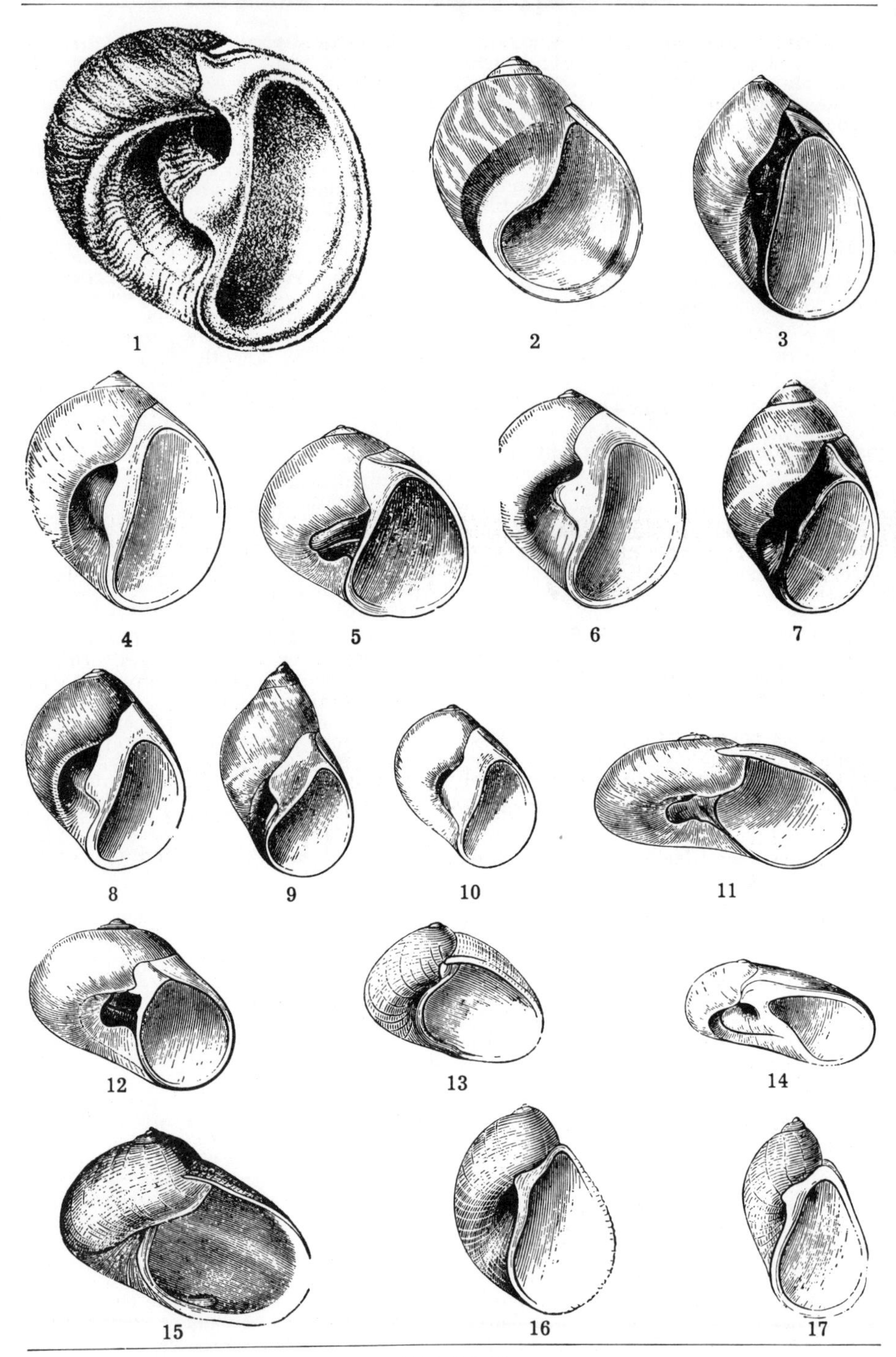

PLATE 56

1. **Polinices albumen,** Lam. New Caledonia. There are few shells with such remarkable umbilical region. It shows a tendency to merge into the next genus Sinum, which also merge from the type of the next genus to this. This species is a yellow shell with white base. 2″ or more.

2. **Polinices fluctuata,** Sow. Mauritius. The apical whorls are very small and most of the entire shell consists of the last whorl. It is of a yellowish-brown with zigzag stripes of white. The white umbilical patch is shaded with reddish-brown. Rather thin and light. 2¼″.

3. **Polinices maura,** Lam. Philippines. It is of a reddish-brown color with wide aperture and dark splash of color in umbilical region. 1 to 2″.

4. **Polinices otis,** B&S. Galapagos Ids. A white shell with white callous around part of aperture and very curious umbilical region as shown. 1 to 1¾″.

5. **Polinices chemnitzi,** Recl. China. A shiny shell with faint trace of yellow, open umbilicus and thin outer lip. Lives on sand bars between tides. An attractive shell. 1½″.

6. **Polinices powisiana,** Recl. Philippines. A very handsome shell of a bright orange-yellow color, the upper whorls white and greenish. Open umbilical region and aperture white. 1 to 1½″.

7. **Polinices bifasciata,** Gray. Gulf of California. A round conical shell of a light shade of brown with three rings of white. The wide umbilical region is deep brown. A fine species. 1½″.

8. **Polinices aurantia,** Lam. Philippines. It has a very smooth polished surface of a deep yellow color of varying shades. 1 to 2″.

9. **Polinices conica,** Lam. Australia. A very conical shell as its name indicates of a horn color, the umbilical region being slightly darker. 1½″.

10. **Polinices mamilla,** Lam. Philippines. A pure white robust solid shell with white enamel. Fairly common over a wide range and there are a number of similar glistening white species. 1 to 2″.

11. **Polinices glauca,** Humboldt. Acapulco, Mexico. A peculiar species with concave base showing a tendency to merge with other forms in a different genus. Not common. 2″.

12. **Polinices petiveriana,** Recl. Samar, P. I. A small white species that much resembles other white species except the dark open umbilicus. 1¼″.

13. **Polinices martinianus,** Phil. West Indies. A thin white species that is much the form of some of the Sinum. Aperture very large and oblong. 1½″.

14. **Polinices mittrei,** H&J. Moluccas. A fine species ranging from light to dark yellow with very peculiar umbilical region. About 1½″.

15. **Sinum neritoideum,** L. Borneo. A remarkable form of this genus which shows a divergence toward the Polinices. It is white as are most of the 60 known species in the world. 2″.

16. **Sinum papilla,** Gmel. Philippines. A narrow upright species with inflated aperture, all white. The mollusk of this genus completely covers the shell when feeding and is usually very much larger than the shell. They live on sand bars usually just beneath the surface. 1½″.

17. **Sinum oblongus,** Rve. Philippines. Somewhat similar to the preceding species, white and radically different from most of the other forms of the world.

The Naticidae consist almost exclusively of the Naticas and Polinices but include such small genera as Sinum (Sigaretes) and Eunaticina. The old genera Natica included 7 subgenera but some of these have been raised to generis rank like the Polinices. One of the reasons for this change was that the true Naticas always have a shelly operculum that fits the aperture perfectly while the Polinices have a horny operculum. In arranging a large collection of the Family it is well to have each group separated.

There are several hundred recent species and over 500 fossil forms have been described from the Silurian. The mollusk is entirely blind. The typical Natica move quickly, are carniverous, mostly living in sandy places where they hide under the surface and burrow after bivalves. Range to 100 fathoms in depth. The colored markings of the recent shells are frequently found on the fossil forms, which is unusual.

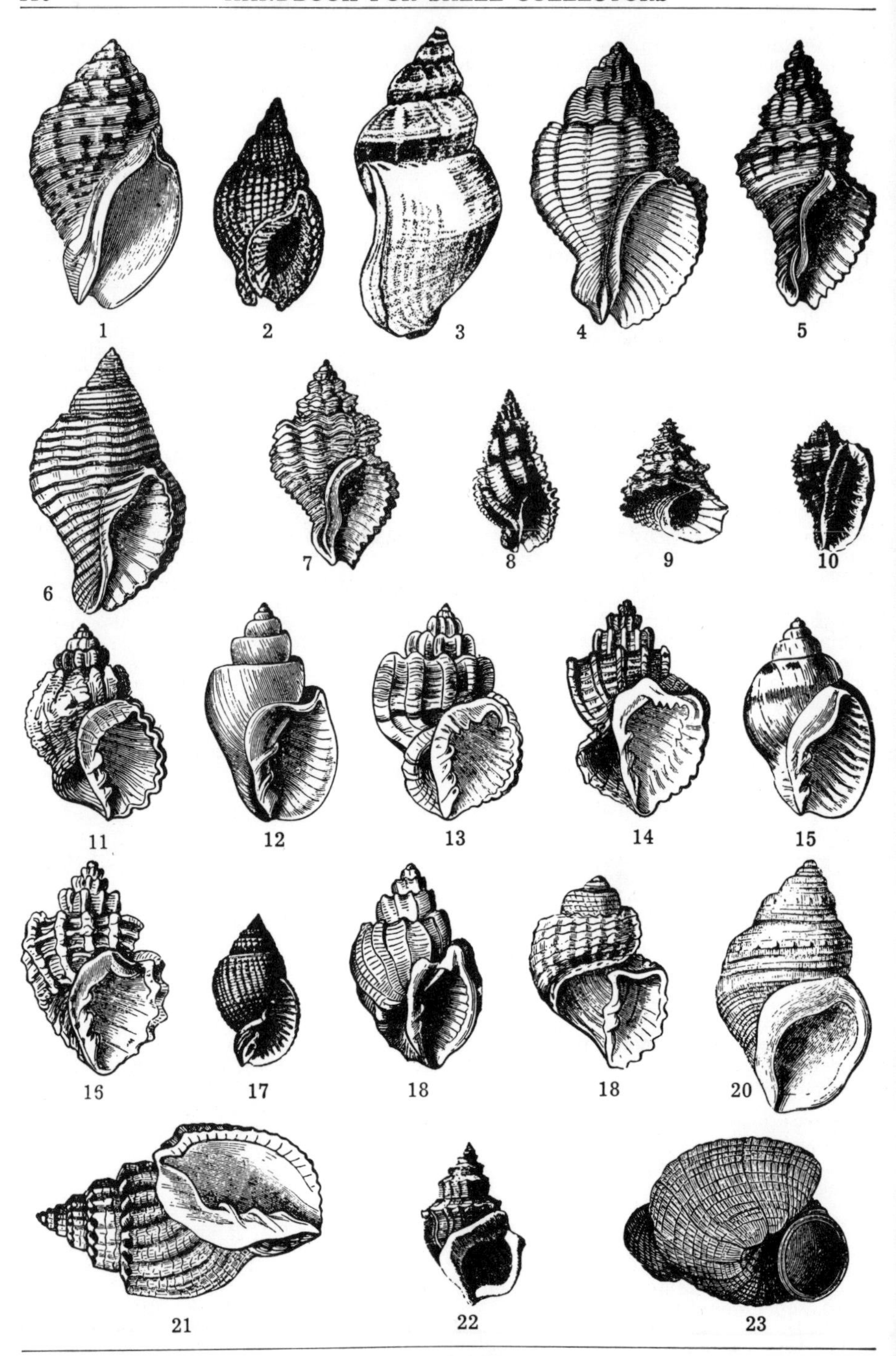

PLATE 57

1. **Cominella lagenaria,** Des. South Africa. Has few ridges, dark brown aperture and black leathery operculum. Quite distinct. There are 50 species in the genus, mostly from the southern hemisphere. 1″.

2. **Cancellaria laticostata,** Kob. Japan. A finely reticulated shell of a light brownish color. 1″.

3. **Cominella adspersa,** Brug. New Zealand. Rather large shell for this genus, of a brownish color, somewhat ridged, with smooth orange aperture. 2½″.

4. **Pollia tranquebarica,** Gmel. Ceylon. A very neat shell finely ridged, the ribs being of a darker color than the ground shade. Not real common. 2″.

5. **Pollia insignis,** Rve. Panama. It is a fine shell well ribbed and notched, of a reddish-brown color. Aperture scalloped. 2″.

6. **Pollia undosa,** Lam. Australia. One of the best known forms of the genus and widely distributed. The numerous ridges are of a deep brown color with white between. Usually covered with a hairy periostracum. Lives on rocks between tides. 1¼″.

7. **Pollia erythrostoma,** Lam. Ceylon. It is finely ridged with lines of a shade of brown, and white aperture. There are over 100 species in the genus. Fairly common. 1¼″.

8. **Phos senticosus,** L. Philippines. A very attractive ridged shell of intricate design as are all of the some 50 species. This is one of the finest and most often seen in collections. 1″.

9. **Tectarias pagodus,** L. Philippines. This species is the largest of the 50 known varieties in the genus. I recently had fine lot from British Solomons. It is a noble shell, completely covered with sharp points or knobs. Really a giant as compared with the other species of the genus. 2″.

10. **Morum (Oniscia) cancellatus,** Sow. China. Its surface is remarkably sculptured, lives under rocks and is about the finest of the genus. There are 12 known species. Rather attractive. 1″.

11. **Cancellaria chrysostoma,** Sow. Panama. A short stubby shell with deep ridges and orange aperture. There are over 160 kinds in the world and very few are at all common. One of the very hardest groups to complete even if most of the shells are small. Largest species is from deep water off California coast. 1″.

12. **Cancellaria excavata,** Sow. Australia. A grayish-white form with white aperture. Excavated whorls, with a tinge of brown near top. 1″.

13. **Cancellaria haemastoma,** Sow. Galapagos Ids. A roundish chubby species with horizontal ribs somewhat marked with brown and orange aperture. 1″.

14. **Cancellaria goniostoma,** Sow. Peru. This unusual species is deeply reticulated with indented aperture. It is one of the large fellows of the genus and rather unique. 1¼″.

15. **Cancellaria laevigata,** Sow. Victoria, Australia. An almost white form with faint ridges on upper whorls and tiny teeth in aperture. 1″.

16. **Cancellaria rigida,** Sow. Peru. It is deeply ridged with flat space at top of each whorl. A grayish-white species not at all common. ¾ to 1″.

17. **Cancellaria reticulata,** L. Sanibel Island. A species fairly common on West Coast of Florida. Pure white specimens have been found. 1″.

18. **Cancellaria scalata,** Sow. Mauritius. A deep brown shell with fine horizontal ridges and white regular marks across same. Each whorl flat at top, aperture nearly white. 1″.

19. **Cancellaria semidisjuncta,** Sow. South Africa. An unusual species finely ribbed, with deep umbilical space. It is uncolored but of unique shape, the last whorl being partly separated from others. 1¼″.

20. **Struthiolaria vermis,** Mart. New Zealand. A white shell with faint traces of brown. It has very small nodules hardly noticeable and white aperture. 1½″.

21. **Cancellaria spengleriana,** Des. Japan. One of the large forms of the genus, finely ridged and one faintly shaded brown. The aperture is white and distinct. A very neat shell, admired by collectors and not rare. 2″.

22. **Struthiolaria papulosa,** Mart. New Zealand. It is white with faint streaks of light brown. Has one row of nodules in middle of whorl. Aperture white and of a peculiar shape not seen on other shells. There are five species in the genus. 2½″.

23. **Vermetus atra,** Rouss. Philippines. This species coils around a piece of rock to which it is always attached. Is of a dark color and every shell will be of a different shape. There are over 150 species of the so-called Worm Shells scattered over the whole world. There hardly is a pair of similar pattern. Very interesting genus. 2″.

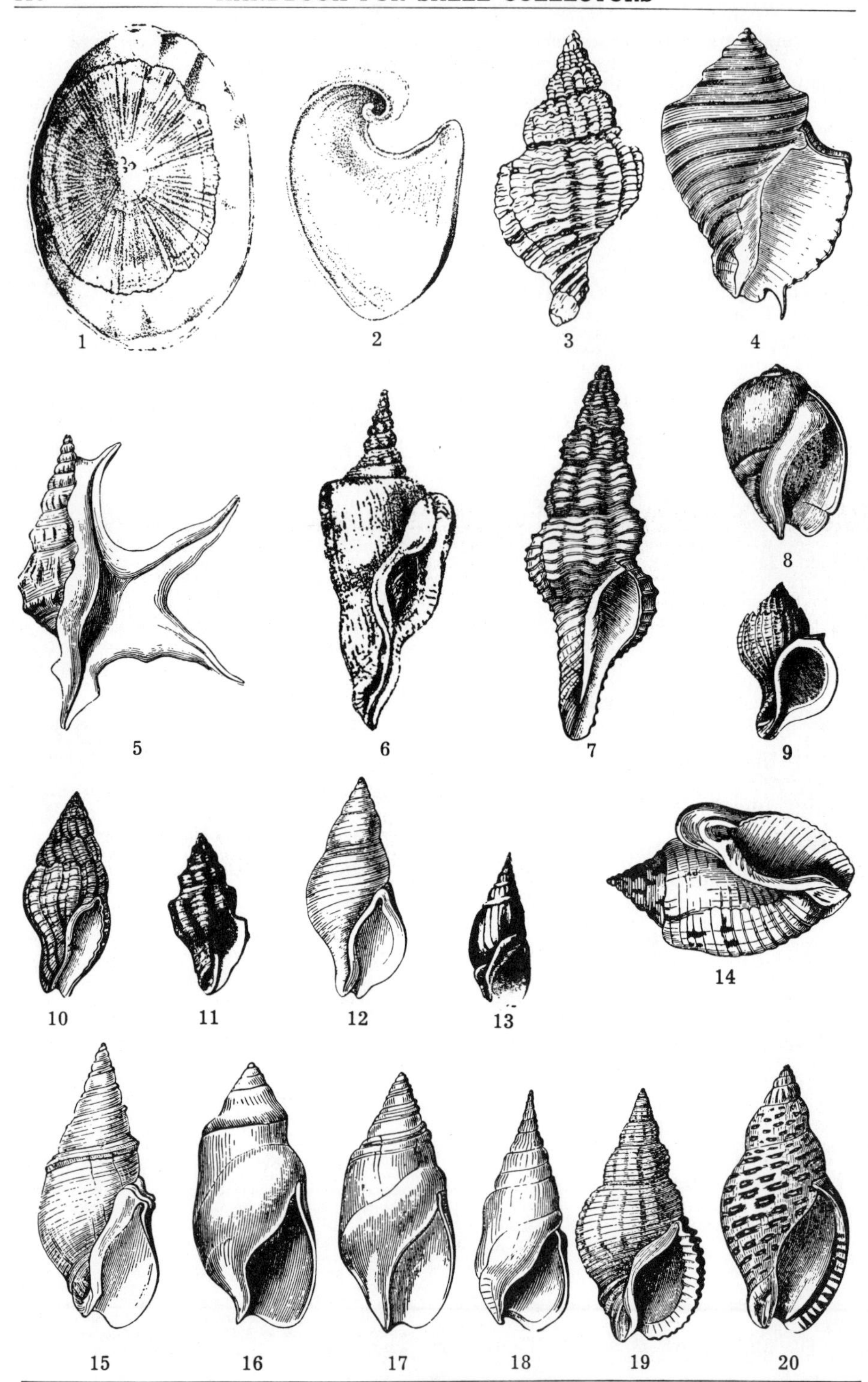

PLATE 58

1. **Umbraculum indica,** Lam. Mediterranean region. A roundish flat shell with sharp edges. It has a thick periostracum. There are 8 known species in the world. Most of them live under rocks at extremely low tide. 3″.

2. **Dolabella gigas,** Rang. Mauritius. Shell is white with yellow periostracum. The mollusk is much larger than the shell, which is really rudimentary. There are 12 species of these strange shells, and all are quite similar to cut, with hardly one whorl. 2″.

3. **Trophon rugosus,** Q&H. New Zealand. A small shell devoid of color and is usually more smooth than cut shows. Has a reddish leathery operculum. 2″.

4. **Latirus cingulatus,** L. Panama. A small, robust, dark colored shell with a sharp horn at base of aperture. It is finely lined with white. The 150 species of the genus are found sparingly over the world but they are a most interesting and handsome lot. Live among the rocks at low tide. 2″.

5. **Chenopus (Aporrhais) pes-pelicani,** L. Pelican Foot. Mediterranean Sea. A very odd shaped shell. The flaring pointed aperture is built on after the shell is mostly full grown. There are seven known species some of which are from deep sea and rare. 2″.

6. **Clavella serotina,** Hinds. Australia. The shell is dull white of the peculiar form of cut. It is a solid sharp pointed species. Only three varieties are known, which are related to the Fasciolaria. Not common. 2½″.

7. **Latirus filosus,** Schm. Senegal. It is covered with ridges of light yellow, and the body of the shell is russet color. A fine 2″ species.

8. **Trophon craticulatus,** Fabr. Greenland. A solid shell of dark color covered with strong periostracum, as are most of the cold water species. 1½″.

9. **Trophon geversianus,** Pallas. Straits of Magellan. A finely fluted shell of typical shape for shells of the genus, most all of which inhabit very cold water. There are 110 species from all the cold parts of the world, some from deep sea. If you ever secure one-third of them for your collection, you will be doing very well. 1½″.

10. **Latirus incarnatus,** Desh. Mauritius. A small shell of a reddish-russet color. It is covered with ridges and cross lines. Attractive. 1¼″.

11. **Latirus aplustre,** Gmel. Philip pines. A typical species of this great group. Ridged and line of dark color. 1″.

12. **Euthria lineata,** Mart. New Zealand. Of a reddish-brown color with regular circular lines of dark brown. More elongated than most species of the genus. There are 32 known forms. 1¼″.

13. **Bullia cochlidium,** Kien. Straits of Magellan. This is one of the largest of the 50 known species of the genus. Without color pattern, it is usually smooth and sharp pointed. 2″.

14. **Cominella (Triumphans) distorta,** Lam. Panama. This unique shell has been switched around from one genus to another in my life time and is now included in a sub-genera of Cominella. It is a heavy thick white shell, with few markings and distorted aperture. Lives among rocks on mud flats. 1¼″.

15. **Bullia annulata,** Lam. Algoa Bay, Africa. This species is smooth with very fine circular ridges and prominent knobs on top of each whorl. It is very faintly marked with brown. 2″.

16. **Bullia callosa,** Wood. South Africa. Usually uncolored, the white callous of the aperture extends over the last whorl in a prominent patch. 2″.

17. **Bullia gradata,** Desh. Brazil. It has a pointed spire and prominent callous patch at aperture. Quite typical of many of the larger forms of the genus. 2″.

18. **Bullia semiplota,** Gray. Senegal. This shell might be called an intermediate form between the robust large species and the slender sharp pointed shells of which there are many species. It is uncolored with modest umbilical callous. 2″.

19. **Bullia tahitensis,** Gray. Karachi, India. There are very few Bullia with ridged surface like this one. It is faintly shaded with brown. Not often seen in collections. 2″.

20. **Cominella alveolata,** Kien. South Australia. It has a yellowish-white ground color, with black dots covering entire surface. Qute distinct. 1½″.

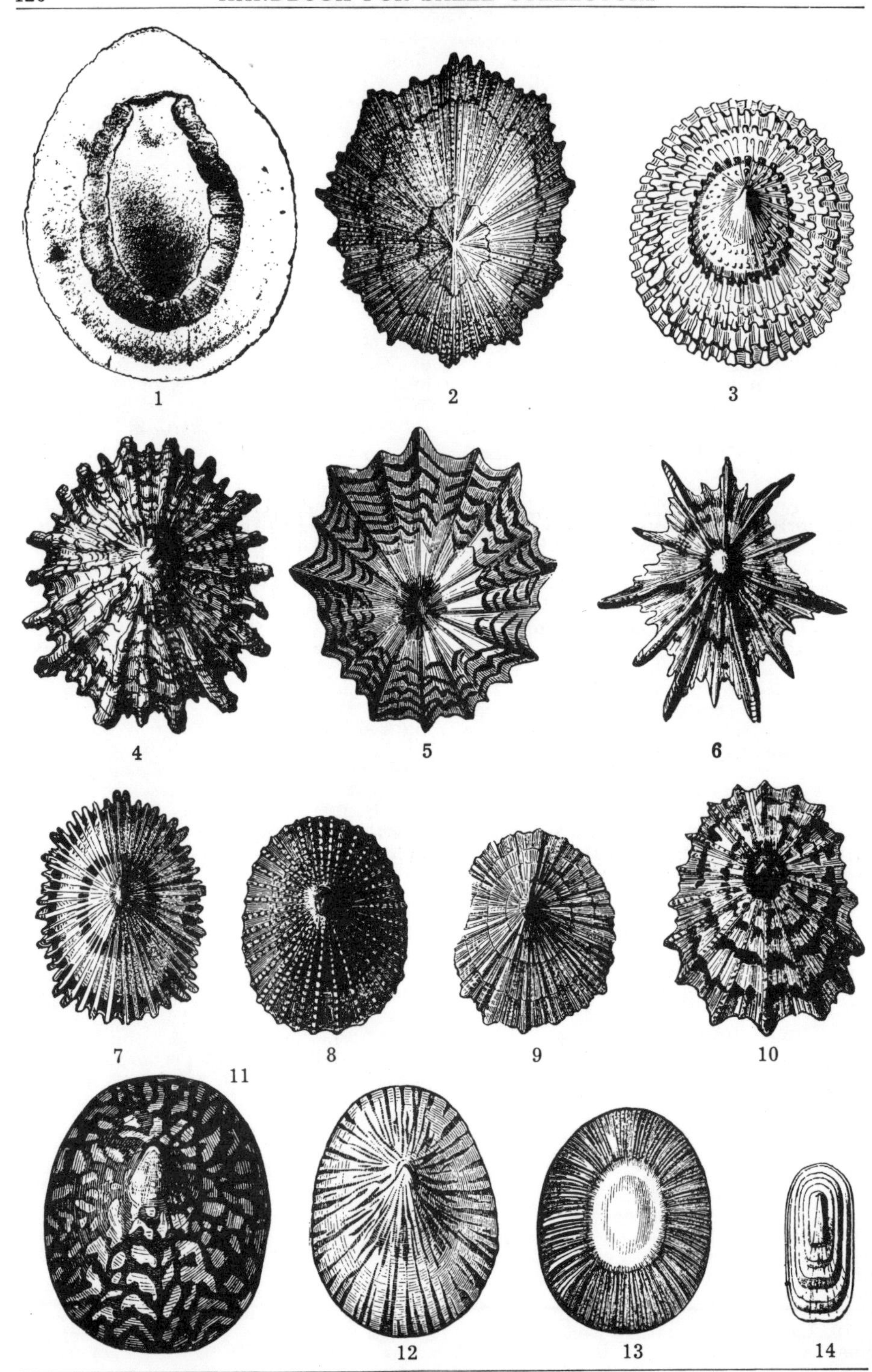

PLATE 59

1. **Patella kermadecensis,** Pils. Kermadic Island, north of New Zealand. A very fine large species, mostly of a yellowish color. It is one of the largest and heaviest forms in the world. There are some 250 known species in the genus but few the size and weight of this one. The shell is hard and takes a fair polish. 5″.

2. **Patella neglecta,** Gray. Australia. Rather flat with few marginal points and many fine ribs. Found in the coral reefs. Not very common. 2″.

3. **Patella nigrosquamosa,** Dkr. Japan. A light colored beauty, as cut indicates. There are many rows of small scalloped edges of a yellowish color with darker circles. 2″.

4. **Patella ferruginea,** Gmel. Mediterranean Sea. The back is of dark color and covered with ridges or ribs that are mottled. The edge terminates in a row of irregular spines. 2″.

5. **Patella granatina,** L. Cape of Good Hope. Rather thin with scalloped edges and finely ringed with dark stripes. This is one of the splendid forms of the region, of which there are many species. 3″.

6. **Patella longicosta,** Lam. Cape of Good Hope. A fine shell with sharp arms extending out like a star. It is white, shaded with darker colors. 2 to 3″.

7. **Patella cretacea,** Rve. Tahiti. A finely ridged shell with dark color around the edge. There are many Patella through the East Indies but New Zealand and South Africa seem to be particularly rich in splendid forms. 1″.

8. **Patella granularis,** Rve. South Africa. A rather small shell of almost black color and ornamented with ribs that are divided into knobs. Fairly common. 1½″.

9. **Patella caerulea,** L. North Sea. A medium sized shell that is quite variable being mostly shades of gray and darker colors. Rather common through the region. 1½″.

10. **Patella oculus,** Born. South Africa. Of rather flat form, thin and has dark circular stripes. It is considered one of the finest of the genus. My collector in Natal had trouble in detaching them from the rocks as the shell would break before it would loose its hold. Finally a thin bladed knife had to be used. 3″.

11. **Patella testudinaria,** L. Philippines. One of the fine large oval forms with dark mottled surface. It is usually rather smooth, thick and an outstanding species. 2 to 3½″.

12. **Patella radians,** Gmel. New Zealand. Rather thin, mottled with brown, the color showing through the shell. It is quite variable some markings being in splashings, others in lines and stripes. 2″.

13. **Patella plumbea,** Lam. Senegal. A neat little shell of rather distinct form. It is ridged around the edges with light smooth space in center. A common species from Indian Ocean region. 1½″.

14. **Scutus ambiguus,** Chem. New Zealand. A pure white plate-like shell with oval back. There are 20 species in the genus most of which are white and of the form of this one, varying in size and slightly in form. 2″.

The Patellidae include the genus Patella. They are universally called Limpets, as the shells are wholly external and spend their lives clinging to rocks or other substances. Dr. William H. Dall, whom I used to visit first when he had a small office in the cupola of the old Smithsonian Institution and later in his fine offices in the U. S. National Museum, extensively studied these shells and wrote many papers in the old American Journal of Conchology and Museum proceedings.

The Epitoniidae consists mainly of the genus Epitonium and six or seven other genera of very small shells which consist of one or two species mainly. These shells have always been called Scalaria and the most famous species of the lot is S. pretiosa, the so-called Wentle-trap. There are more than 200 species in the world and are world-wide in distribution and over 200 species fossil from the Trias. Many of the forms are small but they all are turreted, usually white, many-whorled, and often ornamented with transverse ribs. Operculum horny. I have always found the small forms very hard to name. We have some 60 kinds on the east coast of the USA alone but most collectors have seen only a very few of them as they reach down to 400 fathoms.

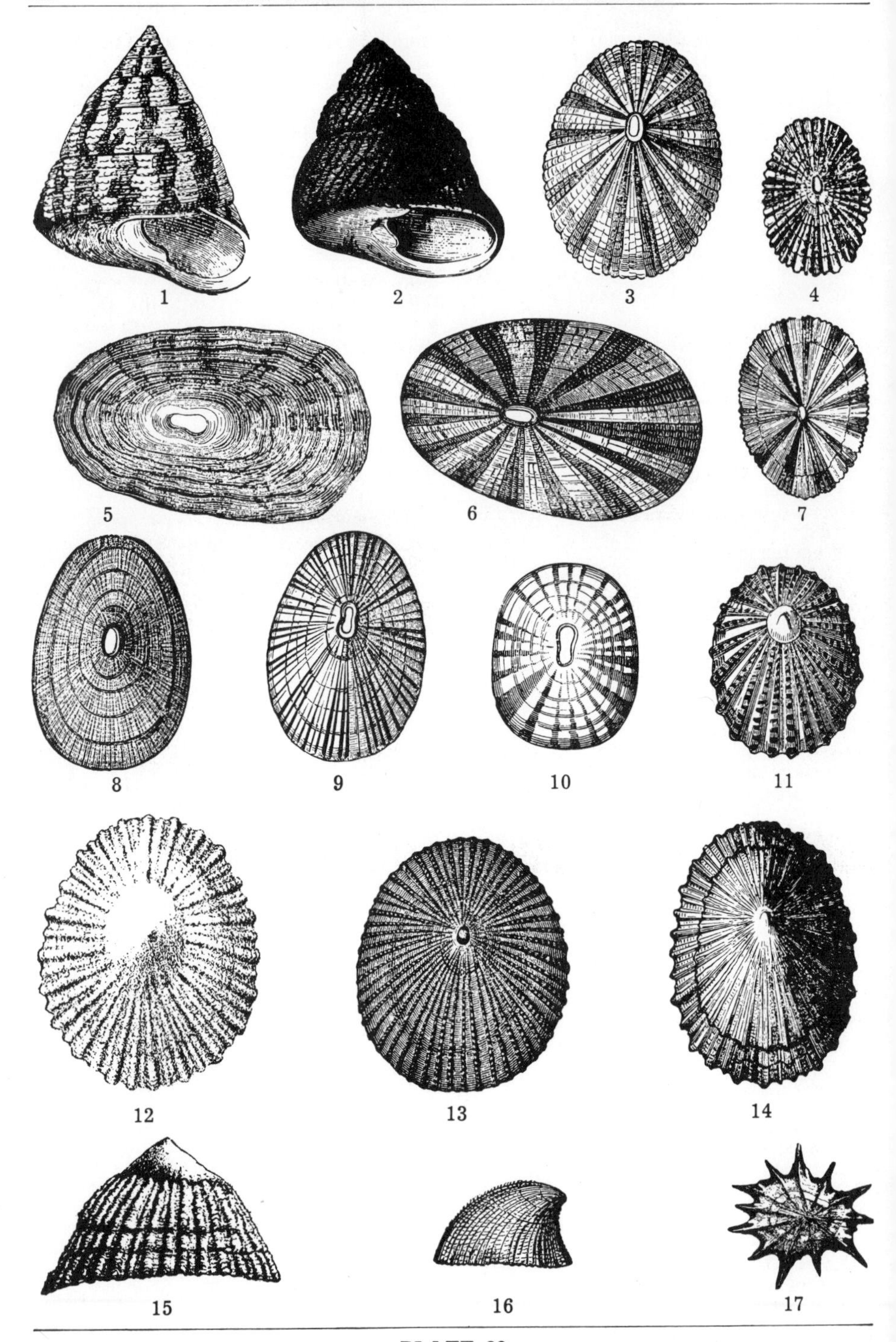

PLATE 60

1. **Trochus maculatus** L. Japan. A handsome conical shell, with red vertical stripes. 1½".

2. **Chlorostoma argyrostoma,** Gmel. China Seas. An almost black shell which is covered with small ridges. It is solid and loves to cling to rocks that are constantly crashed by the surf. 1½".

3. **Fissurella lata,** Sow. Chile. It has a finely reticulated surface with faint radiating streak of color. This part of the coast is famous for its fine shells of this genus. 2 to 3".

4. **Fissurella barbadensis,** Gmel. Bahamas. This Keyhole Limpet is a common shell. Its finely ridged surface and green under side make it an attractive specimen. About 200 species known in the genus. 1¼".

5. **Fissurella crassa,** Lam. Chile. Typical of many other species, the back has radiating shades of light color on white background. Underside pure white. It is a fine large 3 to 4" shell of which there are only a few in the world so large.

6. **Fissurella picta,** Gmel. Peru. A splendid species with radiating bands of brown and white. Along this whole coast there are fine limpets in great variety. 2".

7. **Fissurella peruviana,** Lam. Named for the country where it lives. A fine shell with typical radiating stripe and color markings. It is fairly common as are some of the other local species.

8. **Fissurella latimarginata,** Sow. Peru. A handsome shell with fine radiating ridges and three circular markings. Most of the species are pure white beneath. 2 to 3".

9. **Fissurella nimbosa,** L. Venezuela. It has many fine radiating lines and three faint circular marks or rings. 1 to 2".

10. **Fissurella scutella,** Gray. Torres Straits, North Australia. A neat little shell with natural polish, colored and white stripes. Almost circular. It is not common and only occasionally seen in collections. 1½".

11. **Patella granularis,** Rve. South Africa. There is another illustration of a simliar shell on Plate 59. I had very nice specimens sent in from that rich "Patella territory."

12. **Patella argentea,** Q&G. New Zealand. A yellowish shell of rather flat form and light color. It is a splendid species, is hard and will take a fine polish. 2".

13. **Patella argentatus,** Sow. Hawaii. A very desirable shell, the radiating lines and ridges being covered with tiny ribs. A rich ornamentation. About 3".

14. **Patella compressa,** Lam. Cape of Good Hope. A rather flat thin shell covered with fine striations and a circular dark line. 2".

15. **Patella (Nacella) aenia,** var. Magellanica, Gmel. Straits of Magellan. This fine species belongs to the Patella genus but is more humped than the usual forms and the interior is brilliant with iridescent colors as are most of the section Nacella. 2½".

16. **Emarginula cancellata,** Phil. Mediterranean. All the 114 species of the genus have a slit in edge. Most of the species, many of which are quite small in size, are finely striated with radiating lines. 1".

17. **Patella lanx,** Rve. Japan. A very common shell in this region. It is dark on back, white inside, with a dark patch usually but not always at peak. Used commercially for novelties. 1".

The Aplysiidae include the genus Aplysia, Petalifera, Dolabrifera and Dolabella. In all of the 50 to 100 species the shell is oblong, convex, flexible and translucent. Some of them are called Sea-hares. They live chiefly on seaweed. Are perfectly harmless animals and may be handled with impunity. When molested they discharge a violet fluid from the edge of the mantle which changes to wine-red. The shell is only attached to part of the animal. The Petalifera are small and thin and I only had one species. There are four Dolabrifera quite similar, thin and translucent. The Dolabella has a hard, calcareous shell with a curved and callous apex. Could hardly be called one whorl.

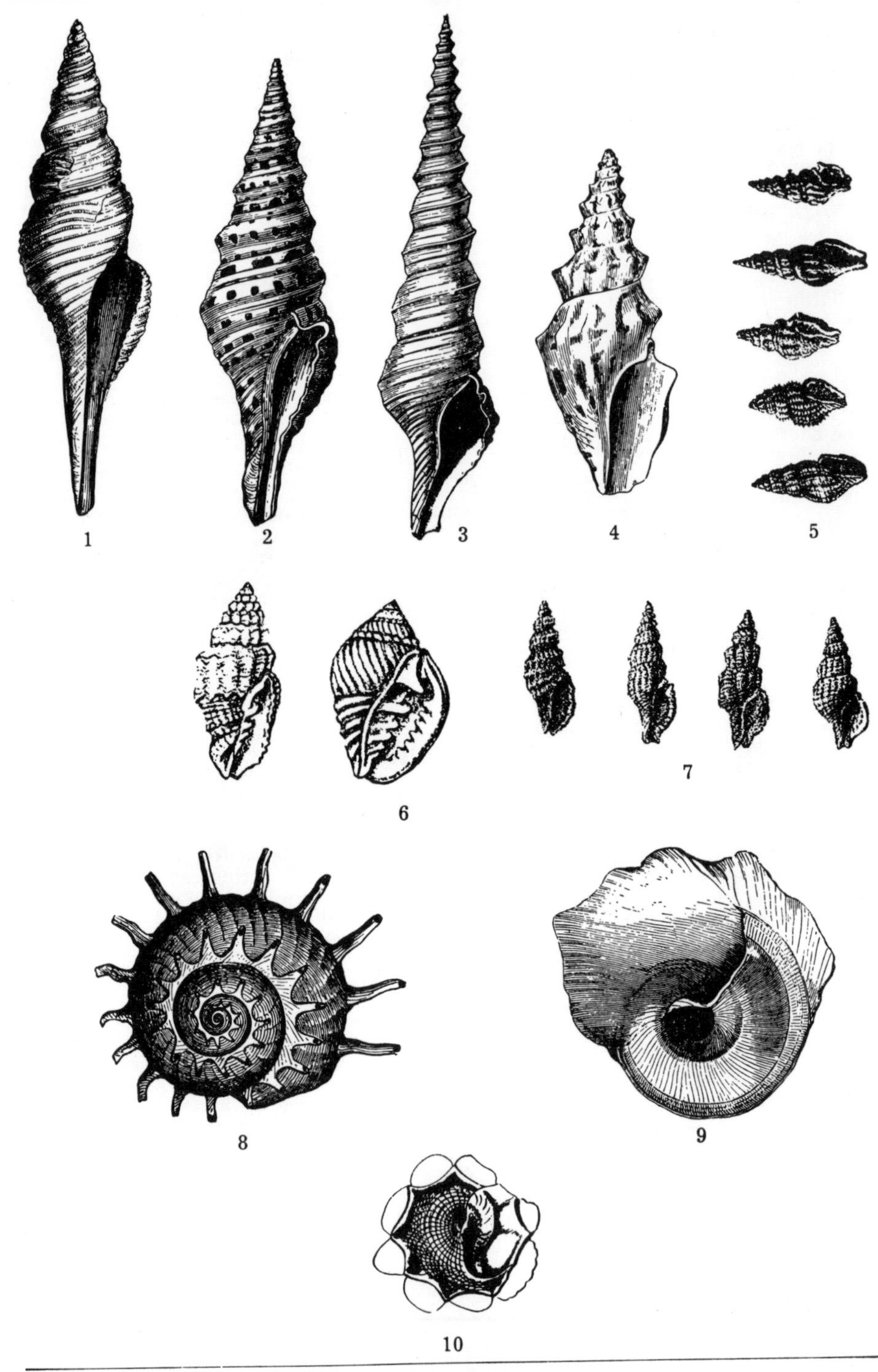

PLATE 61

1. **Turris australis,** Lam. China. An attractive shell ornamented with circular ridges in the form of small dots, colored a light shade of brown. The family of Turridea is a very large one, over 800 species, and there may be hundreds more as they reach down to the deepest seas. There are a number of allied genera in the family. This shell 3″.

2. **Turris babylonia,** L. Mindanao, P. I. The shell is adorned with deep ridges ornamented with brilliant black dots and splashes of color on a white background. 2 to 3″.

3. **Turris woodsi,** Bedd. New South Wales. A tall slender shell usually of a brown color. A desirable species but not often seen in collections. 3 to 4″.

4. **Turris echinata,** Lam. Mauritius. A species with sharp knobs scattered over the surface in one or two rows, the body of the shell ornamented with splashes of dark brown. 1½″.

5. This series of 5 cuts illustrates some of the hundreds of small forms showing their general appearance. They will be found under such genera as Lienardia, Pseudosaphnella, Daphnella, Cythara, etc.

6. Two little Mitra showing shape and peculiar aperture. There are a lot of them in the ocean world, ranging down to about a quarter of an inch high.

7. Another little lot of 4 types usually called Mangilia. There are hundreds of species of them, among some of the daintiest marine shells to be found anywhere.

8. **Xenophora solaris,** Rve. Moluccas. A strange shell of this very curious genus. All of the wonderful species are unusual in form and habits. This form has rows of spines like prongs on the base that are hollow. Their use, if any, is unknown. There are 18 known species in the genus. This one is a rather rare shell in perfect condition. 3″.

9. **Xenophora indica,** Gmel. Japan Seas. In this species the base whorl is spread out into a shell fringe that is rather thin and brittle. Many of the shells of this genus grow other bits of shells to their own as camouflage. This fellow does it in a very small way, usually just small bits of shells. 3½″.

10. **Xenophora pallidula,** Rve. Philippines. It has been found living happily at 100 to 300 fathoms and yet specimens are fairly common in shallow water. It always attaches a choice collection of other parts of shells and I have had specimens with living corals on their backs. How very little we know about these strange invertebrates of the sea. 2 to 4″.

In old collections you will find most of the species, even the small ones, designated as Pleurotoma, the name meaning a shell with a notch. There are a great many fossil forms specially in the Cretaceous. The shells are usually turreted with a straight canal, oval aperture, and outer lip with a notch.

In my own collection I recognized 15 subgenera but some of these are now raised to generic rank. The Family includes such unusual genera as Pusionella, Columbarium (the real Pagoda shell of Japan) and the Halia, a fine unusual rare shell of around 3″ found on the coast of Spain. I have only seen the single species H. priamus which was originally described by Lamarck. But a similar shell has been found in the Tertiary of Italy. My specimen came from off shore near Cadiz, Spain. Dr. Fischer, of France, many years ago examined the mollusk and found its nearest affinity was with the Turris, but the shell has no resemblance to any of the Turridae.

✦

The Xenophoridae contains only Xenophora, which have been divided into three subgenera. Only a few recent species are known and fossil forms have been found in the Devonian. The shells are trochiform, usually flattened to the top of which are often attached shells, corals, stones and miscellaneous broken material which completely camouflages the top of the shell. Some forms only attach minute bits of shells around the upper edge of the whorls. The species usually live in deep water but one form has been found in Lake Worth, Florida, in the shallow water, where it is being studied by local naturalists. They are found very common in some parts of the Java and China Seas.

PLATE 62

This plate was included in one of my earlier editions and is only included here for the reason it illustrates some very nice shells. Some of them are likely duplicated on other plates where they belong.

1. **Terebra strigata,** Sow. Striped Terebra, Panama. Brilliant brown stripes on a yellowish-white background. 3 to 4″.

2. **Terebra robusta,** Hinds. Mottled Terebra, Panama. A mottled shell of brown and white, considered quite scarce. 3 to 5″.

3. **Voluta rupstris,** Gmel. Japan. Zigzag brown stripes on a light brown background. 3 to 4″. See Plate 27.

4. **Cypraea cervinetti,** Kien. Kieners Cowry, Panama, on Pacific side only. Very similar to the Florida "cervus," but usually more elongated, with much darker teeth. A richly colored shell. 2½ to 3½″. See Plate 30.

5. **Cymatium weigmanni,** Ant. Weigmans Triton, Panama. Deep brown shiny ridges, with white between. It is a rather scarce shell. 2 to 3″.

6. **Conus purpurascens,** Brod. The Purple Cone, Panama. Of a rich bluish-purple, with white splashes of color. Very attractive. 2 to 3″.

7. **Voluta hirasei,** Pils. Pilsbry's Volute, Japan. Whorls ridged lengthwise, russet color. 3½″.

8. **Murex troscheli,** Lisch. Japan. One of the largest of the long-spined Murex, ranging 5 to 6″. Lined with brown. Three rows of spines.

9. **Spondylus coccineus,** Lam. Philippines. Thorny Oyster, Philippines. These shells have short spines in great profusion. Colors are orange, purple, red and intermediate shades. 2 to 4″.

10. **Spondylus japonica,** Japanese Thorny Oyster, Japan. Numerous flat spines of a purplish color. 2 to 3″.

11. **Tridacna squamosa,** Lam. Furbelow Clam, Philippines. Illustration is a young shell, which are more attractive than old specimens. Ground color is greenish-yellow, reddish and occasionally lavender. Furbelows are usually white. Old specimens a foot or more long are about smooth. 3 to 4″.

12. **Turbo marmoratus,** L. The Green Snail, China Seas. The illustration is a specimen polished down to the pearl, only the base showing some green. Very brilliant iridescent colors. 3 to 6″. See Plate 42.

13. **Tridacna crocea,** Lam. Baby Giant Clam, Philippines. The common name mentioned is the one used by the natives but it is not really a baby shell of the Giant clam. Colors reddish and white. 2 to 3″.

The Dentaliidae belong to the order Scaphopoda and consist of true Dentalium (Tooth shells, as they are often called). Siphodentalium and Cadulus, the last two genera contain mainly small forms. The shells are all tubular, curved, open at each end. The animal is attached to its shell near the posterior anal orifice. The species are animal feeders, devouring foraminifera and minute bivalves. They are found in sand or mud, in which they usually bury themselves. Two of the largest forms I have seen are D. elephantinum, and D. vernedei, both of which are fairly easy to obtain and attain 4 to 5 inches. There are more than 100 species, some of which were used as money by the Indians of Northwest America before the introduction of blankets by the Hudson Bay Company. In those days a slave, canoe, or squaw, was worth so many blankets, but it used to be so many strings of Dentalium wampum. 25 long shells strung together would make a fathom and at one time such a string would have been worth $50 pound sterling.

✦

The Chitonidae belong to the order Polyplacophora. The Gasterpododa usually are composed of one piece, the shell we love to collect, but the Chitons consist of 8 pieces lodged in a mantle which forms an expanding margin around them and holds them together and in a few unusual cases the mantle covers the whole shell as in the mammoth C. stellari of West Coast of U.S.A. which attains as much as 8″ in length. The whole family includes a large number of genera and the last monographs cover some 550 species. They are particularly abundant on the west coast of the Americas and in the Australian region, but we have some very handsome ones on the east coast of our country. I well remember standing by a sea wall in Panama and watching some fine specimens about 3″ long attached to the rocks very near by on which the surf crashed constantly all the year round. You risk your life to try and get them. The family is so large and so much has been written about them, I will simply refer my readers to books which adequately cover these strange forms of sea life.

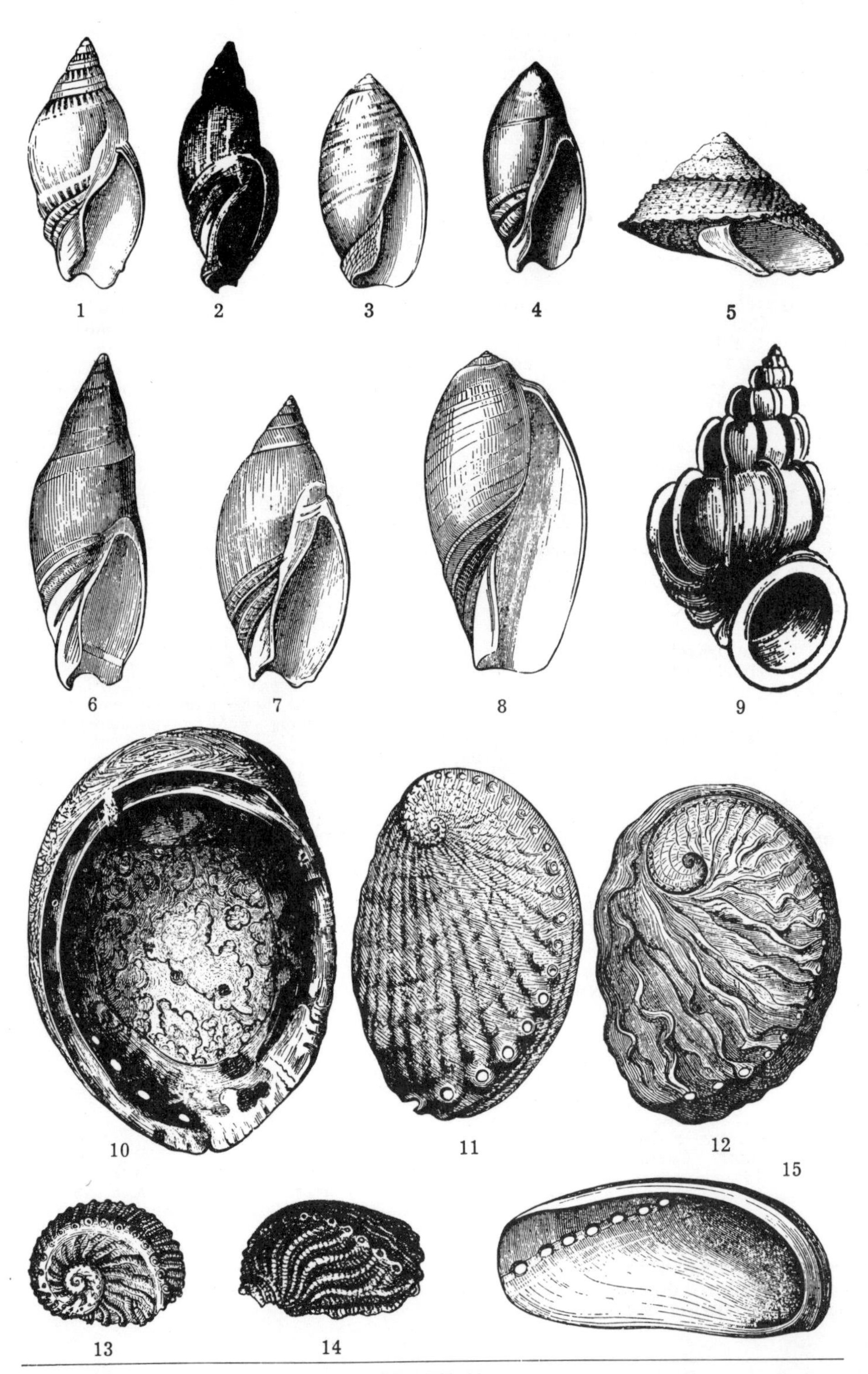

PLATE 63

1. **Ancilla marginata,** Lam. Tasmania. A pointed species of brilliant polish with splashes of chestnut. The aperture is peculiar shape and has high spire. 1½" and not very common.

2. **Ancilla glabrata,** L. Gulf of Mexico to Yucatan. Of a rich golden yellow with an unusual brilliant natural polish that fairly glistens. The aperture has peculiar folds. All of the Ancillas have a neat habit and are closely allied to the Olivas. 2 to 2½".

3. **Ancilla ventricosa,** Sow. Red Sea. It is of a rich golden yellow throughout. Not common. 1".

4. **Ancilla australis,** Sow. Australia. Of a rich shade of brown, high polish, it is an attractive small shell of 1". There are 50 species of this genus in the world.

5. **Astraea rugosum,** L. Mediterranean Sea. Of drab color it is a fairly common species of this region. The surface has tiny ruffles as shown in cut. 2".

6. **Ancilla rubiginosa,** Swain. China Seas. It is of a cinnamon brown highly enameled. Lighter brown at suture. Not common. 2".

7. **Ancilla tankervillei,** Sow. Brazil. A richly enameled high colored shell and is quite rare and only occasionally seen. 2".

8. **Ancilla mauritiana,** Sow. Mauritius. It is of a clear yellowish color, quite highly polished and has a very wide aperture. Only a few other species are similar. 1½ to 2".

9. **Epitonium (Scalaria) pretiosa,** Lam. China Seas. The so-called Precious Wentletrap you will find figured in the oldest shell books as one of the treasures of ye olden times. While it is more common today it has never been called a real common shell, as it lives below the tide lines. Recently I have had numerous fine specimens sent me from round Brisbane, Australia. The collector says he finds many specimens on the beach but only very rarely one alive, with the mollusk. Average size 2".

10. **Haliotis gigantea,** Chem. Japan Sea. One of the largest species of the genus but seldom as thick as some of the other forms, hence it has never known the commercial importance of the California coast forms. Lives on the rocks as do the others. Of very rich dark color inside. Back usually covered with many forms of incrustations. 8" or more.

11. **Haliotis rugoso-plicata,** Chem. Australia. Back is covered with numerous corrugations and interior white. A neat shell not at all common. 2 to 3".

12. **Haliotis midae,** L. Australia. Back is covered with folds. Interior mostly white. It is not common in collections. 3 to 4".

13. **Haliotis pulcherrima,** Mart. Polynesia. A small species with very distinct shell pattern. Interior white. About 1".

14. **Haliotis japonica,** Rve. The back of the shell has numerous folds and nicely marked with shades of yellow and brown. A neat small 2" shell.

15. **Haliotis asinina,** L. Philippines Seas. A narrow shiny shell quite thin and one of the very few naturally glossy forms. It is quite common, and fine specimens can usually be had for a small price.

The Neritidae consist of Nerita, Neritina and Septaria. There are over 500 species of the three genera, and a large number of fossil forms from the Lias on. The true Neritas are solid shells, round, smooth or grooved, epidermis horny, outer lip thickened, inner edge of columella often fantastically grooved as in N. peleronta, the famous "Bleeding Teeth" shell. Only a very few attain 1½", most of the species under one inch in size. They range from white to black and some are ornamented with red and other shades. The Neritina are as numerous as the Nerita but are most of them thin, some ornamented with spines and one form, N. communis, is found in many brilliant colors in some parts of the tropical world. In other places they will be all black or other shades. It just all depends on where you find them. The Navicella come mostly from Pacific area. The shell is oblong, limpet-like, with a posterior submarginal apex, aperture is as large as the shell with a small columellar shelf. They differ radically from the other two genera and like the Neritina are fond of locations in or close to salt water marshes.

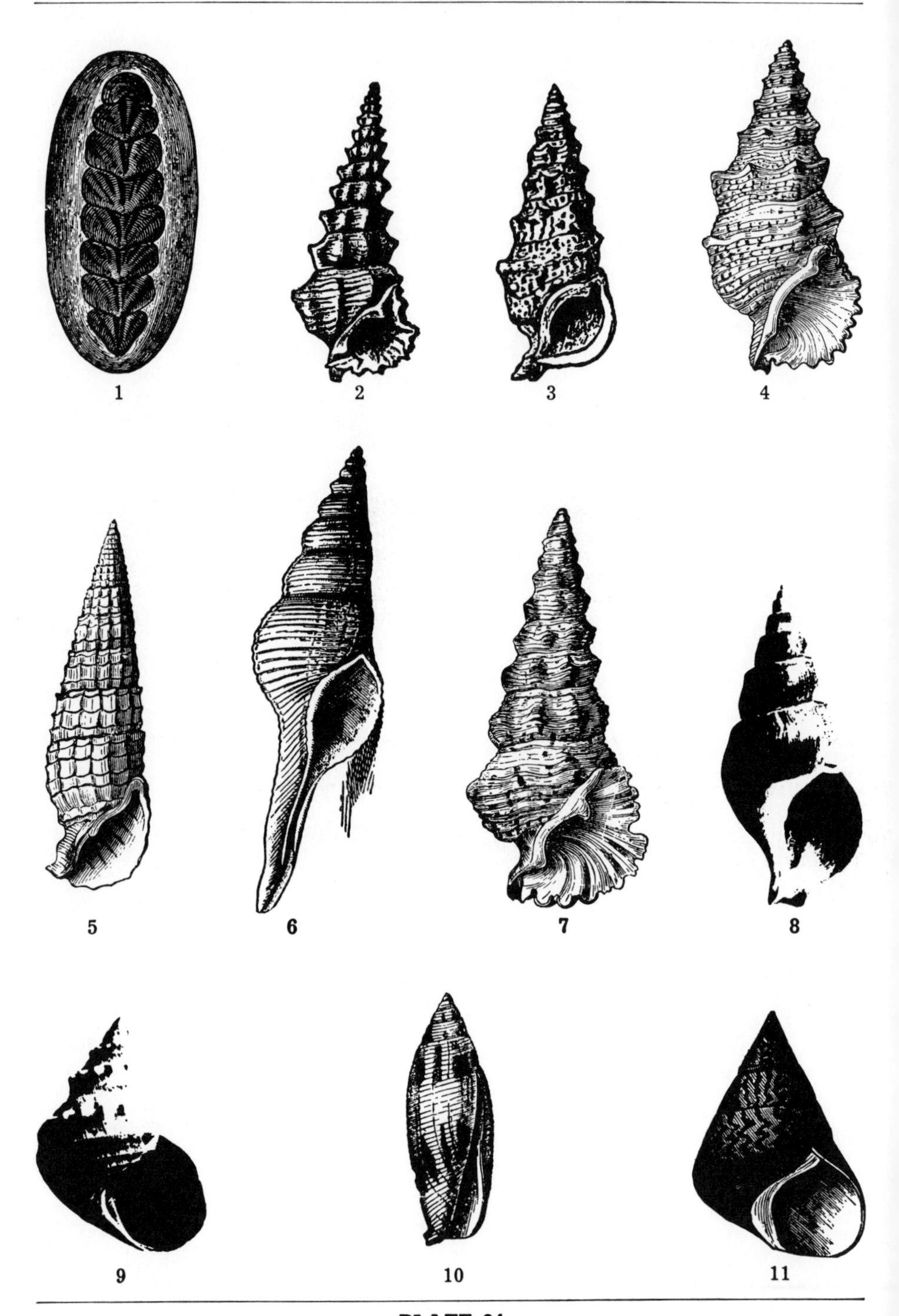

PLATE 64

1. **Ishnochiton coquimbensis,** Born. Chile. A neat shell in which the form of the interior plates are shown on the back. The center is dark and outer rim light. Lives attached to rocks. 2″.

2. **Potamides sulcatus,** Born. Australia. A deep brown shell of varying shades with wide flaring aperture which is highly polished. Completely covered with ridges arranged in circular rows and individual nodules. There are 125 species in the genus. 2″.

3. **Cerithium aluco,** L. Philippines. The shell has a finely mottled natural polished surface much admired by collectors. There are 250 species in the genus widely distributed over the world. 2″.

4. **Cerithium echinatum,** Lam. Philippines. A typical fine shell of this group with prominent tubercles on the whorls. Lives on mud flats. 2″.

5. **Cerithium lineatum,** Brug. Philippines. A very attractive shell for this genus. It has short ridges on each whorl and brown circular lines on a pure white background. Only a few species of this pattern. 2½″.

6. **Fusus colus,** L. Ceylon. The shells of this genus are mostly white, a few are mottled and many are covered with a yellowish-fuzzy periostracum. When this is removed they are often of glistening color. There are about 200 species in the genus ranging from very large 7 to 8″ shells down to 1″. Many are rare and seldom seen. This shell is 6″, white and typical of many others.

7. **Cerithium nodulosum,** Brug. Philippines. A real giant of the tribe. The shell has dark splashes of color with prominent nodules, and wide flaring aperture. One of the largest known species. 3 to 5″.

8. **Neptunea pericochlion,** Sow. Japan. There are 75 species in the genus but this unique shell is hardly typical of the others. It is covered with a gray periostracum and each whorl has an even edge at the top. The true Neptunea do not have it. Rare. 4½″.

9. **Bathybembix argenteonitens,** Lisch. Japan. This is one of the finer deep sea forms. It is of a glistening pearly white, in fact the pearl sheen is distinctly seen over the surface. This sheen is peculiar to many deep sea forms. 1½″.

10. **Dibaphus edentulus,** Swain. Mauritius. A very odd shaped shell the only species in the genus. It is next to the Mitra and a sort of connecting link between that genus and the Fusus. The Vexilla come next to it. It is of a brownish color with darker blotches. 2″.

11. **Cantharidus iris,** Gmel. New Zealand. There are over thirty species in this genus mostly small but very beautifully colored and often with high natural polish. This is one of the largest and fairly rare. It is conical and of a brilliant shade of color, with zigzag markings. I have seen attractive necklaces made of some of the smaller species. 2″.

The Cerithiidae consist largely of the genus Cerithium but they start off with the genus Triforis which are all minute, dark shells, and all sinistral, real little gems, then follows Cerithium, Pleiosotrochus, Bittium which are minute like Triforis but all are dextral. Potamides, which range from large 4″ forms down to 1″ and live largely in brackish salt water, and finally Cerithiopsis which are simply minute forms of previous genera.

There are likely 200 or more species of Cerithium and they range from one large 4″ species C. nodulosum to rather small 1″ or smaller forms. They are world-wide in distribution, the finest forms as a rule from tropical regions. Over 500 fossil forms have been discovered from Triassic and later. The shells are all turreted, with indistinct varices, aperture usually small with tortuous canal in front, outer lip expanded, inner lip thickened. Operculum is horny and spiral. Some of the species emit a bright green fluid when disturbed.

In my collection I recognized four main subgenera which were more or less distinct in form. There are about half as many forms of Potamides and numerous fossil forms from Eocene. The shells are turreted, whorls angulated and coronated, aperture prolonged in front into a nearly straight canal, outer lip thin and sinuous, epidermis thick and operculum corneous. Many of the smaller forms have the habit of dropping off some of the earlier whorls and filling up the space with nacre. You just do not find any of them with full whorls.

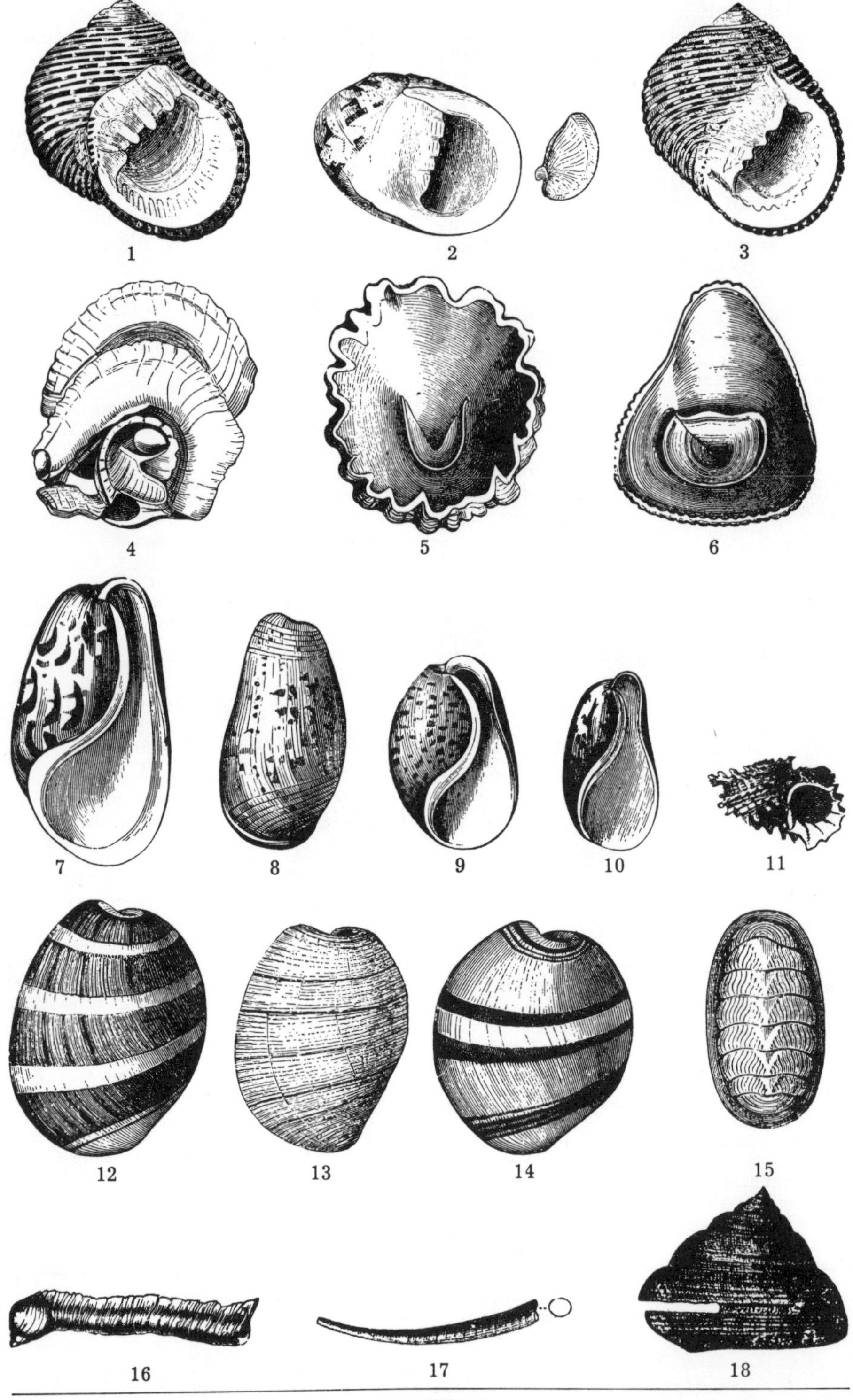

PLATE 65

1. **Nerita ornata,** Sow. Panama. A grand, solid shell covered with black lines on a lighter background. This is really the southern form of **scabricostata.** Lives on wave-beaten rocks. 1¼″.

2. **Nerita polita,** L. Philippines. A natural highly polished species often ornamented with red and other bright colors. One of the most attractive of the 200 or more forms of the genus. The cut shows the operculum, which always perfectly fits the aperture of the shells of this genus.

3. **Nerita undata,** Gmel. Philippines. Somewhat similar to ornata but usually the color pattern is more diffused and clouded. It has dark ridges but a rather distinct species and very common. 1″.

4. **Vermetus centiquadrus,** Val. Gulf of California to Panama. A large heavy species which is common to that region and of which several good varieties have been named. The shell is flatly coiled and attached to piling or rocks. All Vermetus are very strange forms of molluscan life. 2″.

5. **Mitrularis equestris,** L. Pacific generally. A remarkable form of the Cup and Saucer shells as the saucer is of horseshoe pattern. It has a wide range but few shells seem to come in from collectors. About 60 known species live on mud flats. 1½″.

6. **Crucibulum scutellum var quiriquina,** Less. Mazatlan to Chile. The species scutellum is a very variable one, and there are many so-called mutations of which this is one. They are all of triangular shells of the Cup and Saucer variety and cling to rocks. There are 70 species in the genus, all very strange shells. 1½″.

7. **Bulla oblonga,** Ads. Mauritius. A fine species with mottled surface and not real common. It is a solid attractive shell. 1 to 2″.

8. **Bulla striatus,** Brug. Malta. A very common species in this sea. Found everywhere on the shore lines. The surface is dotted. It is claimed to have been found in Florida. 1″.

9. **Bulla aspersa,** Ads. Philippines. The peculiar shaped aperture, and dotted shiny surface, will usually identify this form. Lives on mud flats as do many of the species. 1¼″.

10. **Bulla australis,** Quoy. A very neat little Bubble Shell with peculiar aperture and mottled back. There are about 70 species in the genus which is world wide. 1″.

11. **Delphinula laciniata,** Lam. Philippines. There are about a dozen species in this genus and this is one of the most common. Some shells have short stubby spines, others are long, curved and irregular. Shell is usually black and solid pearl. Some forms are very rare and seldom seen. 1½ to 2″.

12. **Hydatina albocincta,** Hoev. China. A very thin yellowish-white shell with yellowish-brown periostracum. The prominent white bands are distinctive. Rather rare. 1½″.

13. **Hydatina physis,** L. Perhaps world wide. Has been found in China, Mauritius and recently fairly common in Lake Worth, Florida. How did such a thin, delicate shell find its way over the whole world of ocean? There are numerous lines of light brown. 1½″.

14. **Hydatina velum,** Gmel. (Vexillum Ch.) Mauritius. A striking shell with brilliant dark bands. There are 12 species in the genus and all are thin, delicate and attractive. 1½″.

15. **Chiton marmoratus,** Gmel. Barbadoes. A very handsome species with shiny back, quite common in that region. The color pattern is all that could be desired in a Chiton, where so many of the backs are rough and covered with ocean life. There are few others finer. 1½″.

16. **Magilus antiquus,** Mont. A mollusk with very strange habits. The tiny shell attaches itself perhaps to a head of coral or reef. By the time the shell is of full size ¾″ the coral has grown so that it might envelop the little fellow living on its surface and so kill the mollusk. Something must be done at once, so the mollusk turns a trick which seems impossible. It forsakes its shell, gradually fills it with solid nacre, starts to build a tube, in which it lives thereafter. The tube keeps up with the growth of the coral head. To collect the entire shell and tube you must crack open the coral, which is no easy job, and at the same time not crack the shell and tube. Naturally a shell with its tube of perhaps 3 to 6″ is not very common in collections. There are 18 known species in the world all with strange habits.

17. **Dentalium vernedei,** Hanl. Japan. There are over 100 species of the so-called tusk shells and this is one of the largest known ranging from 4 to 6″.

18. **Pleurotomaria.** This cut illustrates a type of very rare shells with a deep notch. There are only a few recent species and all from deep water. Most of the shells in collections have been brought up by real dredging ships. The shells, of which there are only a very few recent species are 3 to 5″ often of a reddish color. There are many fossil forms and the few recent species are the last of a great race of shells very common in geological times. I have sold many specimens in my life time at $50 each.

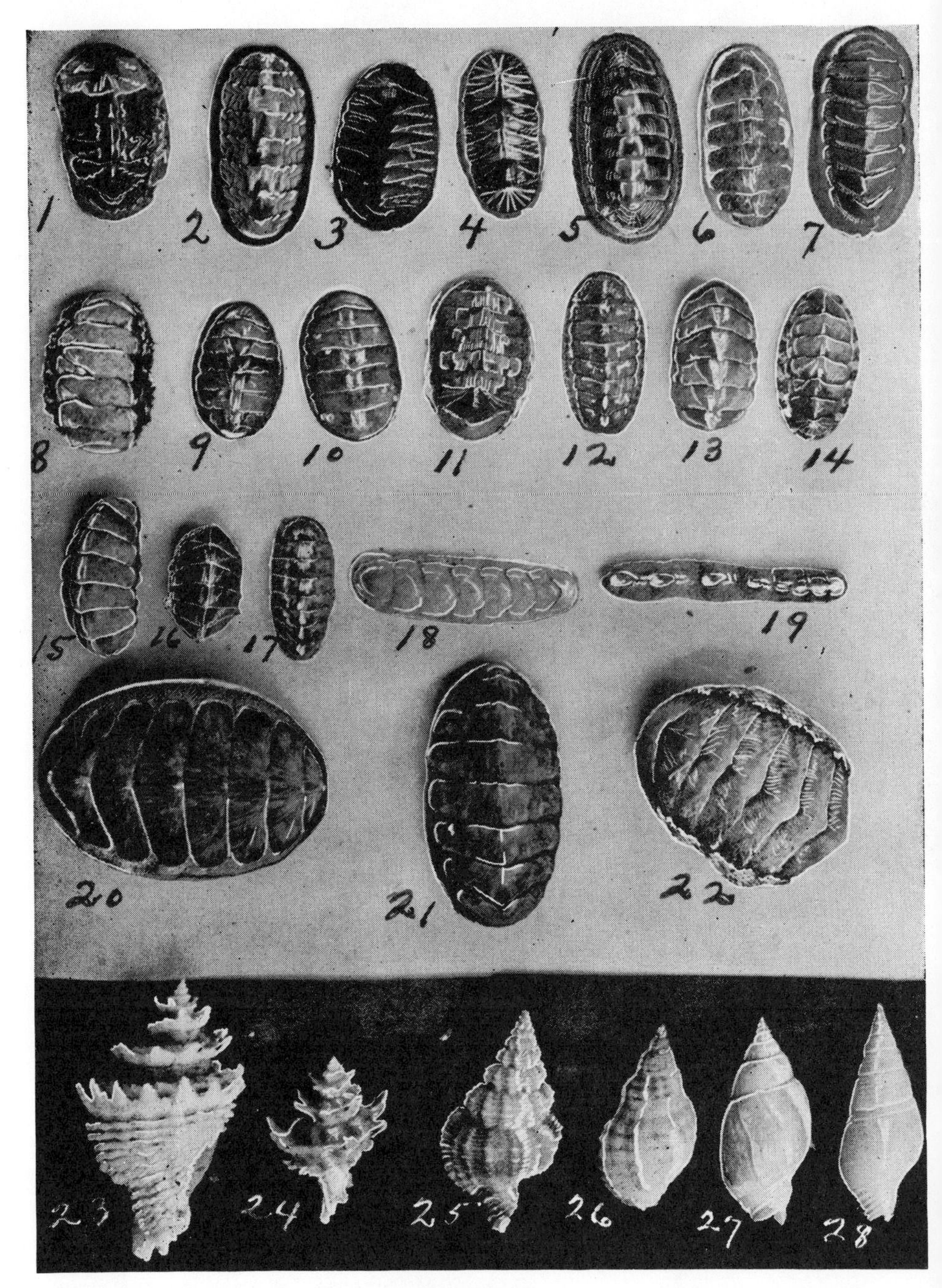

PLATE 66

1. **Ishnochiton alfredensis,** Ashby. Natal, Africa. A finely mottled shell with girdle like serpent. 1¼″.

2. **Tonicella lineata,** Wood. The Red-lined Chiton. All California coast. A brilliant shell of russet-red, with red mantle. 1½″.

3. **Chiton magnificus,** Desh. Chile. Dark black with fine lines. 1½ to 2½″.

4. **Chiton albolineatus,** Sow. Acapulco, Mexico. Brilliantly adorned with dark green and white. 1½″.

5. **Chiton cummingii,** Frem. Chile. Back smooth, with fine circular lines, mantle with fine dots. 1¾″.

6. **Pallochiton lanuginosus,** Dall. San Diego to Lower California. A dark shell with fine lines and fairly smooth mantle. 1¼″.

7. **Mopalia muscosa kennerlyi,** Carp. Alaska to Monterey. A rather dark shell that is quite smooth. 1½″.

8. **Chiton olivaceus,** Speng. Malta. White above with hairy mantle. 1¼″.

9. **Chiton inermis,** Cooper. La Jolla, California. A thin flexible shell, brownish-yellow, with no trace of calcification. 2″.

10. **Chiton marmoratus,** Gmel. Florida Keys. Greenish-brown, rather smooth, narrow mantle. 1¼″.

11. **Chiton opilaris,** Cpr. Monterey Bay, California. Light greenish mottled with white, yellowish mantle. 1½″.

12. **Lepidochiton ruber,** L. Labrador to Conn. A narrow shell, mottled with light black and white. 1¼″.

13. **Ceratozona rugosa,** Sow. Juniper Inlet, Florida to Key West. A handsome shell mottled with shades of brown and white. 1½″.

14. **Chiton stokesi,** Brod. Panama. A narrow brownish shell with white marks on the mantle. 1¼″.

15. **Chiton rhodoplax,** Pils. Japan. A narrow light brownish and white shell. 1½″.

16. **Ishnochiton proteus,** Rve. Port Jackson, Australia. A finely lined shell marked with gray and black. 1¼″.

17. **Ishnochiton scabra,** Rve. California coast. A narrow shell, marked with green and black. 1¼″.

18. **Chiton tuberculatus assimilis,** Rve. West Indies. A very narrow light gray shell. 1½″.

19. **Ishnochiton versicolor,** Sow. Port Jackson, Australia. A long narrow drab-colored shell, much resembling a worm. 1½″.

20. **Acanthochites floridana,** Dall. Cape Florida to Dry Tortugas. A broad gray and black shell, with yellowish mantle. 2″.

21. **Chiton oniscus,** Krs. Natal, East Africa. Of a pinkish-brown color with black mantle. 2″.

22. **Mopalia lignosa,** Gould. Woody Chiton. Alaska to Magdalena Bay. Of light greenish color, the mantle exactly like snake skin. 2″.

23. **Latiaxis japonica, Dkr.** Japan. Usually white. Each whorl is adorned with flat spines, all resembling a pagoda. The genus comes next to the Thais. 1¾″.

24. **Latiaxis spinosa,** Hwass. Japan. A much smaller form of pink color with broad curved spines. 1″.

25. **Hindsia acuminata,** Rve. Formosa. In form it much resembles **Phos senticosus,** being finely reticulated with ridges on each whorl. 1¼″.

26. **Agnewa tritoniformis,** Blv. New South Wales, Australia. Of a yellowish-red color, it has numerous circular ridges. The genus of one species comes right after the Thias. 1¼″.

27. **Pusionella recluziana,** Pet. West Africa. A genus of about 15 species which vary greatly in form. I had ten of the species mostly from the shores of Africa. This is a smooth white shell with very thin yellowish periostracum.

28. **Pusionella nifat,** Brug. West Africa. A slender pointed white shell of about six whorls, with fine lines on last whorl near base. 1¼″.

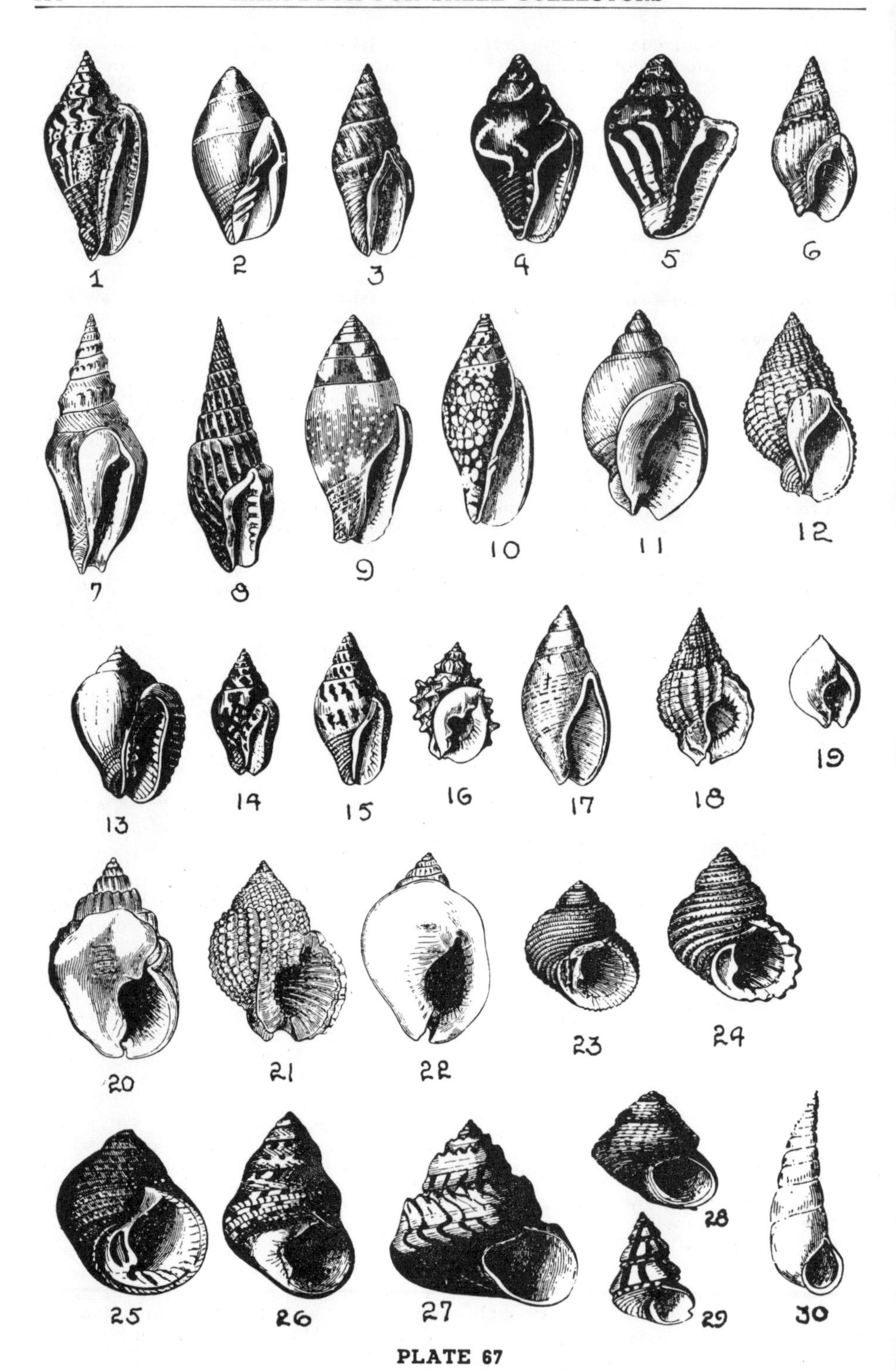
1
2
3
4
5
6
7
8
9
10
11
12
13
14
15
16
17
18
19
20
21
22
23
24
25
26
27
28
29
30

PLATE 67

1. **Columbella philippinarum,** Reel. Philippines. One of the finest of all species in this great group of shells. It is well ornamented.

2. **Columbella unifasciata,** Sow. Coquimbo. A small shiny species with very peculiar aperture.

3. **Truncaria modesta,** Powis. Panama. This elongated shell looks much like some of the Columbellas but the eight species of the genus are placed between the Bullia and Columbella genera.

4. **Columbella fulgurans,** Lam. Philippines. A handsome prolific species which seems to be found over much of the Pacific.

5. **Columbella strombiformis,** Lam. Gulf of California. A shell of angular shape and well colored, they make attractive specimens.

6. **Nassa elegans,** Kien. Torres Straits. One of many forms from this rich molluscan territory. The Nassas are all attractive shells.

7. **Columbella lanceolata,** Sow. Galapagos. The shell is a very large one for this genus and shaped much like a strombus.

8. **Columbella elegans,** Ads. Panama. A really remarkable shell from a territory that is rich in other species of the genus.

9. **Columbella semipunctata,** Lam. Australia. The dots and wavy markings of this shell are very attractive.

10. **Columbella nitida,** Lam. West Indies. A very common neat little species from this territory. It is extensively collected for trade.

11. **Nassa mutabilis,** L. Malta, Mediterranean Sea. A large species that is fairly common and representative of the territory.

12. **Nassa reticularis,** Lam. Mediterranean Sea. This and the preceding species are the two most commonly met with in that region.

13. **Columbella turturina,** Lam. Mauritius. A small distinct species that is uncolored.

14. **Columbella pardalina,** Lam. New Caledonia. I have had this striking shell sent me from many places in the Pacific.

15. **Columbella tyleri,** Gray. Japan. It much resembles the preceding species.

16. **Nassa muricata,** Quoy. New Ireland. A nobby species with glossy aperture which is common in many other species of this genus.

17. **Columbella grana,** Lam. Algiers. A rather rare species not often seen in collections.

18. **Nassa tritoniformis,** Kien. Philippines. There are many beautiful Nassas from the Philippine territory.

19. **Nassa gibbosula,** L. Egypt. In this species the region of the aperture covers the whole side of the shell. There are a few others similar.

20. **Nassa arcularia,** L. Philippines. A large form for this genus, brilliantly white and highly polished.

21. **Nassa gemmulata,** Gray. Torres Straits. A robust strong shell completely covered with small nodules.

22. **Nassa thersites.** Gmel. All over the Pacific. Widely distributed and largely collected for commercial uses.

23. **Euchelus canaliculatus,** Lam. Philippines. A typical representative of this group of 58 species, found all over the Pacific. They are very attractive small shells.

24. **Euchelus denigratus,** Chem. Philippines. Another species of this group which well shows the remarkable ornamentation of the shells.

25. **Monodonta canalifera,** Lam. Luzon, P. I. A neat small species which seems to be quite common in this territory.

26. **Monodonta aethiops,** Gmel. New Zealand. A little more elongated than most forms of this genus and well marked.

27. **Gibbula magus,** L. Mediterranean Sea. One of the largest of the group and one of the most attractive when well cleaned. There are 150 species in the genus with many fine forms in this territory.

28. **Monodonta concamerata,** Wood. Australia. Another fine species of which there are many all over the Pacific.

29. **Forskalia fanulum,** Gmel. Mediterranean Sea. Brilliant small shells all more or less rare and range over Red Sea and Indian Ocean. About one dozen species.

30. **Eulima major,** Sow. Hawaii. A white shell of brilliant polish, the upper whorls of which always bend to the right. There are 165 species in the genus.

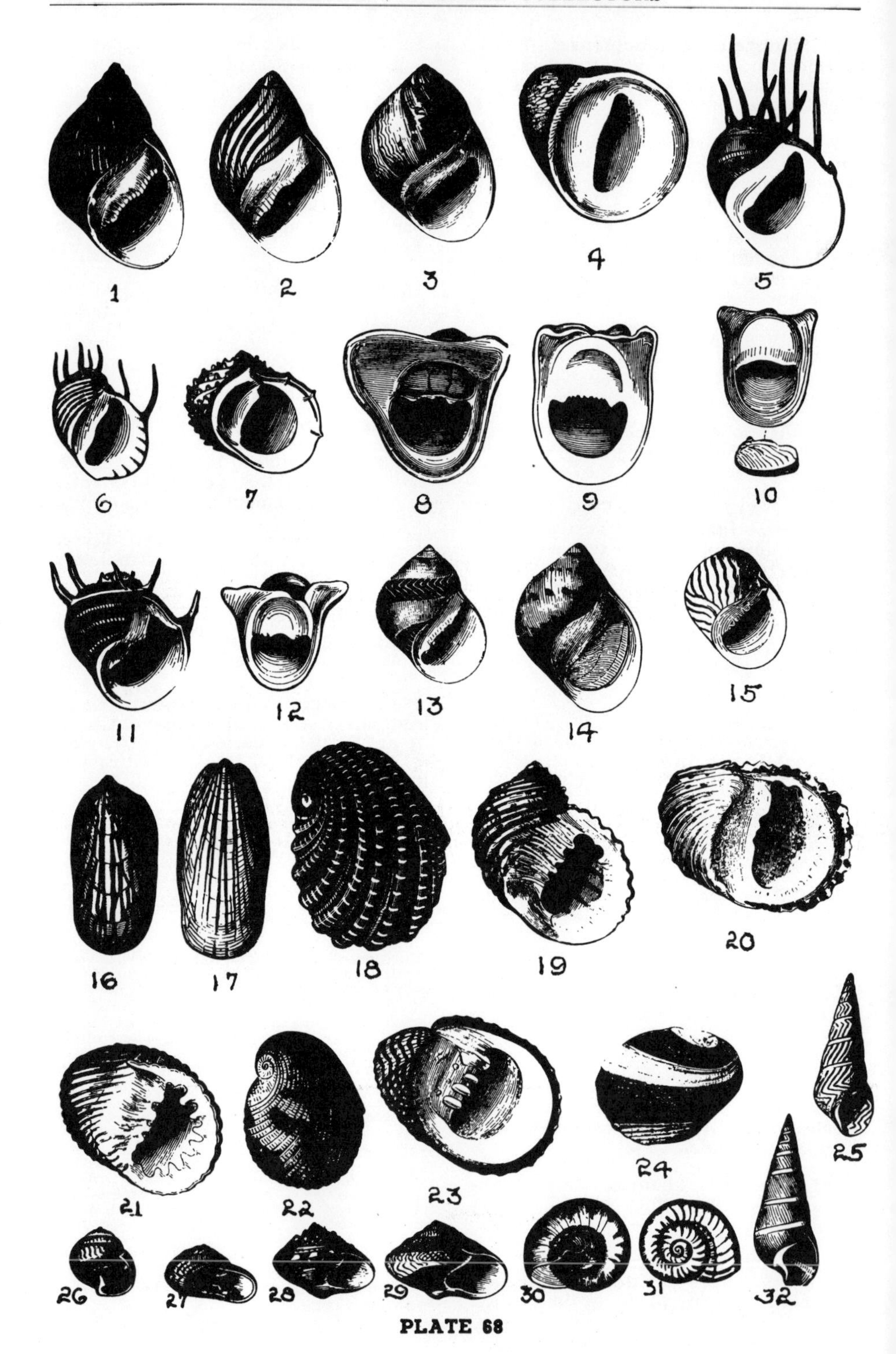
1
2
3
4
5
6
7
8
9
10
11
12
13
14
15
16
17
18
19
20
21
22
23
24
25
26
27
28
29
30
31
32

PLATE 68

1. **Neritina cumingiana,** Recl. Philippines. A fine closely lined smooth species of white and black that is one of the large forms. I have illustrated about 15 species of this genera of some 600 known forms. You will see they are very variable. The shells are found in salt water but often extend up rivers to where the water is low in salinity.

2. **Neritina turrita,** Beck. Philippines. A more widely lined black and white shell. Some writers make the No. 1 a variety of this species.

3. **Neritina smithi,** Gray. India. One of the most beautiful of all known species.

4. **Neritina labiosa,** Brod. Philippines. The largest Neritinas I have seen are of this species. All forms have remarkable operculums, usually of a hard shelly nature and often highly polished.

5. **Neritina longispina,** Recl. Mauritius. There are many spiny varieties but this fellow is the most highly developed of all.

6. **Neritina spinosa,** Sow. Tahiti. Finely lined and covered with short, sharp spines.

7. **Neritina aculeata.** Guild. Andaman Islands. In this species the spines show only faint development.

8. **Neritina vespertina,** Nutt. Hawaii. A type common to this territory and there are several others, somewhat similar over the Pacific.

9. **Neritina tahitensis,** Less. Tahiti. The aperture or front of this form is the most attractive part of the shell.

10. **Neritina bicanaliculata,** Recl. with form of operculum. Fiji Islands. Very common in this group of islands and elsewhere.

11. **Neritina corona,** L. Solomon Islands. A very round, well marked and spiny species that is always much admired.

12. **Neritina auriculata,** Lam. Philippines. A fine horned variety that is quite distinct from other species.

13. **Neritina communis,** L. Philippines. There are very few marine shells in the world that show as wide a variety of markings and colors as this one. It would be possible to secure hundreds of them no two alike.

14. **Neritina semiconica,** Sow. Calcutta. Rather rare and seldom seen in collections.

15. **Neritina pupa,** L. Bahamas. The little zigzag shell always attracts attention on account of its brilliant markings.

16. **Septaria tessellata,** L. Java. All of the 70 species of this genus are closely allied to the Neritina and of somewhat similar habits but the shell is half aperture, hardly showing one whorl. Almost limpet-like in form and often well marked.

17. **Septaria lineata,** Lam. Philippines. Another similar species to the preceding. The various forms reach over the Indian and Pacific.

18. **Nerita exuvia,** L. Philippines. A most remarkable species that has deep ridges over the whole outer surface. Also well marked.

19. **Nerita grossa,** L. Philippines. A finely marked shell which attains very large size.

20. **Nerita plexa,** Chem. Mozambique. Another fine large form which attains much larger size than most of the 200 species scattered over all oceans.

21. **Nerita costata,** Chem. Formosa. You will find this shell most everywhere in this part of the Pacific.

22. **Nerita chameleon,** L. Malacca. I had many forms of this remarkable shell sent in from the Philippines where it is fairly common. Its name well represents its many shades of color.

23. **Nerita atropurpurea,** Recl. Philippines. Seems to be fairly common along the coast of Negros Island, P. I. It shows a wide range of color patterns.

24. **Nerita albicella,** I. Mauritius. I had many of these sent in from the Philippines so it must be common over both oceans. A fine distinct species.

25 and 32 are both patterns of **Bankivia fasciatus.** Mke. Australia. A very variable highly polished shell of which you seldom find two exactly alike. There are four known species.

26. **Isandra coronata,** Ads. Torres Straits. A small curious shell of which only a few species are known. Another form is found in the Red Sea. They are closely allied to the Chlorostoma.

27. **Camitia limbata,** Phil. Sardinia. A neat small shell of which there are only a few species and allied to the Clanculus which they resemble.

28. **Umbonium monilifera,** Lam. Japan.

29. **Umbonium suturalis,** Lam. Japan.

30. **Ethalia zealandicum,** Homb. New Zealand.

31. **Ethalia guamensis,** Quoy. Cebu, P. I. The Umbonium and Ethalia are all highly polished shells, extremely common over the Pacific and in some species an endless number of color patterns. A few varieties hold fairly true to type but others will vary in every group of islands. The sailors call them Button shells as many of them much resemble buttons.

32. See No. 25.

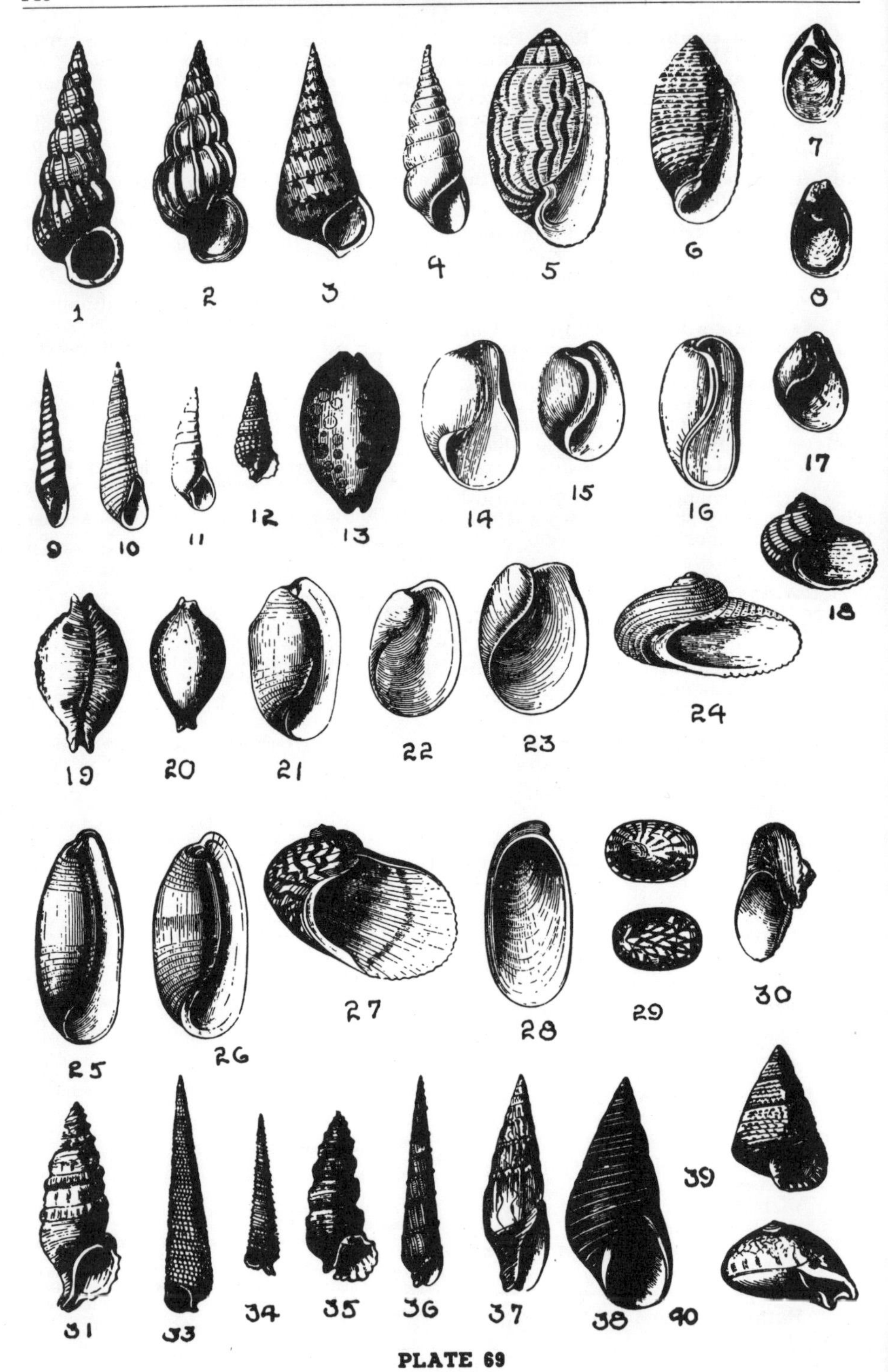
1
2
3
4
5
6
7
8
9
10
11
12
13
14
15
16
17
18
19
20
21
22
23
24
25
26
27
28
29
30
31
33
34
35
36
37
38
39
40

PLATE 69

1. **Epitonium communis,** L. Palermo, Sicily. This genus which has always until recently been called Scalaria, comprises over 300 species. Nearly all are white, but this form is mottled.

2. **Epitonium lamellosa,** Lam. Barbadoes. A very neat white species which ranges over a good share of the Atlantic area.

3. **Niso venosa,** Sow. Karachi, India. There are about 15 species ranging around the Indian and African territory. Most of them are about the size of cut.

4. **Stylifer sublatus,** Ads. Mauritius. Over 300 species are known. Mostly from Indian Ocean but a few from Pacific. They are sharp pointed, neat polished shells.

5. **Solidula flammea,** L. Indian Ocean. This genus used to be called Tornatella and comprises some 300 species scattered here and there over all oceans. They are neat shells, often well marked and ridged, or granular surface.

6. **Solidula solidula,** Lam. Mediterrean Sea. A well marked species somewhat smaller than the preceding. Other forms range down to ½".

7. **Smaragdinella viridis,** Rang. Hawaii. A remarkable small green shell which is allied to the **Atys, Schaphander** and other genera of the Schaphandridae. There are only 6 species known, mostly in the Pacific.

8. **Smaragdinella glauca,** L. Australia. Another form of this peculiar genus which hardly represents more than one full whorl.

9. **Strombiformis subulata,** Don. Ajaccio. A large group of over 40 species, ranging over the world. They are all slender and usually shiny.

10. **Aclis bitaeniata,** Sow. Japan. Slender, finely ridged shells, of which 30 species are known. They are usually rare and hard to secure. Many Pacific localities.

11. **Eulima tortuosa,** A. and R. Pacific. A wonderful group of over 150 species extending over the world. They are all very highly polished, usually white, and the upper whorls are distorted.

12. **Cerithopsis tubercularis.** Mont. Lower California. Typical in form of this great genus of more than 100 species extending over all oceans. They are closely allied to the Potamides.

13. **Cypraea annulata,** Gray, Ebon Island. Pacific. A handsome small species covered with spots. It is not widely distributed.

14. **Haminoea brevis,** Quoy. South Australia. Usually of a greenish color. There are some 60 known species, well distributed over the world.

15. **Haminoea cymbalum.** Quoy. Island of Guam. A white species which is fairly common in this group of islands.

17. **Haminoea virescens,** Sow. California. They call it the Green Bubble shell. The mollusk is much larger than the shell and is usually found on rocks in southern California.

18. **Stomatella baconi,** Ads. Northwest Australia. This genus are all neat shells, like miniature abalones, with very brilliant interior nacre. They are about 50 known species.

19. **Pustularia cicercula.** L. Hawaii. The genus is part of the Cypraea family. Some of the forms are smooth and others pebbly like Triveas. This species has two good varieties.

20. **Pustularia cicercula globulus,** L. Philippines. This is a very good variety of the preceding species and found over much of the Pacific.

21. **Atys solida,** Brug. Philippines. Most of the species of this genus are pure white, some attaining 1½ inch. There are about 36 forms.

22. **Philene aperta,** L. South Australia. This species belongs to the main genus of the family Philinidae. All of the forms have a redimentary shell, as compared with the mollusk. They are thin and white.

23. **Philene coreanica,** Ads. Porean Archipelago. Another of these very queer shells, of which there are some 35 species known.

24. **Stomatella imbricata,** Lam. Adelaide, Australia. One of the most common forms of this genus and is often seen in collections. It is one of the largest of the group and very interesting.

25. **Atys elongata,** Ads. Ceylon. This species is long and slender, pure and white and a very neat shell. It is fairly common.

26. **Atys cylindrica,** Chem. Viti Islands. A fine white bubble form that is fairly common over the Pacific.

27. **Stomatella sulcifera,** Lam. New Caledonia. An attractive shell of this interesting genus, being finely banded with zigzag markings.

28. **Gena planulata,** Lam. Mozambique. A real gem of a shell, typical of the 28 species that form the genus. It is one that is most commonly seen in collections.

29. **Broderipia iridescens,** Sow. Mauritius. A genus of four species from the Indian Ocean, of which this is one of the finest. They much resemble small Limpets and are quite rare.

30. **Stomatia rubra,** Lam. Cebu, P. I. These little fellows differ from their near relatives, by being ridged. There are 16 known species from all over the Pacific. Very neat little shells.

31. **Cerithium torulosum,** L. Mauritius. All of the species of this genus have a general resemblance but cover an immense variety of shape. This form differs radically from either the east or west coast forms of the U.S.A.

33. **Triforis perversus,** L. Mediterranean Sea. This is one of the largest species of the genus, which is a remarkable one. Over 200 varieties are known, all are sinistral and finely ribbed or nodulated. Most of them are dark but a few are red or lighter shades. Some are from deep sea.

34. **Triforis corrugata,** Hinds. New Caledonia. Another form of this great group of small shells. Some of the little fellows from the Philippines are real gems.

35. **Cerithium incisus,** Hinds. Guayamas. A small neat form that seems to be entirely different from other species.

36. **Terebra cancellata,** Quoy. Philippines. All of the small forms of this genus are handsome shells, often sharply pointed, with brilliant natural polish. There are many kinds similar to this one in a genus of over 300 species.

37. **Terebra aciculata,** Lam. Gulf of California. One of the small neat forms from this very rich molluscian territory.

38. **Cantharidus rosea,** Lam. Tasmania. Every one of the many forms that comprise this genus are unusually beautiful shells. This is one of the finest, ornamented with a great variety of colors and make very attractive necklaces.

39. **Thalotia conica,** Gray. Australia. A handsome shell typical of the many species which comprise the genus. Most of them are from the above region. They are closely allied to Cantharidus.

40. **Cyclonassa neritea,** L. A neat little shell which is widely distributed over the Pacific area. It is extensively used by native people for ornamentation. Almost everything they make is lined with rows of this little shell. I know of no other species so widely used in so many groups of islands. They can be worked into a great variety of patterns.

GENERAL INFORMATION FOR CONCHOLOGISTS

The only paper published in this country exclusively on Conchology is called the Nautilus. For a great many years it was issued monthly, but for a decade or more, it has been published four times a year. For information about same write the publisher, care the Museum of Zoology, University of Pennsylvania, Philadelphia, Pa.

The Journal of Conchology is the official organ of the Conchological Society of Great Britain and Ireland. It is now issued quarterly. Can be obtained by becoming a member. American members are desired and must be recommended by some other members of the society. Dues ten shillings a year.

The Malcological Society of England has also issued proceedings since 1893 and is now a quarterly. Dues one pound a year and the proceedings are sent to members. A complete set costing $200 or more has more real information on shells than any other similar publication. I was a member of both societies for 30 years and had complete sets of their proceedings and derived great pleasure from same, as will you.

France and Germany had similar societies, but I have not been informed if they have been continued to date due to war. Their proceedings issued in their own language also contain a world of information. For further information about any of these societies write some bookseller in those countries.

The finest work ever issued on shells in the United States is the Manual of Conchology, by George W. Tryon, Jr., and continued to date by Dr. Henry A. Pilsbry. For information in regard to same and what volumes are now available, write the Philadelphia Academy of Science, 19th and Parkway, Philadelphia, Pa. This institution is the oldest in this country devoted exclusively to Natural Science. Organized in 1812, its annual reports have been issued each year but contain only a small amount of information on Conchology.

The great foreign works on Conchology are as follows: Martini and Chemnitz, Systematisches Concyhylien-Cabinet, a work in German, arranged in monographs of separate genera with good colored plates. It was started in 1841. It costs today several hundred dollars.

Conchologica Iconica or figures and descriptions of the shells of mollusks, with remarks on their affinities and synonymy, geographical description. They comprise 20 big volumes. Written by Lowell Reeve and G. B. Sowerby, 1843-1878. 2,727 colored plates and costs today likely $1,000 when obtainable.

Reisen Archipel der Philippines by Semper and finished by Kobeldt 1870 to 1916. Was issued in parts, as were many of the old and famous works on shells. A complete set binds into 4 large volumes and covers all the then known species of Philippine land shells with fine colored plates. Costs today $200 to $300 or perhaps more.

Kiener, L. C. A splendid work in French published 1834 to 1850. There are several editions. One set covers 9 volumes and some 622 plates in color. Other sets have been bound in 13 volumes or more. They are about the finest colored plates on shells ever issued. Cost usually $500 and up.

Sowerby, B. G. B. Thesaurus Conchyliorum or monograph of the genera of shells. 5 vols. imp. 8 vo, 1847 to 1887 with 532 plates. The plates are superb illustrations. A set will cost today around $600 or more but rarely offered.

Biologica Centrali-Americana. Land and Fresh Water Mollusca by E. von Martens. Royal 4 vo, 706 pp. with 16 plain and 26 colored plates. This book covers the land mollusca of Central America and is the one volume on mollusca of a great series of volumes on the whole flora and fauna of that region. 1800-1901. Used to cost around $35.

During the Twentieth Century there have been very few popular volumes on shells. Most of the great advance in the study has been followed by works of a scientific nature seldom of interest to the shell collector. But if you propose to make a scientific collection of shells, the publications are really necessary to understand what is going on.

You will find the above works and several thousand others of less importance in the great library of the American Museum of Natural History in New York. The Academy of Natural Science in Philadelphia, The Museum of Comparative Zoology in Cambridge, Mass., and also in the New National Museum in Washington, D. C.

I have always advised students to become a member of the American Museum in New York. It costs only $5.00 a year and you get the beautiful magazine called Natural History, well worth the price. This membership will carry with it a membership card, that gives you the privilege of visiting their library and looking over any of their books on shells, which is extensive. I have been a member most of my life and have always considered it worthwhile. You can also have membership in the Academy of Natural Science in Philadelphia and the Chicago Natural History Museum. These memberships carry with them privileges which are really worthwhile to any student of shells.

Tourists to Florida should arrange their trip so as to visit the small museum of fine shells located on the grounds of Rollins College at Winter Park. These shells were donated by the late Dr. J. H. Beal who was a valued friend of mine for over 40 years. Most of the shells come from my collection as did many of the finest works in the library on shells. The good doctor made this grand collection and arranged to have it always on display, so that the people who come to this land of sunshine, could always see them. A real worthwhile monument to a grand student of nature.

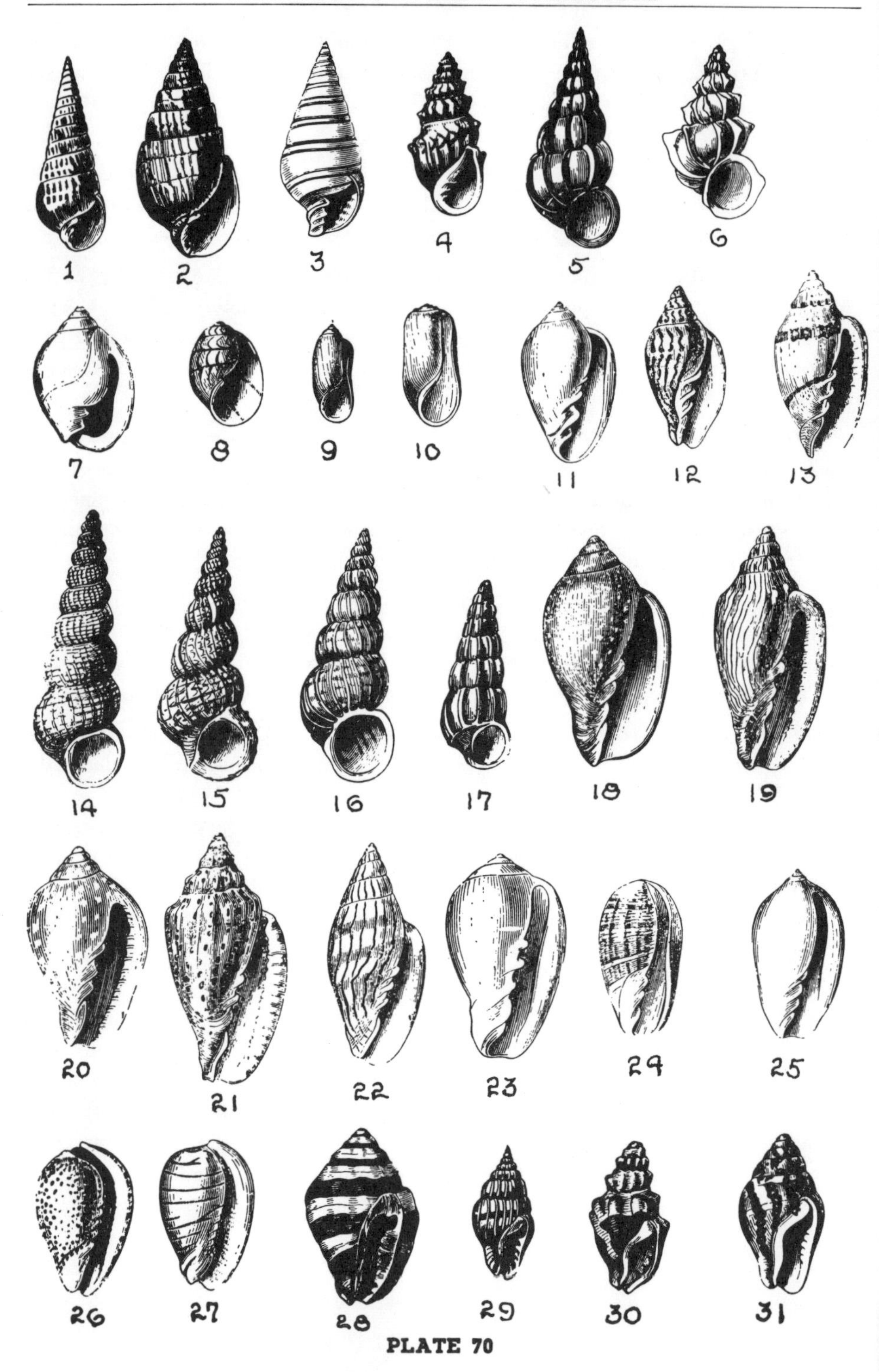
1
2
3
4
5
6
7
8
9
10
11
12
13
14
15
16
17
18
19
20
21
22
23
24
25
26
27
28
29
30
31

PLATE 70

1. **Pyramidella teres,** Ads. Cebu, P. I. A neat little shell, finely ornamented and highly polished. There are about 45 species, a few in all oceans.

2. **Pyramidella ventricosus,** Guer. Viti Islands. A strong robust shell for this genus and finely mottled.

3. **Pyramidella terebellum,** Lam. Mauritius. It seems to be found over a wide territory, as similar shells are fairly common in the Bahamas. It has a brilliant natural polish.

4. **Otopleura nodicincta,** Ads. Philippines. The shells of this genus are closely allied to the Pyramidellas but are of an entirely different form. There are a dozen or more species.

5. **Epitonium pyramidalis,** Sow. New Caledonia. A very neat species from this isolated territory in the Pacific.

6. **Epitonium alatum,** Sow. New Caledonia. This species is ornamented with varices that are uneven, a very unusual type. There are 300 species in the genus.

7. **Ringicula buccinea,** Broc. Capri, Italy. This cut is greatly enlarged as all the shells of this genus are white and quite small. The 44 species range over the world.

8. **Bullina scabra,** Gmel. Australia. A finely ornamented small shell which is related to the **Hydatinas and Aplustrums.** There are five known species.

9. **Tornatina olivula,** Ads. Surprise Island. Typical of some 42 species of the genus, all small and all white but vary greatly in form.

10. **Tornatina coarctata,** Ads. Philippines. Another of this genus of small shells which are widely distributed over all oceans.

11. **Marginella conoidalis,** Kien. Cuba. A very showy white species of high polish and fairly common in Cuban waters.

12. **Marginella scripta,** Hinds. Andaman Islands. An attractive species from this great group of islands in Oceanica.

13. **Marginella bifasciata,** Lam. Senegal. The islands of the Indian Ocean produce some of the finest and rarest shells of this great group of 300 species. This is one of them but not so rare as some others.

14. **Epitonium decussata,** Pease. Cebu, P. I. A very neat finely reticulated variety of which there are few others similar.

15. **Epitonium varicosum,** Lam. New Caledonia. This seems to be quite a rare shell as I have seldom seen it in American collections.

16. **Epitonium coronatum,** Lam. South Africa. One of the finest forms from this territory, which is rich in marine shells.

17. **Epitonium australis,** Lam. Australia. A fairly common small form from this territory.

18. **Marginella glabella,** L. Senegal. A fine large species for this genus and only occasionally met with in collections.

19. **Marginella adansoni,** Kien. Senegal. One of the finest of all forms and seems to be in constant demand among collectors.

20. **Marginella goodalli,** Sow. Senegal. A remarkable shell and one only occasionally met with.

21. **Marginella faba,** L. Senegal. The network of small dots makes this an attractive species.

22. **Marginella cleryi,** Petit. Senegal. Few of this genus have such handsome lines and brilliant polish.

23. **Marginella quinqueplicata,** Lam. Moluccas. A strong robust shell, uncolored but stands out as one of the distinctly large shells of the genus.

24. **Marginella angustata,** Sow. Ceylon. A small neatly colored species that is fairly common. You see a quite similar shell from the Bahamas.

25. **Marginella cincta,** Kien. West Indies. A very strong, robust pure white shell of high polish that is not common. I secured some brilliant specimens from Curacao.

26. **Marginella persicula,** L. West Indies. The brilliant markings of this shell distinguish it from all other species. A real beauty.

27. **Marginella lineata,** Muhl. West Africa. The circular lines make this an attractive species. There are many others in this territory.

28. **Engina mendicaria,** Lam. Philippines. The brilliant black markings of this shell stand out above all others of the 60-odd species of the genus. All of the many forms are attractive.

29. **Columbella lyrata,** Sow. Panama. A very neat small species of this remarkable genus of mollusca covering many hundred species. They are found in all oceans and most of the forms fairly common, but you would have to comb the whole earth to secure all of them.

30. **Columbella rugosa,** Sow. Panama. A well ribbed little species of good color.

31. **Columbella aspersa,** Sow. Japan. A neat well marked species much resembling some of the West Indies forms.

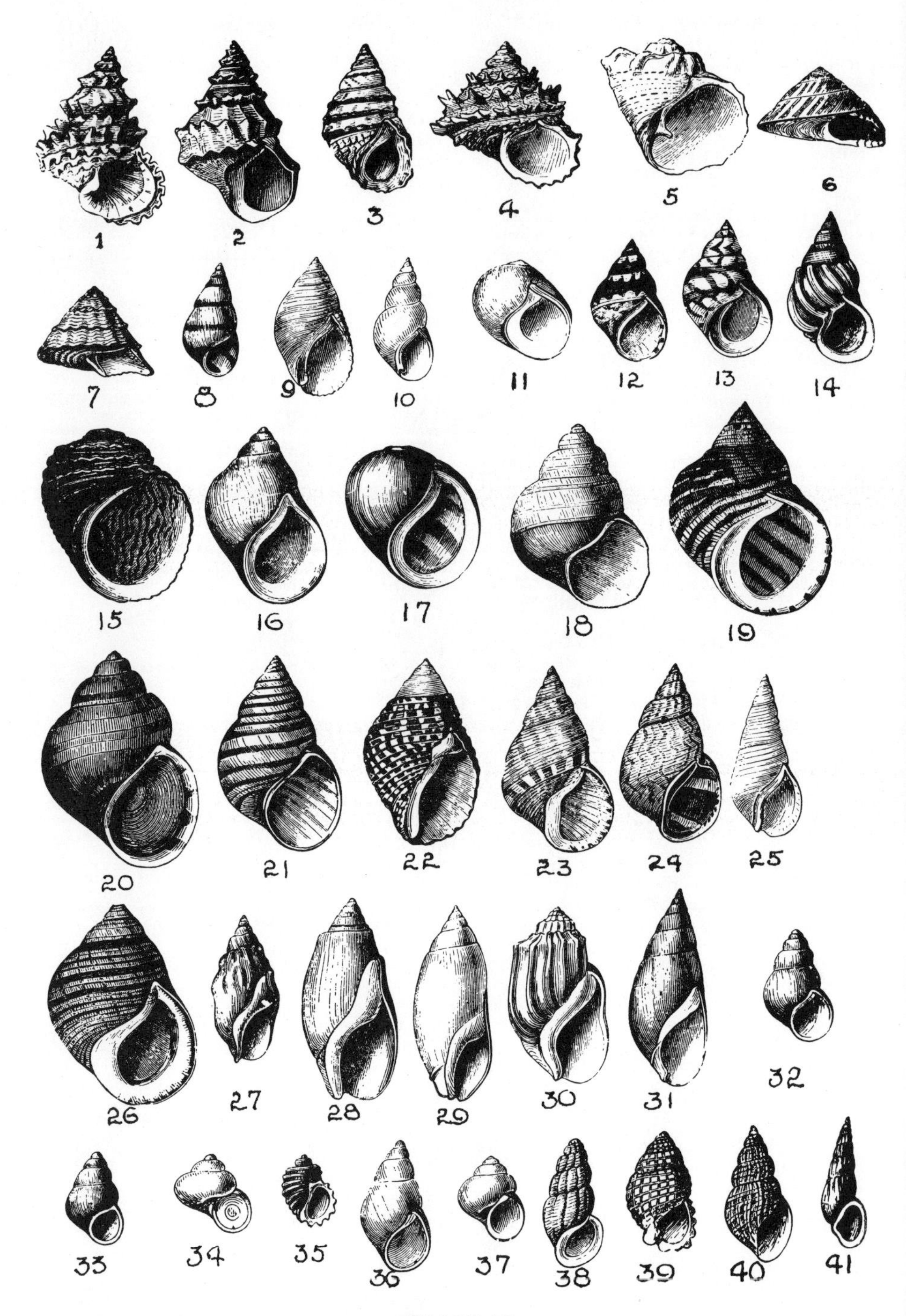
1
2
3
4
5
6
7
8
9
10
11
12
13
14
15
16
17
18
19
20
21
22
23
24
25
26
27
28
29
30
31
32
33
34
35
36
37
38
39
40
41

PLATE 71

1. **Tectarias tectum-persicum,** L. Indian Ocean. One of the finest of this group of shells, all of which are covered with nodules or spines.

2. **Tectarias pyramidalis,** Quoy. Australia. May be common but not often met with in collections.

3. **Tectarias subnodosus,** Phil. Zanzibar. Fairly common over a wide territory and typical of the 50 species of the world.

4. **Tectarias cumingii,** Phil. Paumotus. This species belongs to the section Echinella, all of which are of about this form.

5. **Modulus tectum,** Gmel. Mauritius. One of the largest and rarest of the 20 species that comprise this genus.

6. **Risella melanastoma nana,** Lam. Australia. Typical of the 24 species that comprise this genus. Some of them are faintly marked which show up when well cleaned.

7. **Risella melanastoma,** Gmel. Australia. The typical form of which No. 6 is a variety. There is not much variation in the many species.

8. **Cleopatra bulimoides,** Oliv. Lake Tanganyika, Africa. A small fresh water form from this great lake which is the wonder of the world as far as its forms of mollusca are concerned. They differ from all other lakes and a large percentage of the species resemble marine forms in every way. Many of these marine-like forms are from very deep water and bring $1 or more per specimen, when you are able to obtain them.

9. **Planaxis nucleus,** Lam. Venezuela. It is also found in the Florida Keys. Is a small black species. There are 75 forms in the world of this genus.

10. **Litiopa melanostoma,** Rang (Bombyx). These small shells are pelagic and float around on the surface of the Gulf Stream. There are 60 species in the family arranged into four genera.

11. **Littorina neritoides,** L. Malta. Very common on the rocky shores of this ancient territory.

12. **Littorina undulata,** Gray. Australia. I had many forms of this shell sent in from the Philippines, where it is very common on the rocky shores.

13. **Littorina picta,** Phil. Mindoro, P. I. You always find these attractive shells on the rocks during high tide.

14. **Littorina albicans,** Metc. Sarawak. Everywhere you may travel in the oceans of the world, if the beach is rocky, they are sure to be covered in places with some of the 200 species of Littorinas.

15. **Paludomus gardneri,** Rve. Ceylon. There are about 45 species of these black shells and a good percentage of them are from this island. It is general headquarters for the whole group.

16. **Paludomus conica.** Gray. Ceylon. One of the elongated smooth forms of the genus.

17. **Paludomus globulosus,** Rve. Ceylon. A round smooth species quite distinct from other forms.

18. **Vivipara carinata,** Swain. Bengal. You find Viviparas in all parts of the world, usually in rivers. This is nice carinated variety.

19. **Littorina pulchra.** Sow. Pearl Islands. Considered the largest and finest of the whole genus.

20. **Vivipara vivipara,** L. England. Widely distributed over Europe and typical of most of the other varieties. There must be over 100 species known.

21. **Vivipara bengalensis,** Lam. India. Generally common throughout India and a fine species of its class.

22. **Planaxis sulcatus.** Born. Australia. Most common variety met with, of the large number of known varieties. Also one of the largest. There are 75 species.

23. **Littorina scabra,** L. Philippines. Wherever you may travel in the P. I. you are sure to see some forms of this fine, large shell of the genus.

24. **Littorina zigzag,** Chem. Paumotus. You are liable to find this handsome species most anywhere in the East Indies.

25. **Hemisinus ornatus,** Poey. Cuba. In the few small creeks and rivers of Cuba you are quite sure to find these little fellows on the stones where the water is rushing along toward the sea. There are about 20 species.

26. **Littorina littorea,** L. Massachusetts coast. My first view of the sea most fifty years ago was at Lynn, Mass., and this shell was so common on the rocks, I was tempted to gather many of them.

27. **Melanopsis nodosus,** Fer. Persia. If you traveled over North Africa and adjacent territory you would find many species of Melanopsis in all the fresh water streams. They take the place of the Pleurocera groups in our U.S.A. streams.

28. **Melanopsis dufouri,** Fer. Spain. A fine robust form common in southern Europe. There are about 50 known species.

29. **Melanopsis buccinoidea,** Oliv. Syria. As you travel over ancient Palestine you meet with this neat form.

30. **Melanopsis costata,** Fer. Palestine. Even the Dead Sea has forms like this as well as nearby streams.

31. **Melanopsis laevigata,** Lam. Syria. A smooth species quite distinct from others of the group.

32. **Bythinia tentaculata,** L. Europe and America. On the edge of fresh water streams you may find the stones covered with this fine little operculated shell.

33. **Bythinia leachi,** Schach. England. You will find this small form in the various canals clinging to small stones or grass. There are 50 known species.

34. **Valvata piscinalis,** Fer. There are less than 20 species of this genus but they succeed in finding their way, doubtless through commerce, over all the world. Are fine little fresh water operculates.

35. **Fossorus costatus,** Broc. Sardinia. A small marine shell of which there are about 60 species over the world. Many of them are attractive shells.

36. **Hydrobia ulvae,** Penn. England. Very small shells that used to be called Paludestrina and you will find them so labeled in old collections. They are from fresh water and very common all over Europe where there are 50 varieties or more.

37. **Amincola porata,** Say. Ohio. This genus of small fresh water shells are common in Europe and America and cover many species, all of which are hard to identify.

38. **Rissoa crenulata,** Mich. Corsica.

39. **Rissoa reticulata,** Mont. Mediterranean Sea.

40. **Rissoa aculeus,** Gld. Tronso. These four species illustrate in a small way the over 500 species of the great groups of Rissoa, Rissoina, Scaliola, and other similar genera that are scattered all over the ocean waters of the Eastern Hemisphere. They are small marine forms of great beauty of form and color, worthy of much study. All are quite small, even to minute.

The Anomiidae include the true Anomias, Placunanomia and Placuna. The first two genera are somewhat similar and usually live on back of other shells, assuming the shape of their host. But the Placuna are radically different in form and habits. There are only a few species. The shells consist entirely of subnacreous, plicated laminae, peculiarly separable. The P. placenta is translucent and has been used as window glass for centuries. Many churches of the Orient have windows entirely composed of this shell, which has been slit to the desired thinness and colored as desired. P. sella is the familiar saddle shell well known to many collectors as they are often exported to this country from Japan.

✦

The genus Pectunculus consists of 50 to 100 species and range over most of the ocean world. The shells are orbicular and can always be easily separated from shells of other genera by the semicircular row of transverse teeth. In fact the number of teeth like some of the Arca increase with age by addition to each end of the hinge line. The shells are often well ornamented with colors but some are of a usual shade of brown. There are small genera of allied shells called Limopsis and Lissarcal.

✦

The Limidae are commonly called Lima or File shells. There are 30 or more species and over 300 fossil forms from Carboniferous. The valves are usually white, smooth or ridged, usually with a thin epidermis. The two valves only touch at the hinge and point. The mollusks dart through the water like a scallop, but in a contrary posture. It may be compared to a fish swimming tail foremost. Some species construct nests out of fragments of shell, nullipores, gravel and other material which they ingeniously fasten together by their byssal threads and attach to the roots of large seaweeds. Several young will often be found to be occupying the same nest. The nest is funnel shape, with the larger end contracted and sufficiently wide to admit the Lima freely up and down, but not turning around in it. The case is lined inside with a closely woven net of byssal threads, plastered over with slime. The mollusk has a remarkable grasp with its tentacles. Place your finger near one and he will grasp it and hang on allowing himself to be dragged about. How little we actually know of the life history of the thousands of kinds of shell life found in the ocean.

✦

The Cyrenidae cover many genera which include the fresh water forms of Sphaerion, Pisidium and they range up to the huge heavy Galatea with other minor genera. The largest of the Cyrena that I had sent in were C. buschi from the Solomons, where my correspondent stated the native people were very fond of it as food, it being extremely common. The large 4 to 5" shells being found in great quantities in favorable streams. The genus ranges from true fresh water forms to real salt water species.

✦

The Diplodontidae include Diplodonta and Ungulina. The Diplodontas are almost all of them sphaerical and some of the species build a nest of the sand to live in. If you are where you can collect them be sure to save some of the nests as they are most interesting. The shell is inside and inserts its hollow foot through a hole in its home and brings in the food that it needs. There are about 50 species and quite as many fossil from Cretaceous, etc.

✦

Veneridae. The Venus family contains a large number of genera, several of which have been briefly described elsewhere. These notes refer to Venus proper. There are more than 200 species known and fully as many varieties are found fossil from the Oolite on.

The shells are thick and heavy for their size, one species in our Florida waters attaining at least three pounds without the mollusk. There are 17 or more subgenera, our common Quahog or V. Mercenaria being a typical form. In the tropical seas they are more elegant in design, the V. lamellata with its curved pink fronds and delicate structure one of the most beautiful of the genus. It is found in Australia and Tasmania, not being at all rare. The North American Indians used to make wampum from V. Mercenaria, being particular to get as many of the wampum from the part of the shell that is black on inside. The V. gnidia from West Mexico, while all white, is much desired as specimens.

The Gariidae consists of the genus Gari (Psammobia) Soletellina, Sanguinolaria, Asaphis and Elizia. These genera contain a very attractive lot of shells and are eagerly sought by collectors.

The Gari contain 100 species or more and almost as many fossil in the Cretaceous. They run from 1 to 2½ inch as a rule, very smooth and finely lined and colored. Range from Norway to New Zealand.

The Soletellina are quite similar in form and contain some rather brilliant colored shells. There are 50 or more forms from West Indies to Tasmania and almost as many fossil in the Eocene. There are only a few species of San guinolaria but all attractive. The Asaphis are only a few species not over 10 and extend from the West Indies to Australia. The A. coccinea from West Indies are very colorful and I have seen 100 different shades from brilliant red to purple and many shades between.

A number of the marine univalves, as well as several of the Pteropods and Heteropods are luminous. The light emitted by the mollusks is mostly greenish, or a vivid blue, but some is reddish. Such mollusks live mostly in the open sea, specially in the great sea currents.

If you discover a shell which you believe to be new to science, the first thing to do is to examine all literature on similar forms. If you decide to describe same, the manuscript should be submitted to a scientific society that is incorporated so that it may be published in its proceedings. The type should be fully and accurately illustrated. In no other way will it have a permanent standing throughout the world.

There are many species of Bulla, of which B. Ampulla is the type. A few fossil forms are found in the Cretaceous. The shells are usually spotted, marbled or zoned, spire concave, umbilicated. Aperture as long as the shell, inner margin with columella. They inhabit sandy mud-flats and the slimy banks of river mouths. At low water they often conceal themselves under seaweed.

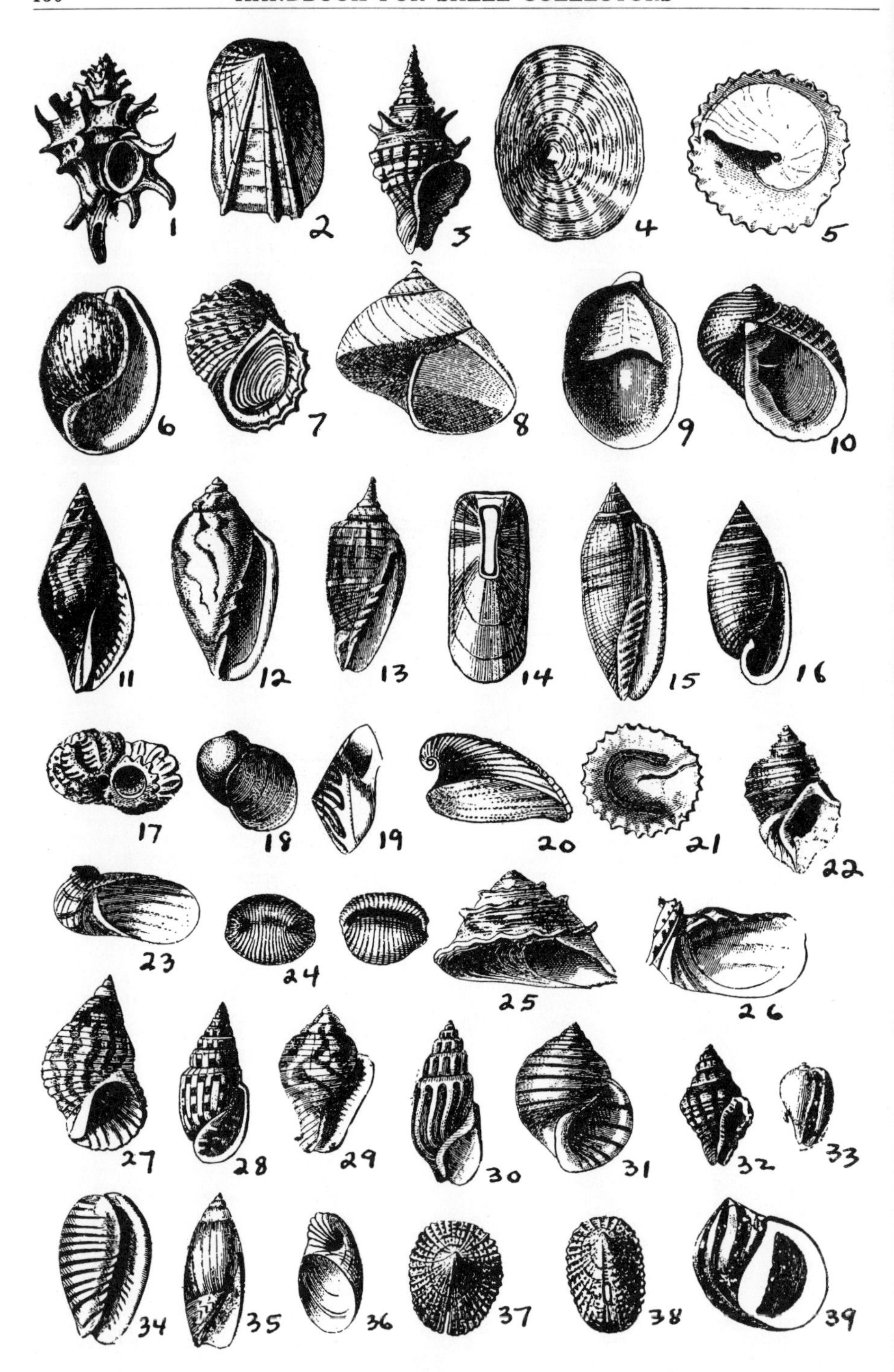
1
2
3
4
5
6
7
8
9
10
11
12
13
14
15
16
17
18
19
20
21
22
23
24
25
26
27
28
29
30
31
32
33
34
35
36
37
38
39

PLATE 72

1. **Murex zealandicus,** Quoy. New Zealand. A rare spiny little fellow about an inch long.

2. **Amathina tricostata,** Gmel. Japan. A small rare form from Japan. A genus of seven species allied to Capulus.

3. **Turris.** One of hundreds of species of a very great group of mollusca.

4. **Umbraculum mediterraneum,** Lam. From Mediterranean Sea. Quite rare. An almost flat form, thickest in middle and has sharp edge. 3 inches.

5. **Calptraea.** About sixty species are known and a few from all oceans.

6. **Bulla ampulla,** L. The Bubble shell and all of its allies, some 60 species, are a great group and most of them are this form or more elongated. None are very rare.

7. **Paludomus.** Typical of a fine group of fresh water shells covering over forty species and mainly from Ceylon. All are black and some are corrugated like cut.

8. **Janthina communis,** Lam. Florida. The Violet Snail is common at some seasons and may not be seen again for many months. It has the habit of forming a float, with a mass of its eggs attached, and swim on the surface of the water, especially in the Gulf Stream. A rough sea may bring thousands of them on the shore. At such times quantities of perfect specimens can be obtained. There are about 30 species.

9. **Crepidula fornicata,** L. Florida. The Slipper shells show a remarkable variety in form in the some 75 known species and other are quite variable in same species. None are rare. They often attach themselves to other shells.

10. **Neritopsis radula,** L. The one species is from Ceylon and other parts of Indian Ocean. At first sight one would be sure was a form of **Nerita** but it is of different texture.

11. **Pisania.** About fifty species are known from all over the Pacific and some in other oceans. They are fine small shells of about one inch, some being smooth and others coated with epidermis.

12. **Marginella.** One of the rare forms from Senegal. Several hundred species of Marginella are known. All are glossy, handsome shells but quite a number are white. The Florida forms called gem shells, find many admirers. They can be had in great variety and rare forms bring from one to five dollars each.

13. **Imbricaria conica,** Schm. Viti Islands. The Imbricaria are simply a section of the great family of Mitra, of which many hundred species are known. This is one of the diverse forms.

14. **Macrochisma.** One of the genus of the great Fissurella or Keyhole Limpet group. About 12 species are known.

15. **Mitra fenestrata,** Lam. Society Islands. Illustrates a type of the great genus of Mitra that assume the form of some of the Conus. There are only a few species of this shape.

16. **Solidula tornatilis,** L. A group of about forty species found to some extent all over the world. Some are finely banded and none over an inch.

17. **Fossorus.** An interesting group of some sixty small species scattered over all oceans. They are related to the Tectarias.

18. **Velutina laevigata,** Penn. England. A small genus of some 20 species with large aperture and very thin shells. Are allied to the Lamellaria.

19. **Umbonium (Rotella).** The button shells are mostly brownish and highly polished. One species gigantea, from Japan, is used to illustrate Violet Rays, as under the rays it is bright red.

20. **Carinaria.** Only a half dozen species of very thin shells like glass and more or less rare. They travel over the surface of the sea.

21. **Siphonaria.** An immense group of limpet-like shells that are about as much fresh water forms as marine. A sort of connecting link. Usually have a hump on one side. Mostly about one inch in size but a few are larger.

22. **Trichotropis.** A widely scattered group of 35 species of small shells from Maine to Greenland and some in the southern hemisphere.

23. **Stomatella.** The fifty species are nearly related to the Haliotis but all are small. Other allied groups of same family are Gena, Broderipia and Stomatia. Mostly from tropical regions.

24. **Trivia.** This large group of 75 species allied to the Cypraea, but all are more or less ridged, whereas the Cypraea are smooth. Many are very common and some rare. Very dainty shells of white, brown, black and other colors.

25. **Risella.** The 25 species are from the southern half of the world and fairly common in Australia. Most of them are of the form of the cut and about one inch in size.

26. **Stomatia.** About 16 species that range from Philippines to Australia. Not common anywhere.

27. **Planaxis sulcatus,** Born. Australia. There are some 75 known species and this is about the largest, attaining one inch. They comprise a family by themselves, there being no other genera closely related.

28. **Pyramidella.** About fifty species scattered over the world and mostly about one inch long. Some are smooth and highly polished and others finely marked like the picture.

29. **Columbella mercatoria,** L. West Indies. This immense group of 600 species covers the world of ocean. Most of the species are small and common. If you find them at all you usually find plenty. All are handsome and much prized.

30. **Melanopsis praerosa,** L. Palestine. A large group of over 100 species of fresh water shells, common through southern Europe to Egypt. They are very distinct forms.

31. **Littorina.** Typical of a great genus of more than 200 varieties of shells, as a marine, but can live great distances inland. They are a fine lot of shells and inexpensive.

32. **Engina.** The 60 handsome forms are all small and from all oceans, but mainly Pacific. None are rare.

33. **Marginella apicina,** Mke. Florida. The little gem shells are common on Florida shores and most pleasant days you will find people gathering them for use in making fancy articles. Always have a high polish, as do all the group.

34. **Marginella cingulata,** Dill. From Senegal. One of the handsome striped forms of this large group of naturally polished forms. There are many others equally pretty.

35. **Olivella biplicata,** Sow. California. A handsome member of this great group is found in all tropical regions in abundance. Over 100 species. Ranges about one inch and always highly polished.

36. **Hipponyx pilosus,** Desh. Hawaii. There are about 30 species of Hipponyx and they are an interesting lot. All small but of curious shapes. Scattered over the world.

37. **Emarginula.** There are more than 100 species of this genus, all small, with a little slit in the shell, which distinguishes them. They are not easy shells to secure.

38. **Rimula exquisita,** A. Ad. Ceylon. Another fine small shell, of which some nine species are known in the genus. This and the preceding species belong to the great family of Fissurella of Keyhole Limpets. There is a marvelous variety of them.

39. **Neritina zebra.** Brug. Oceanica. The great family of Neritinas cover over 500 species of many diverse markings and forms of shell. All are common where found and inexpensive but make a wonderful show, if you secure enough of them. One of the largest forms is from Hawaii.

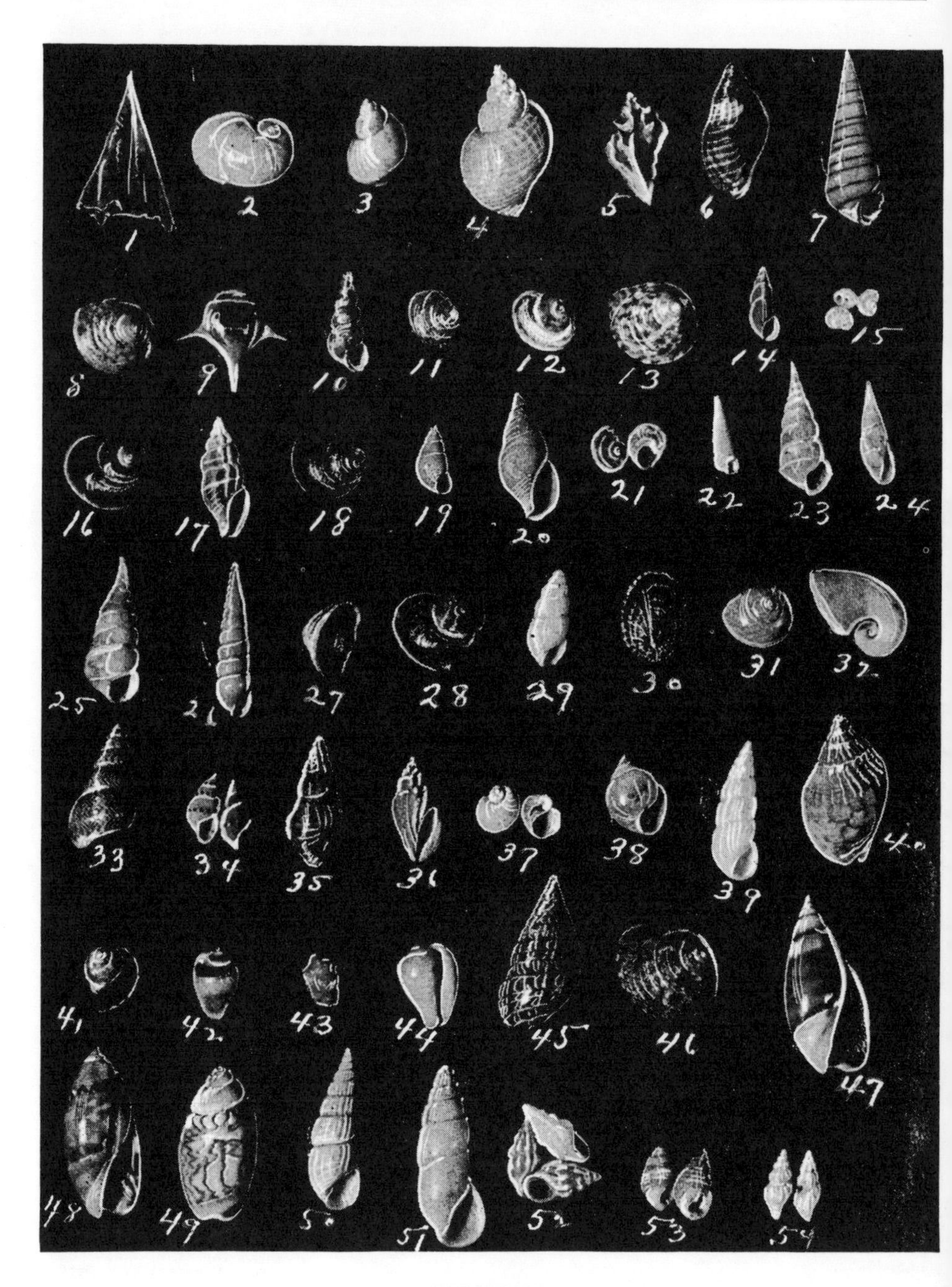
1
2
3
4
5
6
7
8
9
10
11
12
13
14
15
16
17
18
19
20
21
22
23
24
25
27
28
29
30
31
32
33
34
35
37
38
39
40
41
42
43
44
45
47
48
49
51
53

PLATE 73

San Diego Shell Club

NAME	BID
MINIMUM BID	

1. **Cleodora lanceolata,** Chem. Pacific. Pelagic. As thin and clear as glass. All the Pelagic shells are so different from ordinary marine, they attract wide attention.

2. **Lacuna pallidula,** Da Costa. Europe. A small uncolored pointed shell of which there are many kinds in the various oceans.

3. **Lacuna vincta,** Mont. Labrador to N. J. A small pointed uncolored shell, very common over a wide territory of the northland.

4. **Admete viridula,** Fabr. Spitzbergen. A small pointed uncolored shell of which there is the one species in the Arctic regions.

5. **Typhis tetraptera,** Brown. Malta. There are about 20 species scattered over the world but I never had but two of them. All are like fine small Murex but they are more closely allied to the Trophons. This little fellow has fine ridges and curved spines.

6. **Usilla variabilis,** Desh. Reunion Ids. A small smooth black shell with a few fine circular lines. I have only seen this species and one from Japan. It comes next to the Iopas.

7. **Niso venosa,** Sow. Karachi, India. There are about 13 species in the world, but I have only seen two. This little sharp pointed shell glistens like glass and has two tiny brown bands. It comes near Pyramidella.

8. **Minolia variabilis,** Sow. Karachi. All forms are small, usually smooth, occasionally with fine ridges and somewhat mottled. I had 18 species all from Eastern Hemisphere.

9. **Diacra trispinosa,** Les. Cape Verde Islands, Pelagic. I have only seen two species of which this is one. They are real little glistening white butterflies of the sea.

10. **Turritellopsis acicula,** Stimp. Spitzbergen, 40 fathoms. Very small, horn colored shell with fine spiral lines.

11. **Thoristella appressus,** Hutt. New Zealand. Very small trochiform shells with fine circular lines. All the forms I have seen are from the above territory.

12. **Fossarina varia,** Hutt. New Zealand. Very small shells usually white. A genus mostly confined to this country.

13. **Thoristella chathamensis,** Hutt. New Zealand. A larger shell than No. 11 and brilliantly marked with red and white.

14. **Fairbankia quadrasi** Bttg. Manila. I only have a record of three forms of this genus and have only seen one other from Bombay. It is a very small dark colored shell.

15. **Cyclostrema sulcata,** A. Ad. Manila. A very tiny drab or white shell. I have fifteen species, most of them not much larger than the one figured.

16. **Margarella fulminata,** Hutt. New Zealand. They are small reddish-brown shells like a tiny Turbo. I had three species from this territory.

17. **Mangelia sinclairi,** Smith. New Zealand. A tiny white shell finely lined with red.

18. **Cremniconchus conicus,** Blf. Bombay. A tiny dark shell which much resembles a land shell. In fact it is a modified Littorina. There are two forms, both from India.

19. **Mucronalia caledonica,** More. New Caledonia. Very tiny glistening white sharp pointed shells. They are closely allied to the Eulima.

20. **Columbella rosacea,** Gld. Spitzbergen. The small shells are a drab white with hardly a tinge of rosy color and from very cold water.

21. **Risselopsis (Risella) varia,** Hutt. New Zealand. Very tiny flat, rounded shells of the same form as the larger shells of the genus.

22. **Seila terebelloides,** Mart. New Zealand. 10 fathoms. Very tiny, slender white shells of many whorls and deep fine lines.

23. **Scalenostoma carinata,** Desh. Mauritius. A sharp, pointed dull white shell.

24. **Strombiformis metcalfei,** A. Ad. Lifu Ids. A thin, glistening white shell with tiny red band. There are about 50 known species.

25. **Stylifer exaratus,** A. Ad. Mauritius. A round, sharp pointed glistening white shell typical of the more than 30 species known. Most of them are from islands in Indian Ocean.

26. **Syrnola aciculata,** A. Ad. Cebu, P. I. Tall, slender, many-whorled smooth white shell. They belong to the Pyramidellidae and the 25 species are largely from P. I. or near by.

27. **Broderipia eximia,** Nev. Ceylon. Very small iridescent limpet-like shells of which I had three species. They belong to the Stomatidae.

28. **Margarita helicina,** Fab. Greenland. A small round smooth light brownish shell from cold water.

29. **Menestho humboldti,** Risso. Adriatic. There are four species, but I have only seen this one. It is small, white and finely lined.

30. **Notacme helmsi,** Smith. New Zealand. A small conical dark shell like the great Acmaea shells.

31. **Photinula nitida,** A&A. Cooks Id. A small shiny bluish shell covered with tiny dots. There are about 20 species all from southern hemisphere closely related to the Monodontas.

32. **Lippistes grayi,** A. Ad. Algoa Bay, Africa. There are four known species of which this is one. Consisting of one detached whorl they are shaped like a trumpet. Uncolored. They belong with the Trichotrophididae.

33. **Leiopyrga picturata,** H&Ad. Victoria, Australia. A slender white shell completely covered with reddish lines. The one species belong next to the Phasianellas.

34. **Diala monile,** A. Ads. Port Albert, Australia. There are about 25 species in this genus and they belong to the Litiopidae. This form is a small light colored shell with a row of dots.

35. **Jeannea hedleyi,** Iredale. Kermadec Id. It looks like a little Columbella with finely reticulated surface.

36. **Eucythara capillacea,** Rve. Mauritius. This genus is one of the great Family of Turris. Combined with the small Daphnellas there are more than 600 species known. They are small to minute and most attractive. Usually finely reticulated and of elegant form.

37. **Tornus (Adeorbis) patula,** Sow. Hong Kong. They are tiny white shells. The few known species are placed in a family by themselves, the Adeorbilidae.

38. **Apicalia bertsi,** Preston. Ceylon. A small bulbous white shell that is translucent. The genus is a very small one and is placed in the Strombiformidae.

39. **Chrysalida rissoina,** Ads. Japan. A fine slender white shell with numerous lines on each whorl. Belongs to same family as preceding species.

40. **Cyllene oweni,** Gray. West Africa. The entire surface is covered with fine light brown markings, and upper whorls are lined. The species follow the Cominella.

41. **Canthoridella suteri,** E. A. Smith. New Zealand. A well-rounded light brownish shell. The writers on mollusca in New Zealand have erected a lot of new genera for species of shell they have discovered, and collectors who wish to know more about the shells of that country, should send to me for the book they have published on the Shells of New Zealand.

42. **Erato lachryme,** Gray. Queensland. A neat little fellow, with natural polish, brown stripes above, and white through the middle whorl. They belong with the great Family of Cypraea.

43. **Pedicularia pacifica,** Pse. Kingmill Ids. A very small and irregular pink shell. It lives on the coral polyps.

44. **Erato callosa,** A & R. Japan. A very neat little white shell. The shells of this genus have a natural polish like cowries.

45. **Cantharidus bellulus,** Dkr. Australia. A brilliant little fellow with circular stripes of red and perpendicular stripes of white on a brownish surface. Most of the species of this genus are brilliantly adorned.

46. **Gibbula strangei,** A. Ad. Queensland. The shell is finely lined and striped with various colors and naturally smooth surface.

47. **Olivella semistriata,** Gray. Panama. The main body of shell is a grayish-brown and balance above and below white. All Olivella are as beautiful as the Olivas.

48. **Olivella tergina,** Duc. Mazatlan. The apex is white, with orange below, grayish body and base orange. Many Olivellas are adorned with various colors.

49. **Olivella petiolita,** Duc. Mazatlan. It is a short stubby white shell adorned with zigzag stripes of brown.

50. **Rissoa rex,** Pils. Japan. A white shell with numerous lines. It is one of the largest species of the genus.

51. **Rissoina montroizieri,** Souv. New Caledonia. A white shell typical of the larger forms of the genus. The Rissoinas are usually much larger than the Rissoa.

52. **Risso variabilis,** Muhl. Medit. Sea. Attractive little shells which well illustrate the hundreds of species of this great genus. One can easily spend a lifetime on the small and minute shells of the world.

53. **Rissoa cimex,** L. Malta. A finely lined little yellowish shell with pure white aperture.

54. **Asperadaphne dictyota,** Hutt. New Zealand, 23 fathoms. One of the many small attractive shells found in this great island continent in recent years. The genus name is unknown in this country.

The Tellinidae include the very large genus Tellina, one Gastrana, many Macoma, most of which are white. Strigilla, which are smaller and often deep red inside and the very odd shells of Tellidora. There are 3 to 400 species of the Tellinas, most of them quite flat and many of brilliant colors, but the great majority are plain white or drab. There are at least 200 species fossil in the Oolitic, etc. They are found to some extent on nearly all ocean shores and down to very deep water. The most beautiful forms are from Philippines and on to Australia in very warm, shallow water. They have been subdivided into 17 sub-genera. One of the largest species is T. magna found from Florida to Panama but only rarely in the northern part of its range.

✦

The Crassatellites are heavy shells with thick epidermis and are found around tropical rivers where they empty into the sea. There are about 40 species and fully twice as many fossil from the Cretaceous. You do not find very many of them in collections and they should be more intensively studied.

✦

The Ostreidae include only the one genus Ostrea or Oyster. They have been extensively studied due to their commercial importance, and it is easy to get literature that covers many of the species. Some species grow to immense size. There are about 100 forms and more than twice that many fossil, from the Carboniferous. In some places they live in vast quantities, the rocks being covered with them growing one on another. A favorite is O. cristagalli due to the strong, sharp fingers which adorn the two halves and fit perfectly.

✦

The Genus Callista of the Veneridae are about the most colorful shells of the whole family. Most of them are smooth and brilliantly mottled with various colors, specially those from the Australian region and other parts of the tropical Pacific. The Sunetta, while a small genus, are also smooth and very colorful shells ranging from 1 to 2 inch diameter.

The Donacidae consist almost entirely of the genus Donax, only one genera called Iphgenia which contains a very few species are included. They are almost world wide in distribution and often cover the beaches in immense quantity. You will not find them easily unless you follow the receding tide. As the water exposes the sand they dive out of sight quickly. The brilliant colors of the species found on the two coasts of the U.S.A. furnish plenty of amusement for the tourists. You will often see people wading in the shallow water, scooping up a pail full of sand and sieve out the little shells. Then they often take the plunder home, wash it clean and make a delicious soup. As you near the tropics, they attain larger sizes up to one inch or larger and most of them are very attractive. There are over 100 species and fully half as many fossil in the Cretaceous, etc.

The Pteriidae include the very odd forms of Pteria (Avicula), the Malleus which in some cases are called Hammerhead Oysters, the Crenatula and the Melina (Perna). As a family they are about the strangest of all bivalves. There are 50 to 100 species all told and about three times as many fossil from the Lower Silurian on. I would like to have illustrated more of the very strange forms, some of which are 6 to 8" and many are brilliant pearl inside in which are found some fine pearls.

The genus Dosinia are shells almost circular and largely white or drab. Only a few have any color design, mainly from tropical regions. The names of the various subgenera of the many genus in the Veneridae have been changed so many times in the past century you will find them very confusing until you make considerable study of them.

The Fissurellidae is mainly composed of the genus Fissurella which are commonly called Keyhole Limpets as they have a hole in the top and the largest known forms 3 to 4" are in this genus. There are 15 to 20 other genera now included in the family, some without the slit in the top, others with the slit in the edge, some flat and oblong of a shining white. One of the finest is called Lucapina crenulata from the California coast which attains 3" or more.

There are likely over 200 species mostly confined to warm seas and 30 or more species fossil in Carboniferous. They are real homey mollusks, selecting some place on a rock to which they return when away foraging for food. A collection of even a hundred forms including a few of the many genera is a very interesting lot of shells, always greatly admired. They are usually white, gray often finely reticulated, and under surface of some forms is a rich green.

PELECYPODA
BIVALVE MOLLUSCA

The Bivalves differ from the Gastropoda in having no specialized head, though provided with a mouth, organs of sight, etc. They consist almost always of two valves, connected usually by a hinge but there are forms that start as a bivalve and become a univalve when fully adult.

The foot is a fleshy appendage, much used as food, as in scallops, but is used by the mollusk mainly in digging rather than locomotion. The security against many enemies which their completely enveloping shell affords the bivalves, have tended to produce in this class of mollusca a much less active existence. The adults of attached families such as Spondylus, Ostraea, etc., remain for life in one position, without the power to change their residence.

They cannot be said to search for their food but simply select it from such vegetable and minute organisms as the water may float within reach of their mouth. Thus the term "Stupid as an oyster" is justly applied to many of them, still they are supported with organs for the enjoyment of a placid life.

The embryonic shell forms the umbo of each valve and it is often very unlike the after growth. In the Oyster, Anomia, etc., the umbo frequently presents an exact imitation of the surface to which the young shell originally attached.

In the boring shells the carbonate of lime has an atomic arrangement like arragonite which is considerably harder than calcareous spar and which enables them to bore in either hard or soft substances with apparent ease.

Some bivalves continue to increase in thickness long after they have ceased to grow outwardly, and the greatest addition is made to the lower valve, especially near the umbo. In the Spondylus some parts of the mantle secrete more than others, so that cavities, filled with fluid are left in the substance of the shell.

In their native element, the oyster and scallop lie on one side and the lower valve is deeper and more capacious than the upper. Most other bivalves live in an erect position, resting on the edge of their shells, which are of equal size.

The boring shells have a strong and stout foot with which they bore vertically into the sea bed, often to a considerable depth. These never voluntarily quit their abodes and often become buried and fossilized in them. They usually burrow in soft ground and some small forms even bore into Ascidians. They are found in the hardest compact limestone. The holes of Lithodomus often serve to shelter other forms of sea life after the death of the rightful owners.

The boring shells have been called "stone-eaters" (Lithophagi) and the "wood-eaters" (Xylophagi). Between the bivalves having attached valves as in the oyster and those which are free like the Cardium are those which spin a byssus by which the shell is attached to some foreign substance and thus form colonies. Usually the byssus-spinners like Modiolas are inhabitants of shore lines, in situations where at low tide they are exposed above the waters; they are thus subject to the lifting powers of the incoming waves and but for their byssus would be quickly torn away and likely destroyed, as are most free shells in similar situations.

Bivalves are said to be close when the valves fit accurately and gaping when they cannot be completely shut. In many shells both ends are open as in Solen, etc.

All bivalves are clothed with an epidermis which is organically connected with the margin of the mantle. It is developed to a remarkable extent in Solemya, and in Mya it is continued over the siphons and closed mantle-lobes, making the shell appear internal.

The structure of the hinge of bivalves characterizes both family and genera, hence if you wish to learn to identify them, make a special study of same. I have known collectors to make drawings of all the many forms of hinges. The locomotive species generally have the strongest hinges but the most perfect examples are presented by Arcas.

The central teeth, those immediately beneath the umbo, are called hinge or cardinal teeth, those on each side are lateral teeth. Some of the lateral are developed and not the cardinal. In young shells the teeth are sharp and well defined, in old specimens they are often thickened and even obliterated.

The abductor muscles attach the animal to the valves of the shell upon either side and their contraction serves to close them together. At a certain distance within, and parallel with the margin of the shell is an impression caused by muscular mantil-margin and termed the pallial line. It connects the scars of the abductors.

Some bivalves have excellent eyes in the larval state and later lose same, but is partly compensated by numerous ocelli on the mantel-border. These are particularly noticeable in the Pecten. Some Spondylus have as many as 60 ocelli on the right or fixed side and 90 on the left side. In some of the other genera they are placed in different position.

While the bivalves have no real organs of hearing and seem to be insensible to sound, they are very quick to detect motion. Wave your hand over a Pecten in the water and he will quickly respond, doubtless due to the difference in light or shadow.

The mouth of bivalves is situated at the anterior part of the body between two pairs of branchiæ. These lips are equal in dimension, triangular and varying much in size in different genera. From the mouth a short aesophagus opens into the stomach and the long intestine and is terminated by a rectum, opening above the posterior abductor muscle.

The young bivalves are hatched before they leave the parent. The eggs remain in the gill tubes and there retained until the escape of the young mollusks.

PLATE 74

1. **Modiola metcalfei,** Han. Philippines. A light brown shell with reddish-brown periostracum, and usually fine hair on part of shell. Prominent ridge through middle. 2½″.

2. **Mytilis pellis-striatus,** Dunk. Mazatlan. It is rather straight in outline, of light color and darker periostracum. Usual growth ridges. 2½″.

3. **Mytilis viridis,** L. Bankok, Siam. A brilliant green shell, the last growth usually being the darkest color. Rather attractive 2″ or larger.

4. **Mytilis grayanum.** Dke. Japan. Rather wide, pointed and only slightly curved. Of a light shining brown color. 2¼″.

5. **Mytilis pictus,** Born. Algiers. Medit. Sea. Prominent hump through middle, almost straight in outline and of a light brown color. 3″.

6. **Mytilis crenatus,** Lam. Algiers, Medit. Sea. A medium sized very dark shell that is prominently ridged lengthwise. Very slightly curved. 2″.

7. **Mytilis dunkeri,** Dunk. Medit. Sea. A splendid species almost straight below, triangular above, of a handsome shining brown color. 3″.

8. **Mytilis magellanicus,** Ch. New Zealand. A moderately curved species of very dark color and finely ridged. There are about 130 species of this genus in the world. They often grow in clusters, held together by a strong brissus, attached to each other, or some rock, which will withstand the power of the surf. A very interesting lot of shells, most of which are edible.

9. **Ostrea edulis cristata,** Poli. Medit. Sea. The usual type of the oysters, many of which always grow in very irregular shape as they are either attached to one of their own species or some solid foundation. This species is frilled on back and almost circular in outline. Of light shades of brown. 2½″.

10. **Arca fusca,** Brug. Madagascar. An oblong shell with very finely reticulated dark surface which is usually covered with hair. This genus is divided into 11 sub-genera and this species belongs to the Barbatia. Most of the shells of this sub-genus are of this form. It is the most prolific of all the sections, being about 90 species. 2″.

11. **Arca antiquata,** L. (malculosa). Philippines. I believe widely distributed in the Pacific. Somewhat similar to the preceding usually darker color, same fine reticulations and mossy surface. 2″.

12. **Arca tankervillae,** Rve. New Holland. Of the general form of the two preceding but more regular in outline, similarly adorned and the interior of all the three species is a rich satiny dark brown. 2″.

13. **Hemithyris (Rhynconella) psittacea,** Gmel. Washington. A very neat little Brachiapod almost black and very typical of the many fossil forms found in the rocks. The recent Brachiapods are very few as compared with the immense number found fossil. 1″.

14. **Arca decussata,** Sow. Solomon Ids. A very neat smallish shell of the Barbatia section, very strongly reticulated and the fuzzy surface being around the edge. 1½″.

15. **Mesodesma erycinaeum,** Lam. Tasmania. A white shell completely covered with a shiny light brown periostracum. The genus is a large one divided into 7 sections, most of the species from tropical seas and extensively in southern hemisphere. While not at all rare you see very few of them in collections. 1¼″.

16. **Modiola elegans,** Gray. Cochin, China. A shell that is well named as it is really elegant as compared with many of the species in this great genus. It is of the form of cut, of a shining brownish-green with extremely thin hinge. It belongs to a section of the genus that are all thin. 2½″.

17. **Modiola Adriatica,** Lam. Naples. A short prominently humped shell of similar color to the preceding but darker on the umbones. 1½″.

18. **Modiola barbata,** Lam. Malta. As the cut shows the shell is covered with hairy periostracum, being slightly curved and humped. Interior light colored but darker at umbones. 1½″.

19. **Mytilis hirsutus,** Lam. Japan. This is a really hirsute or hairy mussel the hairs being long and wiry. The form is similar to many of the smaller species of the genus, with rich iridescent interior. 1¼″.

20. **Mytilus latus,** Lam. New Zealand. A smooth shining dark brown shell of the form of the cut and rich shining interior. 1¾″.

21. **Modiola rhomboides,** Han. South Africa. A very thin shell of the form of cut, the hinge extending fully four-fifths of the shell. It is a shining reddish-brown color. 2″.

PLATE 75

1. **Modiola auriculata,** Krs. Japan. A typical form of the genus, with hump along middle of whorl, and covered with yellowish-brown periostracum. 1¾″.

2. **Modiola philippinarum,** Han. Philippines. A more oval form than preceding and of light brown color. Rather thin. 1½″.

3. **Modiola sirahensis,** Jouss. Aden. Arabia. A slender elongated shell with shining light brown periostracum. Rather thin. 1¾″.

4. **Modiola arata,** Dunk. Australia. A short, well humped, light yellowish-brown shell of about 1¼″.

5. **Mytilis ovalis,** Lam. Chile. A curved almost black shell with regular lines following the curve. 1¼″.

6. **Modiola japonica,** Dkr. Japan. A slender shining shell of light greenish color, with zigzag rings of brown. 2″.

7. **Modiola penetida,** Verco. South Australia. A small russet-colored shell, coated with a hairy periostracum. ¾″.

8. **Modiola plicata,** Gmel. West Africa. A very smooth, shiny russet-colored shell with hinge the full length of shell and few fine lines. 1½″.

9. **Modiola arborescens,** Chem. Mauritius. A glistening white shell covered in part with an intricate lace work of fine brown lines. Rather unique for a shell of this genus. 1¼″.

10. **Modiola watsoni,** Smith. Bay of Bengal. A small slender greenish-black shell, rather smooth, a little over 1″.

11. **Modiola fluviatillis,** Han. New Zealand. An elongated oval black shell that is rather smooth, little over 1″.

12. **Modiola cinnamomea,** Ch. Philippines. An oval thin russet-colored shell of about 1″.

13. **Modiola senhausi,** Rve. Japan. A slender thin shining shell with zigzag stripes of brown on green surface. 1¼″.

14. **Modiola rodriguezi,** Orb. Argentine. A slender russet-colored shell of about 1″.

15. **Petricola lithophaga,** Retz. Malta. A small pure white finely lined shell that is a finer borer in wood or soft rock. About 1″.

16. **Petricola monstrosa,** Gray. Kurachi, India. A small white frilled shell which with the frills is almost square. It is lined but the extended valves make it a very unusual species. 1″.

17. **Jounnetia globulosa,** Q&G. Philippines. An elongated small white shell with unusual long hinge, valves meeting only at the tip. The wide open space as shown in cut makes it a very unusual species. 1″.

18. **Gastrochaena retzi,** Jouss. Japan. A slender white shell, very thin, typical of the many kinds of boring shells found throughout the world. They are almost a life study, specially if you go into the history of the millions of damage they have done to wharves throughout the world. 1¼″.

19. **Modiolaria cumingiana,** Dkr. New Caledonia. A very bulbous rounded thin shell of a light green, shining color with bands of brown. The few species extend over the world. You often find them living in colonies firmly lashed together with a net like brissus. 1″.

20. **Lithophaga zettelinus,** Dkr. Japan. A large round russet-colored shell that is rather smooth, showing some growth lines. It is a great borer as are all the shells of this genus. You will find whole colonies, each close together and living within a hole that just fits the shell and as smooth as glass. 2½″.

21. **Lithophaga corrugata,** Phil. New Caledonia. A somewhat similar shell but more pointed at one end, of a light brown color, the whole pointed end of the shell covered with chevron lines. 2½″.

22. **Lithophaga teres,** Phil. Philippines. A slender rich dark brown shell. The surface of the shell being smooth. 1½″.

23. **Lithophaga gracilis,** Phil. New Caledonia. A very small almost black form that is only about ¾″.

24. **Perna chemnitzi,** Orb. Peru. The cut well shows the shape of the interior of the valve and the extended part beyond the pearly interior. About 1½″.

25. **Lithophaga cinnamomea,** Ch. Philippines. A slender form, reddish-brown changing to black, at the point of the shell. 1¼″.

PLATE 76

1. **Soletellina violacea,** Lam. Philippines. A very thin purple shell with white radiating rays. The rays show brilliantly on the interior. 1½″.

2. **Gari reevei,** Mart. Tasmania. It is very faintly purple, covered with yellowish shade and very fine rays. Interior uniformly purple. 2″.

3. **Gari ferroensis,** Ch. Ferroe Ids. A very thin greenish-white shell, with chalky white interior. 1½″.

4. **Gari caerulescens,** Lam. Manila, P. I. A glistening grayish-purple shell with many spiral lines. Interior light purple. 2″.

5. **Gari vespertias,** Ch. Malta. A glistening purple shell covered with gray, and tiny reddish umbones. Interior rich purple. 1½″.

6. **Gari amethystea,** Wood. Hong Kong. A white shell with stripes of pale reddish. Interior with shades of brown. 1½″.

7. **Gari lineolata,** Gray. New Zealand. A grayish shell with pink umbones finely lined. Interior rosy. 1¼″.

8. **Soletellina virescens,** Des. Philippines. A dark greenish shell with zigzag markings under the periostracum. Interior the same. 1″.

9. **Soletellina minor,** Des. Philippines. A bluish shell with light brown rays and rich purple interior. 1″.

10. **Nuculana minor,** Sow. Ceylon. A pure white finely lined shell sharply pointed at one end. 1⅛″.

11. **Ungulina alba,** Rang. West Africa. A nearly round dull white shell showing many growth lines, fairly thick and white inside. Only about four species are known. The few forms I have seen were all from above locality. 1″. They come next to the Diplodonta.

12. **Gastrana fragilis,** L. England. A thin dull white shell with uneven surface. Come next to Macoma. 1¼″.

13. **Heterocardia dennisoni,** A. Ad. Philippines. A thin white shell with slightly uneven surface, of which about 4 species are known. This is the only form I have seen. Little over 1″.

14. **Trapezium rostrata,** Lam. Philippines. You will find them in old collections labeled Cypricardia. This shell is white with smooth umbones perpendicularly lined below, with prominent ridge extending to right, forming a point. Interior white. There are about 20 species known, all I have seen being of a white color. With the Corralliophaga they form the family Trapeziidae. 1½″.

15. **Trapezium guinaica,** Lam. Mauritius. A more rounded robust species, white, rounded at both ends with slightly prominent ridge and lined in part, mainly below. 1¾″.

16. **Trapezium solenoides,** Rve. Philippines. A white shell with hinge near one end with irregular surface, showing various ridges. 1½″.

17. **Trapezium vellicata,** Rve. Philippines. A white shell with humped surface showing very irregular growth. Interior white with purple tinge at one end. Near 1½″.

18. **Trapezium oblonga,** Lam. Diego Garcia. A white shell with fairly prominent ridge and reddish-brown marks at one end. Interior white. Near 1½″.

19. **Venerupis crenata,** Lam. Victoria, Australia. A white finely ridged shell with nearly a dozen low frills. Interior white with purple blotch at one end. There are about 50 species in the world well distributed everywhere. 1¼″.

20. **Mesodesma triquetrum,** Han. South Australia. A pure white oval shell nearly smooth and white interior. 1″.

21. **Nucula (Acila) mirabilis,** A&R. Korea. A nearly black shell, finely ornamented with chevron shaped lines. Interior glistening pearl. 1″.

The Lucinidae include a number of genera besides Lucina, such as Loripes Phacoides, Codakia, Divaricella and Corbis. Most of the many species are closely allied. White shades predominate and the interior of the shell is often lined with red, yellow, etc. They are distributed mainly in tropical and temperate seas, on sandy and muddy bottoms, from the seashore to the greatest depth in which you find any shell life. There are more than 100 species and about three times as many fossil starting in the Silurian. The foot of the mollusk is twice as large as the rest of the body but is usually folded back on itself and concealed by the gills. There are 15 or more subgenera showing the great diversity of the shells.

PLATE 77

1. **Lima inflata,** Ch. Medit. Sea. A thin pure white shell, prominently ridged. The hinge in the shells of this genus is rather simple and the valves are held together by a prominent ligament. There are mammoth species up to 8″ in the abyssal sea. 1¾″.

2. **Lima ventricosa,** L. Malta. A similar white shell but much thicker and more bulbous with similar ridges. The live shells can dart through the water very fast. 2″.

3. **Scrobicularia cottardi,** Payr. Corsica. A thin smooth light yellowish shell of 1¼″. The Family Scrobiculariidae contains the shells of this genus of about 16 species. Syndesma about 22 species, Iacra 1 species, Theora 3 species, most all of which are thin and white but have similar affinities.

4. **Scrobicularia piperita,** Gmel. Belgium. A pure white dull shell with fine concentric rings. 1½″.

5. **Macoma nymphalis,** Lam. Gambia. A white shell, thin; early growth smooth, and later covered with dark thin periostracum. 1½″.

6. **Macoma nobilis,** Han. Philippines. A thin smooth yellowish shell, but with brilliant pinkish-red umbones. Interior similar color throughout. It is very close to many of the Tellinas but belongs in this genus. 1½″.

7. **Mesodesma glabrata,** Des. Philippines. A thin dull white shell with some fine concentric lines, and yellowish interior. 1¼″.

8. **Codakia lacteola,** Tate. South Australia. The shell is almost circular, pure white, with prominent concentric ridges. Interior chalky white. About 1″.

9. **Semele carnicolor,** Han. Manila, P. I. An almost round white shell with thin concentric ridges and faint yellow interior. 1¼″.

10. **Semele radiata,** Rupp. Aden, Arabia. It is almost round, rather smooth, white with growth marks, the interior is mottled with reddish markings, starting at the umbones, which show to some extent on the outside. 1″.

11. **Hysteroconcha affinis tortuosa,** Brod. Panama. A uniformly white somewhat pointed shell with prominent ridges and flaky white inside. There are about 25 species in this genus, which used to be called Dione. They vary greatly, many forms being adorned with ridges as well as spines. 1¼″.

12. **Semele proficua,** Pult. Manila, P. I. The shell is almost round with prominent growth ridges and fine cross lines. The umbones are marked with pink and the interior is yellow and pink. 1½″.

13. **Gari suteri,** Gray. New Zealand. A dull white shell with very faint zigzag markings and light pinkish interior. 1¾″.

14. **Tellina staurella,** Lam. Philippines. The shell is white with radiating rays of pink, which show through to the interior of the specimen. 1¾″.

15. **Pectenculus insubricus,** Broc. Medit. Sea. A gray shell rather globular, with bands of a shade of brown. Interior white with brown patch near one end. The entire aperture of the shells of this genus is frequently prominently toothed. 1″.

16. **Pectenculus australis,** Q&G. Victoria, Australia. A white shell brilliantly mottled with splashes of brown. White inside with one small brown patch. Shells range up to 2″.

17. **Corbicula fluminea,** Mull. China. A shiny ridged dark greenish shell with patches of darker color. Interior flesh color. There are more than 200 species of this genus, a few forms found in every part of the world. They love to inhabit salt water marshes, and you will find this species often in the rice fields. Most of the forms are smaller. 1¼″.

18. **Cyrena arctata,** Desh. Maricabo, South America. A rather heavy thick shell finely ridged of a dark black color edged with green. Interior mostly rich purple over 1″.

19. **Terebratella rubicunda,** Sol. New Zealand. A very neat small Brachiapod that is entirely red and fairly common in shallow water. About 1″.

20. **Semele subtriangulata,** Desh. Australia. A slightly triangular, rather smooth shell with very faint ridges, white tinged with purple and faint yellow inside. 1½″.

PLATE 78

1. **Semele australis,** Sow. Kingmill Ids. The shell is white covered with a very thin light colored periostracum, interior white. 2″.

2. **Mactra aspersa,** Sow. New Caledonia. The shell is attractively mottled with lavender-brown blotches. Interior white. 1¾″.

3. **Mactra helvacea glauca,** Gmel. Medit. Sea. It is a shining light brown about half way, and balance of shell seems to have a periostracum and shows lines of growth. Interior white. This group frequently attains 3½″ but shell illustrated was 1¼″.

4. **Arca granosa,** L. Philippines. A solid pure white shell with fine large ridges which are regularly dotted with small nodules. There are a number of species of this family of this type, some attaining weight of several pounds. 1¾″.

5. **Arca emarginata,** Sow. Gulf of California. The hinge you will notice in cut is almost as long as the shell. There are many fine ridges and most of the shell is covered with a short, hairy periostracum. 1½″.

6. **Mactra aequilatera,** Rve. New Zealand. A fine triangular whitish shell that is equally as white inside. Very distinct and desirable. 1¾″.

7. **Mactra sauliae,** Gray. West Africa. A very thin shell and very flat. White with radiating rays of faint reddish. Interior the same. 2″.

8. **Macoma lilacina,** Iredale. New Zealand. A perfectly white shell rather flat on one end, the interior with a faint tinge of yellow. 2″.

9. **Lucina ovum,** Rve. Philippines. A perfectly round thin, white shell that is strongly oval. Interior pure white. 1¼″.

10. **Tellina laevigata,** L. Philippines. A very fine pure white shell, with radiating rays of light russet-red, interior white. A rather thin shell as are so very many of this genus. 2″.

11. **Clementia papyracea,** Gray. Philippines. A very thin dull-white shell, with many growth marks showing. Interior white. Hinge very small for size of shell. The genus comes soon after the Venus. 2″.

12. **Codakia (Lucina) interrupta,** Lam. Philippines. A nearly round shell with fine strae both ways, white with a faint tinge of yellow on the umbones. Interior a rich yellow shade. 1¼″.

13. **Strigilla splendida,** Ant. Manila. A smooth light brownish-yellow shell with fine striae showing over surface. Edge tinged with russet. Interior shaded like outside. 1″.

14. **Mesodesma elongatum.** Desh. Australia. A small horn colored shell shaped almost exactly like a small Donax. About 1″.

15. **Solen roseomaculata,** Pils. Japan. A thin small white shell finely mottled with rosy-purple. 2″.

16. **Nuculana (Leda) polita,** Sow. Panama. A small slender pure white shell.

The Arcidae contain mostly the one genus Arca, but the Cucullea are usually included in the family. There are likely 200 or more species, are world-wide in distribution and specially abundant in warm seas. You usually find them on mud flats and they are brought up from 250 fathoms. Over 400 species are found fossil from Lower Silurian.

The shells are equivalve or nearly so, of many shapes. The animal has a long pointed foot deeply grooved. Some of the forms with close valves often have the left valve a little larger than the right. Some species secrete themselves under stones at low-water, in crevices of rocks and the empty burrows of boring mollusks.

About a dozen subgenera are recog- tortuosa has twisted valves and A. Noae which has an opening in the middle of the two valves. There are numerous other forms similar. Many species are clothed with mossy hairs and these should be left on as they add to the attractiveness of the shell, but where the valves are finely marked a few can be cleaned to show up the design. A. tortuosa has twisted valves and A. pexata is called the bloody ark as it emits a reddish fluid. It would fill a large book if one were to describe the many fascinating characteristics of this great family.

PLATE 79

1. **Paphia (Tapes) Graffei,** Dkr. Japan. A shell with many prominent ridges and brilliant natural polish. Of a light yellow color, faintly marked with brown. Interior white. 2¼″.

2. **Paphia rotundata,** Gmel. Japan. Rather narrow and elongated with regular prominent ridges, smooth surface of a yellowish color with few faint blotches of brown. Interior drab white. 2½″.

3. **Paphia textile,** Gmel. (tetrix Chem.) Australia. Same shell as textile, Gmel. Of yellowish color finely marked with crossed zigzag lines of pale brown. Interior white. 2″.

4. **Paphia japonica,** Desh. Japan. A bulbous shaped shell with regular lines and near edge of whorl zigzag lines of brown. Interior white. 1¾″.

5. **Paphia philippinarum,** A&R. Japan. Finely lined perpendicularly of a drab color and few blotches of brown. 1½″.

6. **Paphia virginea,** L. Philippines. A rather thick rounded shell of a yellowish-brown color, rosy apex and white interior. 1¾″.

7. **Paphia senegalensis,** Gmel. Senegal. A rather oval shell finely lined and reticulated of a drab color with few rays of light brown. Interior shaded with yellow. 1½″.

8. **Paphia deshayesi,** Han. Aden, Arabia. A small elongated shell with fine lines and mottled brown surface. Interior white with splash of purple at one end. 1½″.

9. **Paphia striata,** Ch. Philippines. A small light brown shell with fine regular lines over the surface. Interior white. ⅛″.

10. **Paphia kochi,** Phil. Red Sea. An oval shell pointed at one end with natural polished surface. Mostly yellowish-white with few irregular blotches, which usually vary in each shell. Interior white. 1⅛″.

11. **Paphia marmorata,** Lam. Malabar. A small yellowish shell with chevron markings of brown. Interior white. 1⅛″.

12. **Paphia aequilaterale,** Desh. Japan. A yellowish-brown shell with, very smooth surface, adorned with numerous small brown marks. Interior white. 1¼″.

13. **Dosinia variegata,** Gray. Loo Choo Ids., near Japan. A circular, deeply ridged shell of a drab color with milky white interior. 1¼″.

14. **Dosinia histrio,** Gmel. Persian Gulf. A circular, deeply lined shell, the ridges ornamented with brown and one or more distinct perpendicular stripes of brown. White interior. 1⅛″.

15. **Dosinia juvenilis,** Chem. Mauritius. A circular shell with few fine lines and numerous blotches of brown. White interior with faint traces of brown. 1⅛″.

16. **Trigonia strangei,** A. Ad. Tasmania. It has all the usual ridges and dots described elsewhere, peculiar hinge and rich iridescent pearl interior. They are most remarkable works of nature. 1″.

The genus Paphia (Tapes as we used to call them) belong in the great Family of Veneridae. Tapes means tapestry and many of the species surely have that type of design. There are a hundred or more recent species and a very few fossil in the Cretaceous. The mollusk spins a brissus but this seldom ever comes with the shells. Along the Mediterranean Sea, both sides, the people eat the mollusk. You will often see the inhabitants digging for them at low tide almost everywhere. I had the shells divided into 7 subgenera and they look much better in the cabinet.

The Trigoniidae is a Family, almost exclusively Trigonia, of which there are only a few species but of great beauty. I never had more than 3 or 4 species but there are over 100 fossil forms known from Devonian. The shells are thick tuberculated, about one inch, the hinge teeth so intricately placed it is very difficult to connect them together when once separated.

The shells are almost entirely nacreous. Many collectors call them Jewell shells. Australian collectors are very fond of them, although they are never very common on their ocean beaches.

PLATE 80

1. **Spondylus japonica,** Sow. Japan. A well formed medium sized shell, the spines and body being shaded with reddish-brown and white. Interior pure white with reddish-brown edge. 2¼″.

2. **Pecten nobilis,** Rve. Japan. It is of a rich purplish-red color strongly ribbed and white inside. There are various color forms all of which are scarce as usual. 2″.

3. **Pecten nipponensis,** Kuroda. Japan. Much thinner than preceding species; finely ribbed with few fine points, of a rich shade of brownish-purple. 1¾″.

4. **Pecten layardi,** Rve. Solomon Islands. A rather oblong deeply ridged shell of a rich dark brown color. The ridges have numerous small sharp points. Dark interior. 1¾″.

5. **Pecten grandis,** Sol. Newfoundland to Long Island Sound where, in latter locality it is widely fished for the meat. It attains 8″ but the smaller 3 or 4″ specimens are always finest. The valves do not evenly meet. Surface appears smooth but is really very finely ridged of a faint reddish-purple color. You will find it in old collections under the name of **magellanica** and **tenuicostata.** Found to 100 fathoms.

6. **Meretrix planulata,** B&S. Mazatlan, Mexico. A triangular shell with smooth surface. A light yellowish color marked faintly with brown, interior white, lined with light lavender. The Meretrix you will find also labeled Cytherea. There are nearly a hundred species known and they are all attractive shells. 2″.

7. **Callista pacifica,** Dill. China. It is of a light yellowish shade with concentric rings and only faint markings. Interior white. All the shells of this genus are very attractive. 2¼″.

8. **Solen brevis,** Gray. Borneo. Ground color pink, covered with a greenish periostracum. Interior pink. Perfectly straight. 3″.

9. **Donax carinatus,** Han. Mazatlan. The color is white and purple with the appearance of being finely ridged but the shell is brilliant and smooth. Interior light purple. 1¾″.

10. **Solen guineensis,** Gray. Guinea. Of a shining light yellowish color with bars of pink. Interior same. 2½″.

11. **Solen exiguus,** Dunk. Sarawak, Borneo. A white shell covered with light green periostracum, and interior white. There are over 50 species in the world and they are a strange lot indeed. Range from very thin and small to gigantic. The true Solen are mostly straight and the subgenus Ensis curved. 3″.

12. **Glaucomya chinensis,** Gray. China. You seldom see the shells of this genus in any collections and yet they are not at all rare. Usually light greenish in color and of form of cut. There are about 16 species mostly from Philippine or China region and range up to about 3″. They form the family Glaucomyidae. 1¼″.

13. **Solen vaginoides,** Lam. Tasmania. A shell that is slightly curved and belongs to the Ensis subgenera but some species are very much more curved. It is a white shell banded with pink and the usual very thin greenish periostracum. 2¾″.

14. **Pharella acutidens,** B&S. Philippines. A very thin white shell with light green periostracum. There are about 8 species known, all similar to this one, but varying in size. They belong to the Solenidae. 2½″.

15. **Perna maillardi,** Desh. Mauritius. A neat little chap with curved shell, usual prominent hinge, pearly interior, and thin shelly fringe around edge. 1″.

16. **Perna acutirostris,** Dkr. Viti Islands. A smaller form than the preceding of usual shape, with thin fringe extending beyond the usual pearly interior. Nearly 1″.

17. **Callista squalida,** Sow. West Mexico. A finely mottled shining shell with shades of white and brown. There is often usually a very thin periostracum. Interior light colored. I found them fairly common at Panama. 2 to 3″.

18. **Callista impar,** Lam. Australia. A rather thick, heavy shell for this genus with shining surface, concentric rings of light brown, and diagonal stripes with the usual very thin periostracum. The Callistas as a group, are all remarkable shells. Interior white with prominent purple blotch near the nose of the shell. Around 2″.

19. **Pecten ventricosus,** Sow. Panama. I selected a small shell to illustrate. It comes in a variety of colors, too complicated to attempt to describe. You seldom find two exactly alike. Are closely allied to circularis of California but the latter shell is seldom so highly colored. 2 to 2½″.

20. **Trigonia margaritacea,** Lam. Victoria, Australia. This cut illustrates the interior of the shell and on Plate 91 is a cut of the outside. The interior is iridescent. All of the shells of this genus are real Jewels and much admired by collectors.

21. **Crenella nigra,** Gray. Iceland. One of the largest of the 13 known species, of a black color with some brown periostracum. Very thin and pearly inside. They are placed in the Mytilidae but many of the species have no resemblance to the shells of that great family. 1″.

PLATE 81

1. **Mactra carneopicta,** Pils. Japan. A thin light yellowish shell striped with rays of brown and white, and partly covered with thin periostracum. Interior white. You find the valves of some species of this genus on most of the shell beaches of the world. There are more than 200 species and many of the shells are flakey and easily break with a little pressure. 2¼″.

2. **Mactra maculata,** Desh. Japan. It is rather oval, of yellowish color, mottled with flakes of brown. Interior white. Many of the shells of this genus are white with no other coloring. 2″.

3. **Mactra reevei,** Desh. Philippines. A thin white shell somewhat triangular and mottled with light brown. The mottled surface shows on the inside of the shell. 2″.

4. **Diplodonta zelandica,** Gray. New Zealand. There are more than 40 species in this genus and most of the shells are white with concentric lines of growth and shaped like a bullet. This little fellow is about 1″.

5. **Phacoides columbella,** Sow. Liberia. A real bullet-shaped Lucina; it is rather unique. I can think of no other species in the genus just like it. A solid shell, it shows the usual characteristics of the genus. 1″.

6. **Mactra achatina,** Ch. Philippines. It is thin, white and finely mottled similar in coloring to No. 2, but more flat and entirely different shape. Interior white. 1¾″.

7. **Mactra antiquata,** Speng. Japan. A delicate white shell with fine radiating rays of brown. Interior purple. 1½″.

8. **Mactra nitida,** Speng. Senegal. A white shell, with extremely thin periostracum of drab color. Interior white. Quite typical of many species in this genus. 1¾″.

9. **Cyrena turgida,** Lea. Borneo. A rather solid, white shell covered with dark greenish periostracum. There are more than 100 species in this genus and some attain massive size. I had a box of gigantic shells come in from Solomon Islands, where the natives make a delicious dish of the meat of the mollusk. The shells are mostly found in fresh water streams close to the sea, where the water is less saline than usual. 2″.

10. **Arca ventricosa,** Lam. Solomon Ids. A solid bulbous shell of white color with prominent brownish-black ridges. Interior white. About 2″.

11. **Arca scapha,** Chem. Singapore. A shell of somewhat similar shape to the preceding, but more twisted in outline, ridges prominent and the periostracum mainly between the ridges. Interior white. 1¾″.

12. **Arca radiata,** Rve. China. A white shell with ridges perfectly arranged to conform with the shape of the shell and some brownish periostracum. If you keep these shells in the cabinet in their natural condition they show the difference in the species better than if cleaned down to a perfect white. 1½″.

13. **Cardita difficalis,** Desh. Victoria, Australia. The shell is almost round but the prominent ridges are curved and covered with crossed lines. It has a thin periostracum and white inside. 1¼″.

14. **Plicatula muricata,** A.Ad. Japan. A small thin reddish-white shell, covered with short spines. There are about 25 species in the world and this shell well illustrates why scientists include them with the very spiny Family Spondylidae. Most of the species are knobby but few have sharp spines. 1″.

15. **Iphgenia altior,** Sow. Panama. A large, rather thin shell of white with dense brownish-yellow periostracum. The hinge is placed about middle of shell. Interior purple. You will find it labeled with genus name of Capsa in old collections. They belong with the Family of Donax and about 8 species are known in the world. Some of the Donax are similar in shape. 2¼″.

16. **Pteria mauritii,** Jameson. Mauritius. A rather stubby variety quite different in form from most of the shells of this genus. Irregular in shape, it is of a yellowish color. 1⅛″.

17. **Anomia humphreysiana,** Rve. Peru. It is thin as are most of the more than 60 species in the world. The back is reddish-purple. The species of this genus are a strange lot and worthy of more study than usually given them. 1¼″.

18. **Glaucomya rugosa,** Han. Borneo. A thin light greenish shell of the form of cut and about 2″. Seldom seen in collections.

19. **Anomia cytaeum,** Gray. Japan. A rather solid shell of this genus. The specimen shown in cut had a fine Vermetus living on the valve with the aperture showing it had not been attached to any other shell. It is rather round, the first half of the shell rather smooth but later growths are ridged. 1¾″.

20. **Perna isognomum,** L. Solomon Ids. I have purposely shown the interior of the shell, in this cut. The pearl part is prominent and the shelly part extends outward 1¼″. Back has scales as usual. 3½ to 4″.

21. **Pecten varius,** L. Medit. Sea. One of the commonest species of the genus, usually dark, almost black, finely ridged and rather purple inside. Usual specimens 1½″.

PLATE 82

1. **Venus foveolata,** Sow. Martinique. A round drab white deeply lined shell with white interior. 1½″.

2. **Venus yatesi,** Gray. New Zealand. A shell that is round about two-thirds and balance flat. The whole deeply lined white apex purple. Interior white with large splash of purple. A rather unusual shell. 1¼″.

3. **Venus subrugosa,** Ant. Mazatlan. Rather rounded, pointed at one end, of a gray color, with prominent blackish stripes. Interior white. 1½″.

4. **Venus tiara,** Dill. Philippines. A rather small white shell with deep rosy frills. About ½″.

5. **Venus affinis,** Sow. Senegal. A small white shell with six prominent spiral ridges, tipped with brown, interior white, shaded with purple at one side. Under 1″.

6. **Venus scarlarina,** Lam. Victoria, Australia. A very finely ridged shell, which is marked with brown, interior all white. 1¼″.

7. **Venus gyrata,** Sow. West Mexico. A shell that is finely lined each way, white marked with shadings of light brown. Interior pinkish-white. 1¼″.

8. **Venus corrugata,** L. South Australia. A rather flat shell with numerous faint concentric ridges, white color, the interior orange with splashes of black. 1½″.

9. **Venus spissa,** Q&G. New Zealand. A small globose finely lined shell of gray color with two stripes of brown. Interior white edged with rich shade of brown-purple. 1″.

10. **Venus flexuosa,** L. Yucatan, Mexico. A bulbous shell pointed at one end, with thick prominent vertical ridges of a yellowish-white color. Interior with dark splash of purple at pointed end. 1½″.

11. **Venus donacina,** Ch. Japan. A rather flat, smooth shell, white with zigzag markings of light brown. Interior white. 1½″.

12. **Venus subimbricata,** Sow. Panama. A small triangular shell, with concentric ridges and mottled with white and black. Interior white. 1⅛″.

13. **Venus toreuma,** Gld. Japan. A bulbous round, finely lined shell with faint rose markings. Interior with pink space in umbilical region. 1″.

14. **Venus flumigata,** L. Victoria, Australia. A rather smooth grayish shell of light yellowish color and interior similar. 1½″.

15. **Venus impressa,** Han. China. Very similar to No. 10 except that it comes from an entirely different part of the world. 1″.

16. **Venus macrodon,** L. Brazil. A rather smooth white shell, the wide edge being brilliantly adorned with blue and orange lines and stripes, interior white with deep purple splash at tip. 1¼″.

17. **Venus undulosa,** Lam. Australia. A rather small shell, deeply mottled with yellow and orange. Interior a creamy white. 1″.

18. **Venus laevigata,** Sow. Tasmania. A small rather smooth shell the umbilical region and lower being russet color, with an edge of gray. Interior russet and white. 1″.

19. **Venus alta,** L. Mindoro. A small shell covered with circular ridges that turn over toward the umbones. Of a gray color the interior is white. 1″.

20. **Terebratella cruenta,** Dill. New Zealand. A fine brachiapod of a clear horn color.

21. **Laqueus blanfordi,** Dvds. Japan. Another fine horn colored brachiapod being more rounded than preceding species as are all of this genus.

22. **Lima dunkeri,** Sow. Japan. A thin white nearly smooth form, white inside. 1″.

There are over a hundred forms of Sinum, some of which closely resemble the Naticas and a few fossil forms have been found in the Eocene. The shells are mostly ear-shaped, usually white, very large aperture and a thin epidermis. The mollusk has a very large mantle which usually entirely covers the small shell, the foot lobe being enormously developed. They live on muddy sand flats and the mollusk can frequently be seen moving just under the surface of the sand, when they can very easily be dug out. It is interesting to place them in a globe of sea water when the mollusk will stretch out to 6″ diameter where the shell is only one inch.

PLATE 83

1. **Caryatis inflata,** Sow. Australia. A bulbous white shell, with streaks of light brown. Interior a flaky white. A genus of about 80 species mostly in the Eastern Hemisphere. 1½″.

2. **Caryatis rufescens,** Deb. Philippines. A bulbous pure white shell and interior the same. 1½″.

3. **Caryatis obliquata,** Han. Mauritius. A bulbous glistening white shell and interior is the same. These white shells usually vary enough in shape to be easily distinguished. 1½″.

4. **Plicatula imbricata,** Mke. Queensland. A very small shell with narrow hinge and flaring valves that are deeply ribbed. White less than one inch.

5. **Caryatis subpellucida,** Sow. Loo Choo Ids. near Japan. A bulbous shell, white and well marked with russet. Interior white with a brown patch on posterior end. 1½″.

6. **Caryatis citrina,** Lam. Mindanao. A bulbous shell with hinge to left of middle. The color is yellowish white and darker near edge. Interior the same with dark brown patch on posterior end. 1½″.

7. **Caryatis tumens,** Gmel. Senegal. A bulbous grayish-white shell with similar interior. 1¼″.

8. **Caryatis sulphurea,** Pils. Japan. A small chalky white shell with interior of pale orange. 1″.

9. **Hysteroconcha (Dione) multispinosa,** Sow. Gulf of Mexico to Panama. A small deeply ridged shell with row of nodules at one end. The color is a light rose and white inside. 1¼″.

10. **Tellina truncata,** Sow. Philippines. A pure white typical shaped species and white inside. A large share of the shells of this genus are of this color. 1¼″.

11. **Tellina rubescens,** Han. Panama. A brilliant ruby-red shell with rays of pale white. Interior of same color. There are the usual color forms in a shell of this nature. 2″.

12. **Venerupis siliqua,** Desh. New Zealand. A small dull white well ridged shell and white inside. About 1″.

13. **Tellina linguafelis,** L. Philippines. The entire surface of the shell is covered with fine dots arranged in regular rows. Color white with rays of pink but not always. Umbone pink. Interior yellowish-white. A most remarkable shell and there are very few like it in the world. 2″.

14. **Tellina interrupta,** Wood. Surinam. A rather bulbous elongated shell with fine lines and shades of light brown. Interior yellowish-white. 2½″.

15. **Venerupis rugosa,** Desh. Cape of Good Hope. A pure white lined and corrugated small shell. The interior white, with pink patch in center. Umbones pink. About 1″.

16. **Tellina semiplanata,** Speng. Brazil. An elongated rosy-red shell with growth lines of white. Interior same. 2½″.

17. **Tellina spenceri,** Suter. New Zealand. A pure white pointed shell with growth lines. Interior white. 2″.

18. **Venerupis reflexa,** Gray. New Zealand. A rather square shaped shell with numerous small ridges, white outside, interior white with brown patch at one end. About 1″.

For years I had a correspondent by name of Herbert N. Lowe, of California, who used to visit me when he made trips east. During the Springs of 1929, 30, and 31 he made trips to West Mexico and Central American coasts usually during the months of January to April. The reason he selected these months was due to favorable climatic conditions at that season of the year, the extreme low tides and that many species come into shallow water to spawn at that season. He had a power dredge that he usually used up to 20 fathoms and those trips brought up an unusual lot of very fine shells and he secured over 100 species new to science. These are described in a booklet he sent me of 143 pp. and 17 full page plates of shells. His discussion of species is vastly interesting as in so many cases he tells in what situations the shells were found. If on shore he described the territory well and in other cases mentioned if taken by dredge in deeper water. It is that class of literature that the average shell collector needs, as without it, you are doomed to failure on most of your trips. I found it the same in my trips to Cuba on winter vacations, that if you did not know where to go you might as well stay at home as far as getting good shells is concerned.

PLATE 84

1. **Cardita semiorbiculata,** L. Philippines. A perfectly mammoth shell for this genus. The exterior is black and finely lined lengthwise. Interior a rich purple shade. Hinge very strong. 3″.

2. **Pecten hindsii,** Carp. Puget Sound. A very thin and rather attractive shell with fine small ridges, which are a shade of pink which contrasts well with the white surface. Interior white. Rather common as it is fished for meat. 2″.

3. **Donax deltoides,** Lam. Australia. A pure white, very large shell for this genus, and the interior is mostly white. 2¼″.

4. **Iphgenia rostrata,** Romer, New Guinea. A pure white shell something like the preceding, but of a slightly different shape. White inside. A shell very seldom seen in any collection. 2″.

5. **Terebratella jeffreysi,** Dall. Puget Sound. A fine reddish Brachiapod flat on side with the opening, convex on other, all finely ribbed. 1¼″.

6. **Codakia punctata,** L. Fiji Ids. A fine small rounded bulbous shell with very fine circular lines, mostly white, the interior yellowish and edge lined with red. 1¼″.

7. **Pectenculus flagellatus,** T. W. Bass Strait. A small shell with broad ribs that are covered with fine circular lines. Interior white. 1″ but may come larger.

8. **Macoma cayensis,** Lam. Cuba. A small white shell, flat at one end as are many of this genus, the lower part of the valves covered with a fine yellowish periostracum. Inside white. 1½″.

9. **Cardium subrugosum,** Sow. Philippines. A medium size shell with broad perpendicular ribs, the whole surface being a light shade of yellow. Interior a deep purple. 1¾″.

10. **Mactra discors,** Gray. New Zealand. A symmetrical elongated shell, shell, covered with a yellowish periostracum. Inside white. 1½″.

11. **Mactra quadrangularis,** Rve. Japan. A small rounded shell mostly white with traces of gray periostracum. Interior white. 1¼″.

12. **Mactra dissimilis,** Desh. Australia. A gray shell with fine circular lines on lower half of each valve. Interior a rosy purple. 1¾″.

There are likely 200 or more species and you will find one or more on about every ocean beach in the world. The Marine univalve shells bore holes in the umbones of the Mactra and the waves bring them up on the beach, usually in single valves, but you occasionally find them fully hinged.

A few fossil forms have been found in the Lias. The recent species are fond of burrowing just beneath the surface. The foot can be stretched out and moved about like a finger. It is also used for leaping. The sea stars are fond of them and must destroy great quantities. I have them divided into 18 subgenera and you often see these subgeneric names in books and on labels but they should never be used unless the section has been raised to generic rank.

✦

The Sea Stars are very fond of many forms of shell life. They will open a Scallop or Pecten by slowly dragging the valves apart by means of small sucker-like feet on the lower side of each of the five rays or "arms". When the valves are forced apart the sea star rolls out its stomach, which envelops the soft body of the scallop. Digestive juices are poured forth and the food is digested outside the body of the sea star. They are the most destructive enemy of the scallop fisherman, and do vast damage to the oyster beds in the same manner.

✦

Always save the brissus of the Pinna as it has a great history. In the Mediterranean region it has for 25 centuries been woven into choice silken objects. In fact the Kings of Ancient Greece had made elegant dresses for their wives. It has been said that some of these dresses were so fine they could be pulled through a finger ring.

✦

On the muddy flats you will find Mya (Soft shell clam), Petricola, Macomas and many others. Wherever the salt water grasses check the motion of the water, you find hosts of Astyris, Cerithopsis, Triforis Bittium and very small forms of Odostomia and Mangilia. A little further out where the water rarely leaves them, are the oysters (Ostraea), Anomia, Crepidula and many other small gasterpods. Each class of Mollusks have their own particular habitat, depending somewhat on the ocean floor.

PLATE 85

1. **Tagelus constrictus,** Lam. Japan. A thin elongated white shell covered with light greenish periostracum. Interior white. 3″.

2. **Tagelus coquimbensis,** Sow. Coquimbo. A thin shining white shell with rays of greenish and yellow. 1½″.

3. **Plicatula australis,** Lam. Australia. A small white shell which is attached to coral or rock. Faint dots of brown on upper surface. About 1″.

4. **Cultellus attenuatus,** Dkr. Japan. A shining white shell partly covered with green periostracum. One edge straight, other convex. Interior white. 3″.

5. **Cultellus lividus,** Dkr. Philippines. A thin shell, more curved than preceding species, and finely mottled with brown. 2″.

6. **Petricola bipartita,** Desh. Philippines. A small white finely lined shell under 1″.

7. **Tagelus dombeyi,** Lam. Peru. Lower edge nearly straight but hinge is short, outer surface shining light green. Interior white. 3″.

8. **Solecurtis antiquatus,** Pult. England. A symmetrical elongated shell, small hinge near one end. Periostracum horn-colored, interior white. 1½″.

9. **Petricola lapicida,** Chem. Aden, Arabia. A bulbous small white shell with heavy ribs along the small end. Interior white. About 1″.

10. **Barnea australasiae,** Gray. South Australia. A real white Pinna-like shell with circular striae and a fine dots. Interior white. 2″.

11. **Pholas candida,** L. Queensland. Of similar form to preceding, the large end slightly curved, all with lines and dots. 1½″.

12. **Pholadidea suteri,** Lamy. New Zealand. The cut shows the exact form of the shell which is white. A great borer in wood or soft rock. 1½″.

13. **Barnea similis,** Gray. New Zealand. The two valves of the shells of this genus only meet about half way. Shell is white as are most all of the genus. 2″.

14. **Petricola nivea,** Gray. Nicobar Ids. A small white short shell, rather globose and faint lines. Interior white. 1″.

15. **Pholadidea tridens,** Gray. New Zealand. A very thin fragile white shell that bores in soft rock for its home and thus escapes the fury of the waves. About 1″.

16. **Petricola rubiginosus,** A&R. Tasmania. A thin rather smooth white shell shading to brown near umbones. Interior shaded yellow. About 1″.

17. **Parapholas quadrizonata,** Speng. Japan. A white shell of the form of cut. Rather round and large at one end, quickly tapering to rounded point. The hinge and base is covered with plates. 1¼″.

18 and 19. **Gastrochaena dubia,** Penn. Adriatic Sea. A thin white shell, the valves of which only touch on one side and one end, leaving a large oval opening for use of the mollusk. Cuts show side and front view.

20. **Venus squamosa,** L. China. A rather thick finely reticulated shell, white outside and dull white interior. About 1″.

21. **Arca imbricata,** Brug. Bahamas. A small white sharply angulated shell, edges leaving opening in middle of whorls. Interior white. 1″.

22. **Arca costata,** Rve. Mauritius. A white shell with slightly irregular valves, finely lined and cross lined, white inside. 1¼″.

23. **Arca solida,** Sow. Panama. A small dark deep shell that is finely lined. Under 1″.

24. **Arca reticulata,** Ch. Reunion Ids. A strong light colored finely lined shell with shadings of brown. About 1″.

25. **Arca cornea,** Rve. Mauritius. A rather thin small shell, finely lined with mossy periostracum, white inside. 1″.

26. **Loripes picta,** H. Ad. Mauritius. A nearly round rather smooth thin shell, white outside and inside. 1″.

The Caryatis (called by some authors Pitaria) are part of the Family Venerdae. They are most all very round yellowish or white shells and I had many species from Philippines.

PLATE 86

1. **Meleagrina margaritifera,** Rve. Mother of Pearl Shell, Pacific generally. The shell belongs to the genus Pteria but this sectional genus name has been used so long and is so well known I dislike to use the new designation. In the East Indies if you mention Pearl Shell, every one believes you mean this species. All the finest pearls are found in this variety. Attains 12" and takes a high polish. Is a very rough shell when the native divers bring it up from its sandy bed at 3 to 6 fathoms.

2. **Crenatula phasianoptera,** Lam. Philippines. A very peculiar formed shell. Fairly thin, the valves are crumpled and vary much in form in different shells. 3 to 4".

3. **Crenatula, mytiloides,** Lam. Australia. While there are 15 known species of this genus you only now and then see a shell in collections. They live in sponges, are flat, thin, dark color and usually have a crumpled surface. 3".

4. **Meleagrina margaritifera,** Rve. Pacific. This cut shows a natural shell before it is trimmed or polished. The other cut of this shell on this plate has a finely polished back. In the Jolo Sea, North Australia, off the coast of Ceylon, and in many Pacific islands, thousands of people depend on the Pearl Fisheries for a living. When Pearl brings a low price by the ton, times are hard. Supply and demand of this shell regulates their lives.

5. **Vola (Pecten) dentatus,** Sow. Mazatlan. Lower concave valve is reddish-brown. Upper convex valve is yellowish-brown or russet. A very attractive shell. 3 to 4".

6. **Vola (Pecten) Jacobaeus,** L. St. Jacobs Shell. Mediterranean Sea. The Volas are always interesting with one flat and one convex valve. There are 25 species scattered over the world. Some are real common and are of considerable commercial importance. Usual mottled color. 3 to 4".

7. **Vulsella rugosa,** Lam. Red Sea. While there are 40 known forms of this genus, the shells are always rare in collections. They live in a sponge, is perhaps the reason, and the sponge fishermen usually throw them back in the sea when they trim their products. This form is 2 to 3" and uncolored.

8. **Pecten foliaceus,** Quoy. Australia. A rather scarce form which much resembles some of the common Japanese species. It comes in various colors and of the form of cut. 3".

9. **Pecten pes-felis,** L. Mediterranean Sea. A small shell ornamented with ridges and two dark bands. Fairly common. 2".

10. **Pecten Swifti, Bern.** Japan. A wonderful shell with swollen prominent ridges that are humped in three places. The prevailing color is pink. One of the most attractive of the genus and rather large. 4 to 5".

11. **Pecten purpuratus,** Lam. Peru. A very brilliant 3" shell in which rosy-purple is the main coloration. Holds mainly to this rich shade, the ridges of the shell being deeper color than the spaces between. Most specimens are dredged, as they live in deep water, but the natives and fishermen usually find plenty of them to eat.

12. **Vola meridionalis,** Tate. Tasmania. A neat species from the southern part of the world. The flat valve is reddish and convex valve a faint purple color. Not rare. 3".

The word Pecten means a scallop and there are 200 to 300 known recent forms and more than twice as many fossil forms from Devonian, etc. On the north and west of Ireland you find immense beds of P. maximus, and you always find them in the Lower East Side of London for sale in December. The convex valve of P. maximum which ranges to 5" is much used for scalloping oysters and serving them in the shell.

While most all Pectens are attractive only a very few forms are of unusual colors. Often times the unusual colored shells are simply rare patterns only found about one in a thousand. Scarcely more than 20 species show a good varity of color that can be obtained at modest prices. The P. Jacobaeus is often called the St. James shell, as it was worn by pilgrims to the Holy Land and became the badge of several orders of knighthood. Both Maxmus and Jacobaeus now belong in the Vola. I never had over a dozen species of Vola and none could hardly be called brilliantly colored.

PLATE 87

1. **Vola (Pecten) Maximus,** L. The Great Scallop. North Sea. A very large 6″ shell of the usual shape with fluted back. The convex side has been shipped into this country in quantity for baking fish or oysters and serve in shell. On the East Side of London you will find quantities of this shell in the market during the month of December of each year where it is sold for food. Of a russet-red color.

2. **Amussium Japonicum,** Gmel. Japan. The Sun and Moon shell. Fairly common and is shipped into this country for commercial use. One valve is red and the other yellow. Thin almost circular. The two valves only meet at top and bottom. There are 22 species in the genus. 4 to 5″.

3. **Pecten tigris,** Lam. Philippines. This handsome species is finely ridged and adorned with splashes of reddish-brown. Very attractive. 3″.

4. **Pecten tegula,** Wood. Australia. A thin shell ornamented with ridges and spines. One valve is more flat than the other. The shades of rosy-brown are in waves of color. 2″.

5. **Spondylus crassiquama,** Lam. Lower California. A solid shell with stubby spines that come in all shades of color such as white, red, purple, orange, etc. Not common. 4″.

6. **Spondylus avicularis,** Lam. West Indies. In the young stage they come in shades of color, but as they attain a large size they are white. I have had specimens up to 10″, one of the largest species of the family. There are over 80 species in the world. All attach themselves to rock or coral, where they remain for life and are able to only move one valve, the upper, during life.

7. **Spondylus hystrix,** Bolt. (Nicobaricus, Sow.) Nicobar Ids. A small flat species of reddish-yellow color covered with short sharp spines. It is always quite difficult to detach such a shell from the coral. 1¼″.

8. **Spondylas ducalis,** Bolt. Philippines. A fairly round fat shell of a brownish color. The spines are always short and stubby. Quite variable in form and you seldom find two alike. 2 to 3″.

9. **Spondylus gaederopus,** L. Mauritius. A very variable species of purple or other shades. The spines are short or none at all. Not as heavy as many other forms. 4 to 5″.

The Spondylidae consists entirely of the genus Spondylus and the small genus Plicatula. There are likely over 100 species and about as many fossil from Carboniferous.

The shells are usually attached by the right valve to rock or coral. The valves are held together by a cartilage in a central groove nearly or quite covered. The hinge consists of two interlocking teeth which when separated are not so easy to get together again. Frequently the lower valve is most spiny and least colored. The inner shell layer is very distinct from the outer and entirely wanting in fossil specimens. Water cavities are common in some forms. I have had large specimens from Bahamas a foot long with a very large water cavity in lower valve.

Back in 1893 one could secure the fine colored Spondylus from West Mexico in any quantity. Many had 2 to 3″ spines in perfect condition and the shells ran from red to orange, yellow, white, etc., but of recent years few come on the market. I had 80 species of Spondylus in my collection and they were about the most colorful group of shells, that run from 2 to 8″, that I owned.

✦

There are only about a dozen species of Plicatula but over 100 forms found fossil in Trias, etc. The shell is usually attached to a rock by the right valve and specimens found on the shore often in quantity have been detached by the power of the waves. The mollusk resembles the Spondylus, is reason they are included in that family.

✦

The Tridacna, bivalves are always of interest as the genus contains the largest bivalve in the world, T. gigas. Other smaller forms are obesa, Sow., which is smooth. The T. squamosa grows to a foot long, but the finest specimens for the cabinet are 4 to 6″. These sizes are wonderfully fluted and if carefully collected and shipped are superb but the fluted plates break very easily. T. elongata comes in various sizes with narrow fluted ridges, and T. crocea, my collectors always thought in the Philippines was a "Baby Giant Clam" as it is exactly the shape of the big brother. Most of the Philippine shore lines have some Tridacna and there are places where the bluffs are honeycombed with burrows, each containing a nice specimen. Some species prefer such a situation.

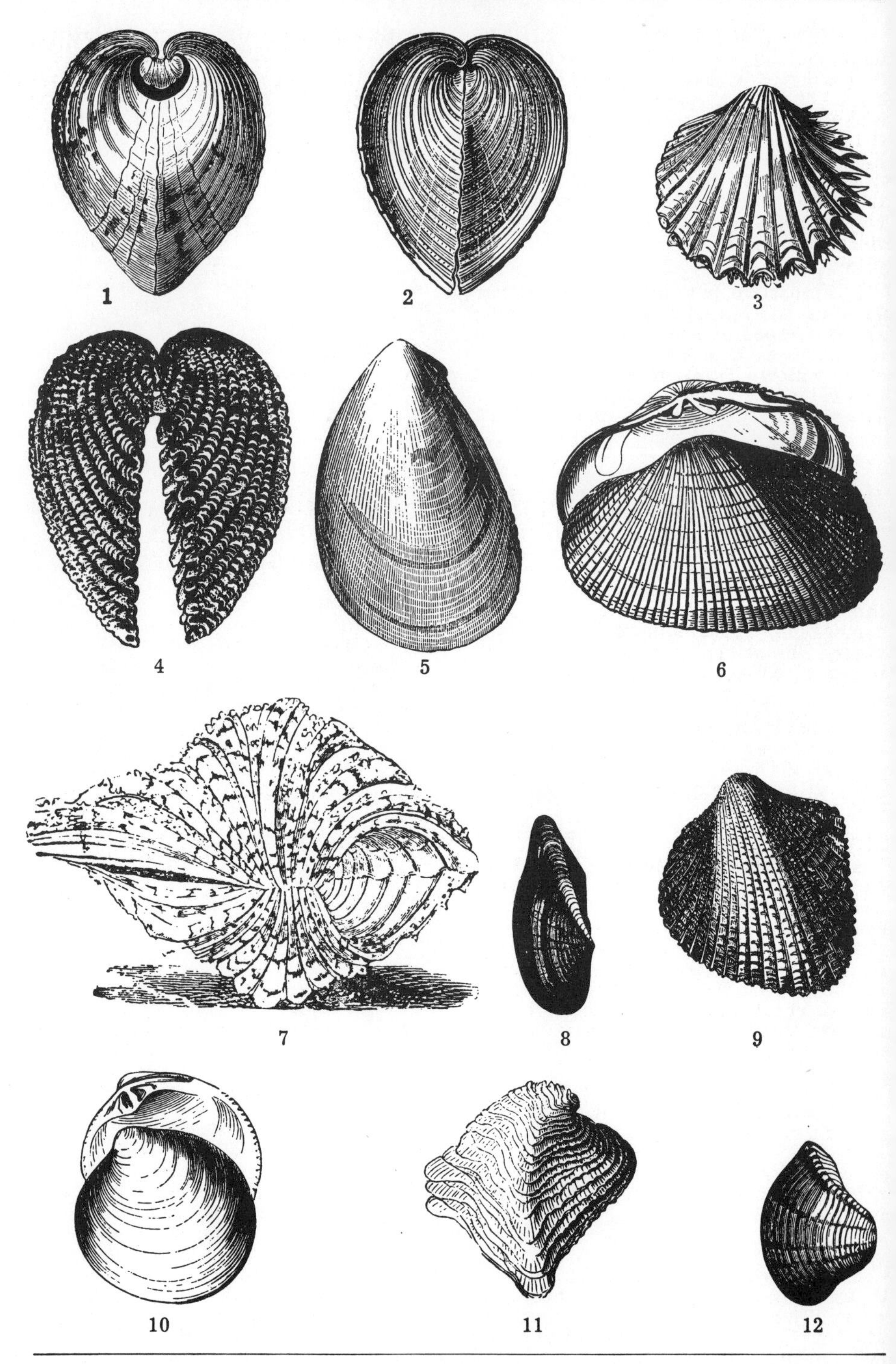

PLATE 88

1. **Cardium auricula,** Forsk. Philippines. A very peculiar shell of a grayish-white color, with ridges. The formation of the hinge and curved cavity around same is very unique. 1½".

2. **Cardium cardissa,** L. Nicobar Ids. A remarkable shell, pure white, concave on one side and convex on other. The form of the valves is exactly the reverse of the usual shell of this genus. 1½ to 2".

3. **Cardium ringens,** Chem. China. This shell is remarkable in that while the valves fit perfectly at one end, at the other they barely touch at their points, which are extended about one-eighth of an inch, leaving small openings. It would be interesting to know the practical use of same. 1½".

4. **Cardium consors,** Sow. Panama. A real Heart Shell of remarkable form of sculpturing. It is completely covered with little cup-shaped ridges arranged in regular rows. Color a rich pink shade. Common on mud flats. 3".

5. **Cardium oblongum,** Chem. Greece. The shell is of a yellowish-white and has a very thin periostracum over part of the surface. It is common and much used by the inhabitants who live along the shores for food. 2½ to 3".

6. **Asaphis deflorata,** L. Bermuda and other parts of the world. The prevailing color is white tinged at one end with purple but as all of the shells of this genus are very variable, there are usually other color shades. There are only seven known species but they include hundreds of color combinations. 2".

7. **Hippopus maculatus,** Lam. East India Clam, Philippines and East Indies generally. It is a finely mottled, heavy shell and has been shipped into this country in vast quantities for several decades. It is usually sold in various sizes ranging from 2 to 9". Old and large shells are devoid of color, being usually white or drab, but the younger shells of 3 to 4 years are beautifully shaded with brown.

8. **Soletellina biradiata,** Wood. South Australia. It is of a yellowish color and highly polished. The young shell is of a darker color. There are 50 species in the genus and they are a very colorful lot as a whole. 2½".

9. **Cardium unedo,** L. Philippines. Here we have a handsome shell that seems to be a sort of intermediate between other species. It is deeply ridged with yellowish-white, the ridges being barred with red. 1½ to 2½".

10. **Cyclina chinensis,** Chem. Japan. Although this shell would be classed as smooth it shows concentric ridges which are adorned with a bluish shade. Interior pure white. There are 12 known species but I have seldom had more than one or two. 1½".

11. **Chama lobata,** Brod. West Indies. A small species of a drab color of very unique form, and when well cleaned makes a fine specimen. 1½".

12. **Corbula sulculosa,** A.Ad. Hong Kong. A trim little shell, thick, smooth, shiny and well mottled with brownish. There are some 90 species scattered over the world and they are all as interesting as this one. Live under rocks at low tides. 1".

Now we have had a world war and millions of our boys have had a taste of tropical life and seen no end of shells, both war-like and sea-like, we will have more beautiful shells brought here for sale. The East and West Shore of U.S.A. contain some very fine species. Those from Artic regions are usually horn color or no color at all and our boys must have seen some beauties when they were located with MacArthur on Torres Straits in North Australia.

✦

The Brachapoda are not true mollusks but the recent species of which one to two hundred forms are known, are usually included in the shell collection. They are largely deep water bivalve shells and they differ from true mollusca by having an internal bony skeleton, proceeding from the hinge in connection with the dorsal valve, for the support of the arms; and being extremely variable in structure, it affords excellent characters to the collector for the distinction of genera. To the paleontologist they are of the greatest interest, as there are thousands of species found in the rocks. You will find much literature on them scattered through the proceedings of scientific societies for the past 150 years.

PLATE 89

1. **Meretrix (Cytherea) petechialis,** Lam. China. The species has a natural polish and covered with chevron markings. Ranges from shades of light brown to white. Very common edible species. 2 to 3″.

2. **Lioconcha castrensis,** L. Ceylon. It is richly adorned with splashes of rosy-brown on a white background. There are about 28 species in the genus and all have brilliant colors and markings. 2 to 2½″.

3. **Hysteroconcha lupinaria,** Less. West Indies to Panama. It is a true Venus-like shell adorned with spines at one end. A very unusual species. 2″.

4. **Heteroconcha rosea.** Brod. West Coast Central America. A rather flat shell with fine concentric ridges and smooth base. Of a light shade of rosy-pink. 2¼″.

5. **Venus lamellata,** Lam. So. Australia and Tasmania. It is pure white and ornamented with at least six frills which curve backward and have distinct pink shade on under side. Very attractive. 2½″.

6. **Sunetta scripta,** L. Ceylon. A small shell of brilliant polish and many shades of color. It is impossible to find two pair exactly alike. Some are pure white, others lavender, gray, brown with zigzag markings. ½″.

7. **Lioconcha picta,** Lam. Viti Ids. This species is smooth and covered with splashes and zigzag markings of a shade of light brown. Very attractive. 1½″.

8. **Lioconcha tigrina,** Lam. New Caledonia. It is rather triangular in shape with markings of different shades of brown. Some shells are much darker than others. 1½ to 2″.

9. **Dosinia juvenilis,** Chem. Indian Ocean. Usually quite round, and adorned with shades of light brown. The Dosinies are a large group of shells covering 140 species and ranging from the big white **D-ponderosa** which attains 5 to 6″ down to little fellows of 1″ or smaller. They are world-wide in distribution. Usually white, there are some forms fairly well colored.

10. **Dosinia circinaria,** Desh. Victoria, Australia. It is of a drab color with fine lines over the entire surface. Almost circular. 2″.

11. **Circe divaricata,** Chem. Philippines. A solid shell with concentric ridges. The color pattern is splashes of different shades of brown with dark patches at hinges. 2″.

12. **Meretrix tripla,** L. West Africa. A small triangular species not so highly colored as some of the other of the genus, but a fine smooth natural polished shell. 2″.

13. **Paphia (Tapes) litterata,** L. Philippines. The Tapes as we used to call them are fairly common over the tropical world. This shell is of a faint reddish-yellow color and covered with lines. It is very variable and lines are not regular in form. The shell of Tapes is quite brittle. 2½ to 3½″.

14. **Circe scripta,** L. Philippines. It is not so thick as some other species but it has the concentric ridges and splashes of reddish-brown. 2″.

15. **Paphia (Tapes) laterisculca,** Lam. Ceylon. A reddish-brown shell with deep lines. Glossy and attractive. Most of the genus are well ornamented with lines and color. 2″.

16. **Paphia papilionacea,** Lam. Indian Ocean. One of the rich fine species of the group. It has the usual ridges and is smooth and natural polish. Color pattern different shades of brown. 3½″.

17. **Venus guidia,** B and S. Lower California. An attractive white shell adorned with ridges that have serrated edges. One of the noble species of the genus. The young and medium size shells are the finest. Lives in mud in quiet water of small bays. 3 to 5″.

18. **Lioconcha hieroglyphica,** Conr. Hawaii. A small shell with striking color pattern of light and dark shades of brown. It is thinner than the usual shells of this genus. 1″.

The genus Circe (Now often called Gafrarium) follow the Lioconchas but are of an entirely different pattern and much more numerous in species. You can usually buy around 25 or 30 varieties. The shells vary greatly in sculpturing, some being grooved on the surface and others covered with tiny nodules, etc. The beaks are often flattened. C. divaricata is typical of the group.

PLATE 90

1. **Spondylus imperialis,** Chem. China. A real imperial shell with round spines, smooth and glistening. Some are very short but there are usually a few 1 to 2″. Usual color pink, and shells are rather scarce. 3 to 4″.

2. **Spondylus pictorum,** Sow. Lower California. One of the finest of all known species, coming in brilliant colors of white, rose, red, salmon, deep purple, and intermediate shades. Has been scarce of late years and never enough shells on the market to supply the demand. I have been told the native Indians are fond of the mollusk and have depleted the shells in shallow water. There are likely as many ever in 5 to 10 fathoms. 4 to 6″.

3. **Cardita pectunculus,** Brug. Madagascar. A fine fluted shell of a dark color, typical of several species in this genus. Lives on mud flats. 3″.

4. **Pinna sacrata,** L. New Caledonia. One of the very odd shaped forms of this genus all of which go under the general common name of Fan Shells. There are 28 known species. Very variable and I suspect at least 100 names have been applied to them in the past two centuries. This species is thin, valves only meeting here and there, with crumpled and irregular surface. 4″.

5. **Cardita calyculata,** Lam. Mediterranean Sea. The deep ridges are scalloped and adorned with brownish spots. Rather odd shape. The silky brissus with which many species of this genus are adorned is shown in cut. This brissus is used to more firmly attach the shell to its permanent location. 1½″.

6. **Aethera caillaudi,** Fer. Leopoldville, Congo, Africa. A very strange shaped shell that lives in the estuaries of large rivers. The natives use them for food and roast the shells in kilns to make lime. Of a dull color. 4″.

7. **Cardita sulcata,** Lam. Malta. A round solid shell of a russet color with deep ridges. There are several forms of the type. 1½″.

8. **Isocardia moltkiana,** Speng. China. A small white shell typical of the genus, all of which have curved umbones. All of the known species are rather uncommon. 1¼″.

9. **Crassatella japonica,** Dunk. Japan. It is rather solid and heavy, well mottled with dark color. You seldom see the shells of this genus in any collection although they are all desirable specimens. 2½″.

10. **Cardita bicolor,** Lam. Ceylon. A splendid shell with white ridges barred with brown. All of the Carditas are strong and solid shells and usually quite common. There are 75 known species. 1½″.

11. **Crassatella antillarum,** Rve. West Indies. A heavy shell which must be fairly rare as you seldom see it. One end is pointed as are many of the species, of which 55 kinds are known. The shell is well mottled and rather attractive. You will work hard to secure 15 species of this genus. 3″.

12. **Isocardia vulgaris,** Rve. China. A real gem of a small shell, white and rather scarce. I have never been able to secure information as to the scarcity of the various species of this genus. All I know is, you only occasionally secure any of them. 1¼″.

13. **Corbis fimbriata,** L. Philippines and Pacific generally. An unusually beautiful white shell with fine sculpture rarely equalled in any bivalve. The surface is smooth and glistening. There are only five species. This is one of the finest and I have had it sent in from various localities. 3″.

14. **Cardita laticostata,** Sow. Panama. A fine little well-marked shell that is in demand. Systematists put this genus next to the Pinnas, perhaps because they are so very variable in form. You must see a real collection to get a true idea of their great variety. 1½″.

The Carditidae consist almost entirely of the genus Cardita, but there are a few small forms under the generic names of Thecalia, Milneria, etc. There are between 50 and 100 recent species found in all seas everywhere. Only a very few are found fossil in Lower Silurian. The shells are usually oblong, radiately ribbed. A typical form would be C. antiquata. The name Cardita means a heart and there are a few real heart shaped forms among the 11 subgenera in which the genus is divided.

The Corbis are really beautiful shining white ribbed shells and are classed with the Lucinidae.

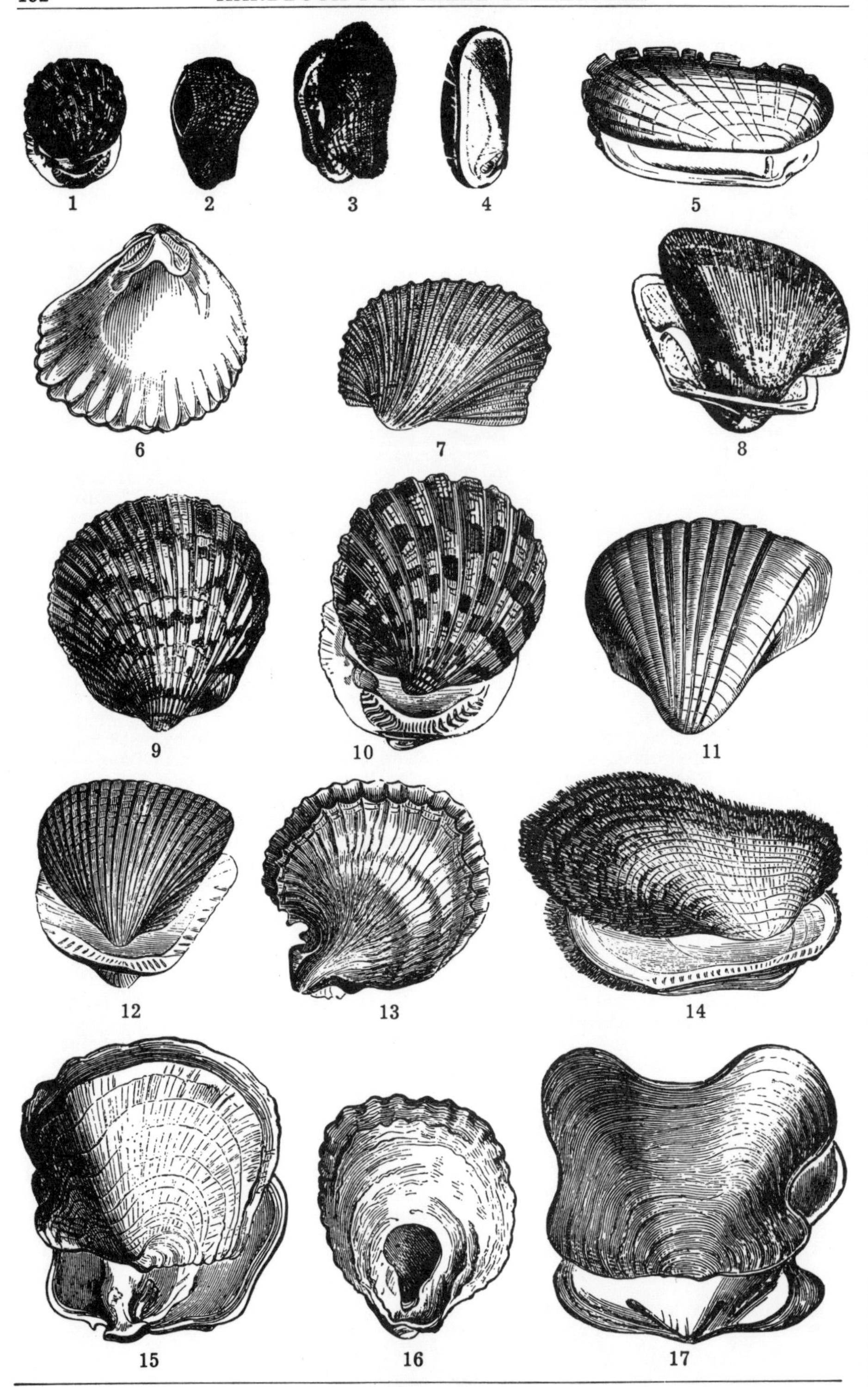

PLATE 91

1. **Pectenculus pectenculus,** L. Philippines. This is considered one of the most typical of the 90 species of the genus. Some are small of about 1″ but others are large, very thick and heavy. This form is ridged, finely mottled with dark color and 2″.

2. **Arca zebra,** Swain. Philippines. A well marked species in which the two valves fail to meet in the middle. There are some quite similar forms in Florida. 2½″.

3. **Arca barbata,** L. Italy. It is covered with a hairy coating as are many others of the great Ark Family. One must see about 100 species from the tiny little fellows to the large 4 to 5″ specimens to realize the tremendous range of the forms of this genus. 2″.

4. **Solemya borealis,** Tott. Nova Scotia to Connecticut. The shell is covered with a highly polished periostracum, that reaches beyond the aperture and breaks up into segments. There are three other forms on the east coast. About ten species in the world. They should be mounted in glass-top boxes. 1½″.

5. **Solemya australis,** Lam. Australia. The polished periostracum is light brownish and extends beyond the aperture, breaking up into segments, as is usual with this genus. All tropical countries seem to have one or two forms of these interesting shells. 1½″.

6. **Trigonia margaritacea,** Lam. Australia. It is called the Jewel Shell and it is in fact. The outside is covered with beautiful ridges and the interior is of the finest colored pearl. The hinge is set at angles and the teeth are so close that once the two valves are separated it is almost impossible to get them together again. 1¼″.

7. **Arca auriculata,** Lam. China. An elongated shell, the ridges of which are covered with fine cross bars. 2½″.

8. **Cucullea concamera,** Brug. Japan. Only three living species are known but there are hundreds of fossil forms, showing one genus which has almost disappeared from the earth in living form. The shell is light brown and has an interior plate on each valve which is very distinctive. 3″.

9. **Pectenculus aurifluus,** Rve. Japan. A fine large shell which is well marked with ridges and zigzag markings. Fairly thick and heavy. 3″.

10. **Pectenculus delesserti,** Rve. Red Sea. One of the finest forms of the genus which is well marked with splashes of color. It has the usual ridges and the peculiar hinge design is shown, which locks the shell tight. 3″.

11. **Arca senilis,** L. Senegal. A very thick solid shell and one of the heaviest of the group. Has deep ridges and is of a light brownish color. 2 to 4″.

12. **Arca reversa,** Sow. Mazatlan. A solid well ridged shell that is fairly common on the muddy beaches. It has much the appearance of some oriental species. 2½″.

13. **Placuanomia Zealandica,** Rve. New Zealand. A very strange and oddly formed shell. Rather thin and the two valves do not exactly match. Horn color. 3 to 4″.

14. **Arca velata,** Sow. Japan. Quite typical of many of the 300 known species of the genus. There are some 28 forms on the East Coast of U. S. alone. A few are from deep water, 500 to 2000 fathoms. This is a fine large species, hairy coated and 3″.

15. **Anomia elyros,** Gray. Japan. Of a greenish-yellow, lower valve concave and wrinkled, upper valve convex and also crinkled and humped. There are 60 known species. Usually lie attached to other shells. 1½″.

16. **Placuanomia patelliformis,** L. Britain. The Ribbed Saddle Oyster. It has 20 to 30 waved ribs, radiating from the beak. Usually found 10 to 60 fathoms and rarely on beaches. There are 20 known species. 1½″.

17. **Placuna sella,** Gmel. Saddle Shell. Philippines. The shell is strong and the two valves measure about ½″ in thickness and about 8″ long. Lives buried in mud. In its early stages it is quite flat but as it grows older, it assumes its saddle shape. There is another species flat and translucent about 5″ diameter. They have been polished and used as window glass from the Philippines thru the East Indies. They were called Window-glass Shells. The natives call our glass crystal.

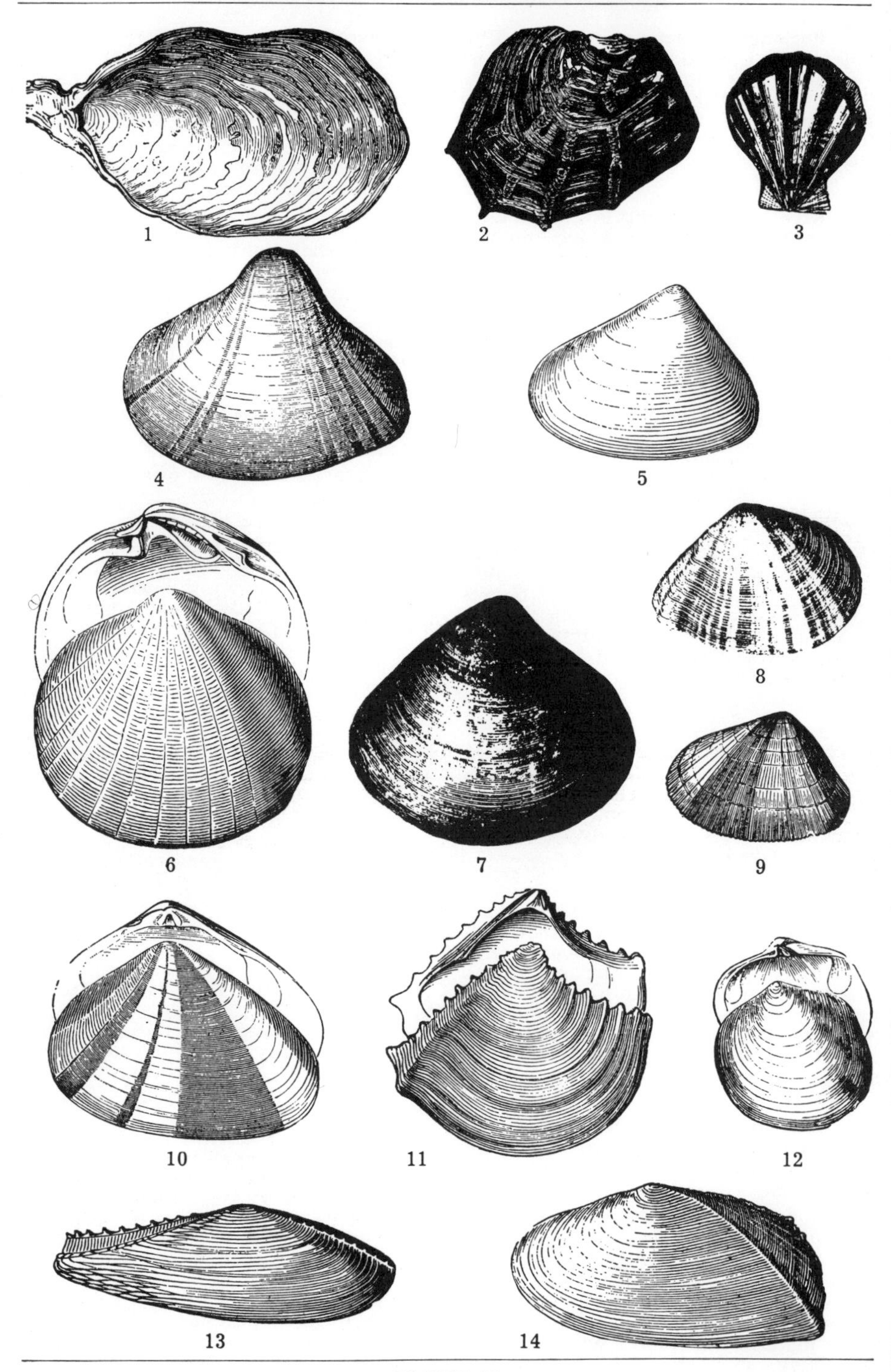

PLATE 92

1. **Mulleria lobata,** Fer. Amazon. Similar in color to our native fresh water clams but often covered with a mineral deposit. The two valves are of different shape. One is more convex than the other. There is no hinge that is perceptable and the shells are rather hard to secure. 3″.

2. **Aethera elliptica,** Lam. Senegal. Old shells are strong and heavy, attaining 5″ and 2″ thick. Are without color and live in the estuaries of large tropical rivers. The natives dig deep pits, fill them with shells, build fires over them, and make a very good grade of lime, which they use in building operations. Are often found in great quantities where water is shallow.

3. **Pecten plica,** L. Ceylon. The Plicated Pecten is adorned with humped ridges which instantly separate it from other species. Of drab color, it is fairly common. 1½″.

4. **Galatea radiata,** Lam. West Africa. There are about 15 species in this genus and they live in estuaries of great tropical rivers. Usually heavy, triangular of a reddish-brown color, pointed hinge and white inside. Not common. 3 to 4″.

5. **Donax compressus,** Lam. Ceylon. The shells are of a yellowish shade with two faint bands of light lavender. There are the usual color combinations. 1½″.

6. **Semele solida,** Gray. Coquimbo. This genus used to be called Amphidesma and you find the shells in old collections, so labeled. There are some 85 species. This is a solid, thick shell, one of the largest of the group. White, almost circular and finely lined. You will find some Semeles on almost every ocean beach in the world. Live on mud flats. 3″.

7. **Mactra violacea,** Chem. Borneo. One of the very few fine colored species of the genus. Of a violet color, it is a rich looking shell as compared with so many of the white colored forms. 3″.

8. **Tellina jubar,** Hanley. Amboina. One of the prettiest shells of the genus which is a very large one. They seem to be widely distributed over the ocean world. I have had this shell sent me from several localities. It is finely ridged and ornamented with pink and white. 2″.

9. **Donax denticulus,** L. West Indies. It is a small shell of the usual shape of other species of the genus, adorned with white and dark bands. Quite variable but easily identified. Lives on sandy beaches between tides. 1″.

10. **Donax cuneatus,** L. Philippines. A very neat shell of reddish-russet color with flaring white bands. 1½″.

11. **Tellidora burnetti,** B&S. Colombia, S. A. There are only five species all white and shaped like cut. The valves are flat and somewhat triangular. Usually brought up on the beaches in storms. 1½″.

12. **Diplodonta rotundata,** Turt. Aden, Arabia. The genus comprises some 45 species, many of which are as round as a bullet and almost as solid. Look like balls in the cabinet. This species is round and white. Some of the forms make a nest of sand so they are completely hidden from view, but the nest always has tubular openings for the siphon of the mollusk. 1″.

13. **Tellina rostrata,** L. Sarawak, Borneo. A brilliant reddish-pink shell, one of the prettiest of the genus. Of the peculiar form shown in cut. 2″.

14. **Tellina foliacea,** L. Philippines. A thin shell of a handsome reddish-yellow. Does not seem to be very common. Most all of the species live on mud flats. 2½″.

If you have a hunger for nice big oysters, just travel down to the west coast of America from California to Panama and you will find the big Ostraea chilensis 10 to 12″ broad and the Indians will likely offer you all you wish of them at 1 ct. each. My friends, who write me from there, say they are delicious. Then there is Ostraea iridescens which lives on the side of wave-washed rocks. You may think it a little coarse, as indeed it is, but if you have not had any oysters for some time you will be satisfied. The Ostraea mexicana will be found by the hundreds on mangrove roots and also the small Ostraea tubulifera, which Prof. Dall described some years ago and gave it this name for the small tubes on the upper valve.

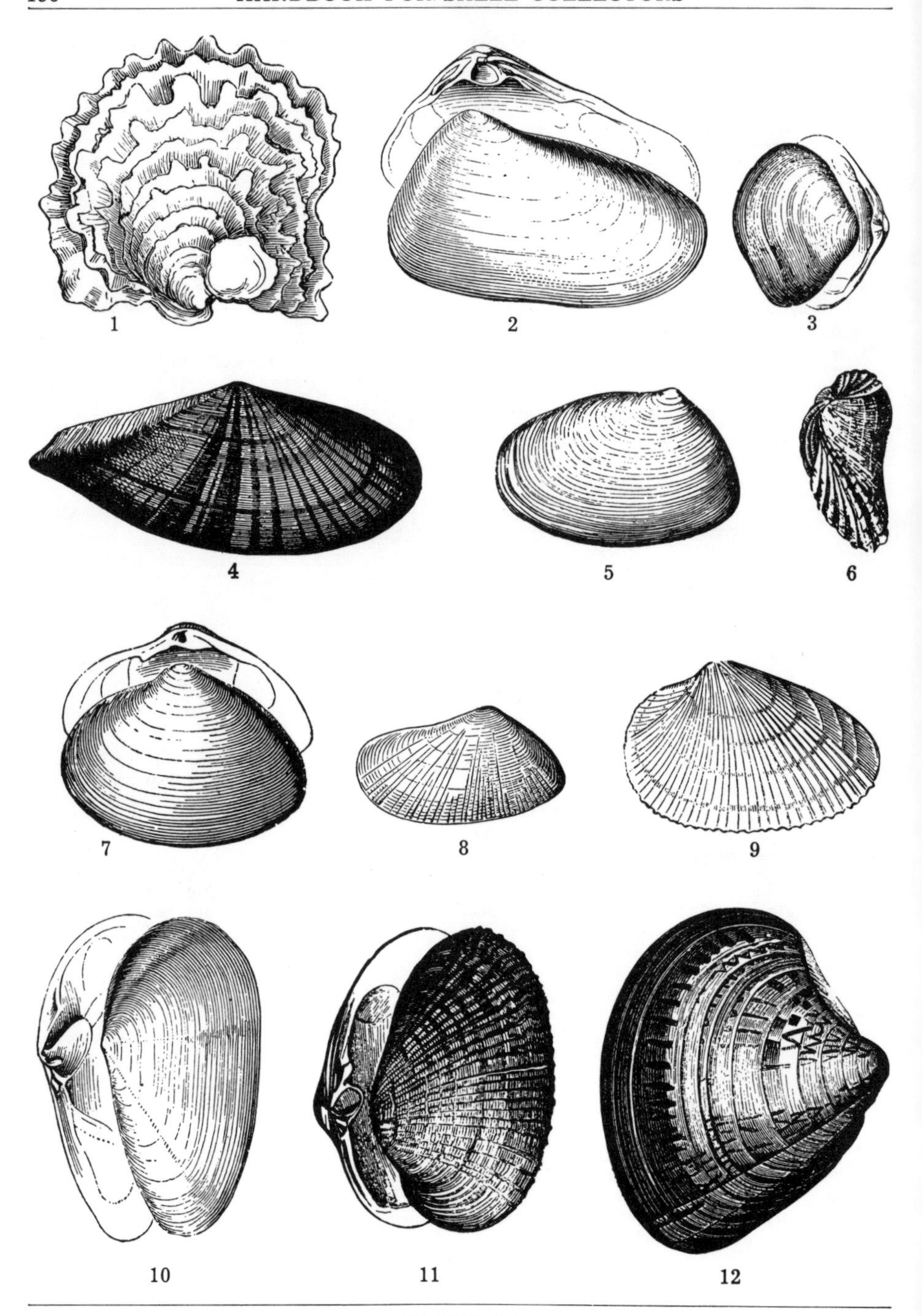

PLATE 93

1. **Chama cristata,** Lam. Java. One of the finely frilled forms of this genus, that is pure white. The cut shows the top valve. The lower valves is plain and always attached to coral or rock. 2″.

2. **Mesodesma donacium,** Desh. Chile. A white shell of peculiar oblong shape, fairly thick and solid. This genus comprises 50 species which follow the Donax. Many of the shells are of similar form to the shells of that genus. 2½″.

3. **Fisheria delasserti,** Bern. West Africa. A genus of only three species. Shell is nearly white and quite scarce. ½″.

4. **Tellina virgata,** L. Viti Islands. One of the several fine rayed species that is reddish-pink and white. It is quite variable. 2½″.

5. **Tellina punicea,** Born. Gulf of California. A dainty little shell of bright reddish color. Many of the shells of this genus are white, red, salmon, pink and rayed. But the great majority of the species are uncolored. Some are wonderfully sculptured. 2″.

6. **Cardita affinis,** Rve. Panama. One of the fine little shells of this region where so many very attractive shells are found. Many years ago Mr. Adams, of Mass., wrote a very good work on the marine shells of this region and his shells are still preserved in his native state. There are later works more complete. 1½″.

7. **Mactra solida,** L. France. A solid white shell of the usual type of the genus. There are about 125 or more species known and most of them are white or yellow and fairly common. Almost all forms live on mud flats. They follow the Mesodesma. 2½″.

8. **Donax trunculus,** L. Sicily. A yellowish-white species, the interior being a deep purple. Very common. 1½″.

9. **Standella (Eastonia) nicobarica,** Gmel. Japan. A finely ridged white shell. The two valves meet in the middle only. There are 26 species in the genus, scattered over the tropical world. 1¾″.

10. **Vanganella taylori,** Gray. New Zealand. A white shell of which there is only one species in the genus. My friends over in New Zealand say it is quite rare. It is placed near Lutraria. 2½″.

11. **Standella (Eastonia) rugosa,** Heib. Portugal. It is deeply ridged and faint color. The shell much resembles similar sized shells of the Asaphis group. 2½″.

12. **Meretrix (Cytherea) lusoria,** Chem. Japan. One of the common edible clams of that territory. It is ornamented with various shades of brown over a natural highly polished surface. The valves are thick and hard. 2 to 3″.

Collectors always like to have positive records on shells of unusual size and I give herewith some measurments of specimens of the Tridacna gigas. A specimen in the British Museum weighs 310 pounds and was 36″. Dillwyn in his work mentions a specmen that weighed 517 pounds and was 54″. Dall in the American Naturalist Vol. 16 and page 698 listed a specimen that weighed 528 pounds and was 36″. Fisher records in Manual of Conchology, page 1035, one that weighed 550 pounds and was 34½″. There used to be, years ago, a specimen in the window of Rule's restaurant in Maiden Lane, Covent Garden, London, a 40″ shell that weighed 434 pounds. These big fellows are now scarce near shore line even in the Barrier Reef section. In the Philippines one of my correspondents wrote me they went down in 20 ft. of water to get them and it took six men some time to get ropes around the specimen and aboard a boat. In fact it cost usually $25 to get one out of the water, and add further cost of packing, freight and some profit to the seller and you are lucky to get even a decent size specimen in good order for $100.

✦

Hysteroconcha come just before the Caryatis. Mostly shells 1½″ or smaller and adorned with fine sharp spines. If you can get them with perfect spines, they are real gems, but they only rarely come that way.

✦

The Family Mactridae consists mainly of Mactra but a number of other genera, consisting usually of one to a dozen forms, include such as Raeta, Standella, Heterocardia, Labiosa, Vanganella, Zenatia, Lutraria, Caecella, Tresus and others.

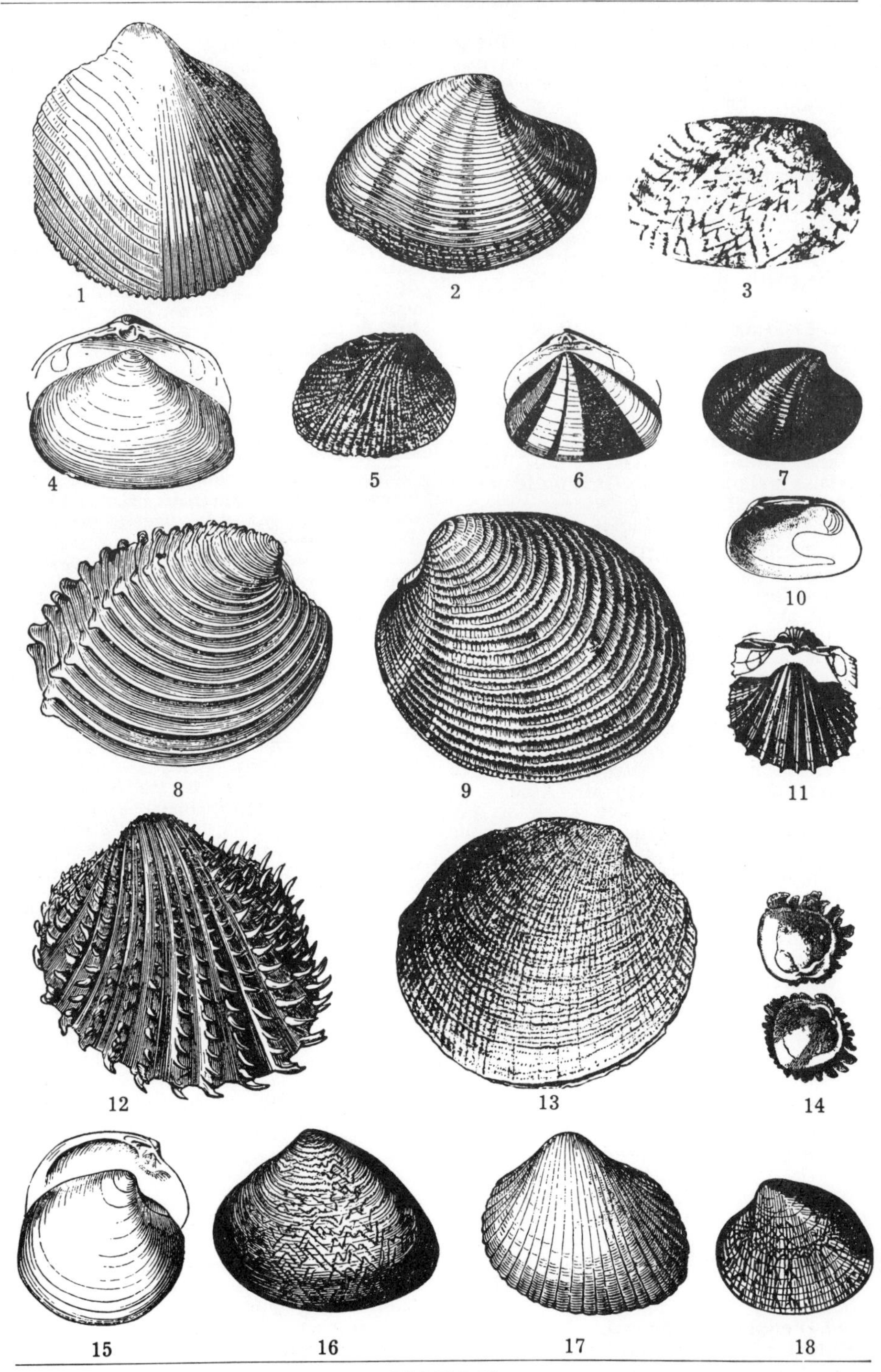

PLATE 94

1. **Venus lyra,** Hanley. Sierra Leonne, Africa. A handsome species with fine ridges and mottled with russet-brown. It is fairly rare. There are 300 species in the Venus family and they are a most remarkable lot of shells when you see them all. 2½″.

2. **Paphia malabarica,** Chem. Indian Ocean. A fine solid species that is well ridged and of shiny surface. One of the desirable shells of the genus. 2 to 3″.

3. **Paphia sulcaria,** Lam. West Africa. A handsome smooth shell finely ornamented with reddish-brown lines and scroll work. Makes a fine cabinet specimen. 3″.

4. **Caecella turgida,** Desh. Aden, Arabia. A small thin shell of a russet-brown color and pure white inside. There are 11 species of this little known genus. They are allied to the Lutraria. 1½″.

5. **Circe gibba,** Lam. Philippines. It has deep ridges across the shell and only coloring is different shades of yellow. 1½″.

6. **Paphia geographica,** Gmel. Mediterranean Sea. A small species, smooth and polished with white and brownish bands as shown in cut. Quite common over a wide territory. Most all species live on mud flats. 1½″.

7. **Callista erycina,** L. Ceylon. One of the fine and showy species of the genus. It is ridged with white but the bands of color are brown with shades of red. There are 50 species in the genus and they are practically all very attractive shells, that usually have a fine natural polish. 2½″.

8. **Venus plicata,** Gmel. West Africa. One of the very distinct forms of this great genus. It has deep ridges that terminate in two rows of small knobs. It is a real plicated shell, as its name indicates. 2½″.

9. **Venus puerpera,** L. Loo Choo Ids. A very solid shell adorned with concentric ridges with tiny cross sections which cover the entire shell. It is shaped and rayed with light brown. 2½″.

10. **Paphia pullastre,** Mont. Coast of France. A very common shell that is considered good eating by the local residents. Uncolored, it is gathered in vast quantities. 2″.

11. **Cardium costatum,** L. China. It is a large pure white shell covered with tall ridges which meet alternately rather than together, as would be expected. The shell is thin, and nearly circular. One of the gems of the genus. 3 to 4″.

12. **Cardium aculeatum,** L. Naples, Sicily. A real prickly Cardium which is quite uncommon in this genus. Most of the species are fairly smooth and usually ridged. Of a drab color. 2½″.

13. **Venus reticulata,** L. Pacific generally. Usually a massive heavy shell completely covered with fine ridges and cross bars. It has rays of light brown. 3 to 4″.

14. **Chama lazarus,** L. Mauritius. One of the most beautiful shells of the genus, which covers over 100 species throughout the world. This variety lives on coral reefs, where it is always solidly attached and cannot move the lower valve. The upper valve is adorned with numerous slender fronds, of a most delicate white color. A real gem of a shell. 2 to 3″.

15. **Mysia undata,** Penn. France. A very thin shell of which the upper part of the valve is yellowish, balance white. It used to be called **Lucinopsis.** There are six species in the genus, and they come next to the Venerupis. 1″.

16. **Venus undulosa,** Lam. Australia. An attractive shell with fine zigzag markings which is rather uncommon in shells of this genus. Not real common. 2″.

17. **Cardium edule,** L. England. The Common Cockle. A small white shell unusually uncolored. It burrows in the sand in great numbers and is much used for food and bait. 1¼ to 2″.

18. **Venus verrucosa,** Lam. Mediterranean Sea. It has heavy ridges which terminate at the lower end into small knobs, close together regular rows. Of a light reddish color. 2 to 3″.

The genus Lioconcha belongs in the family Veneridae. They are thick solid shells from 1 to about 2½″ adorned with hieroglyphic markings of red color. One of the finest is L. castrensis, very common around Ceylon. A fine tray of all the Lioconchas, none of which are rare, make a dazzling array in any collection.

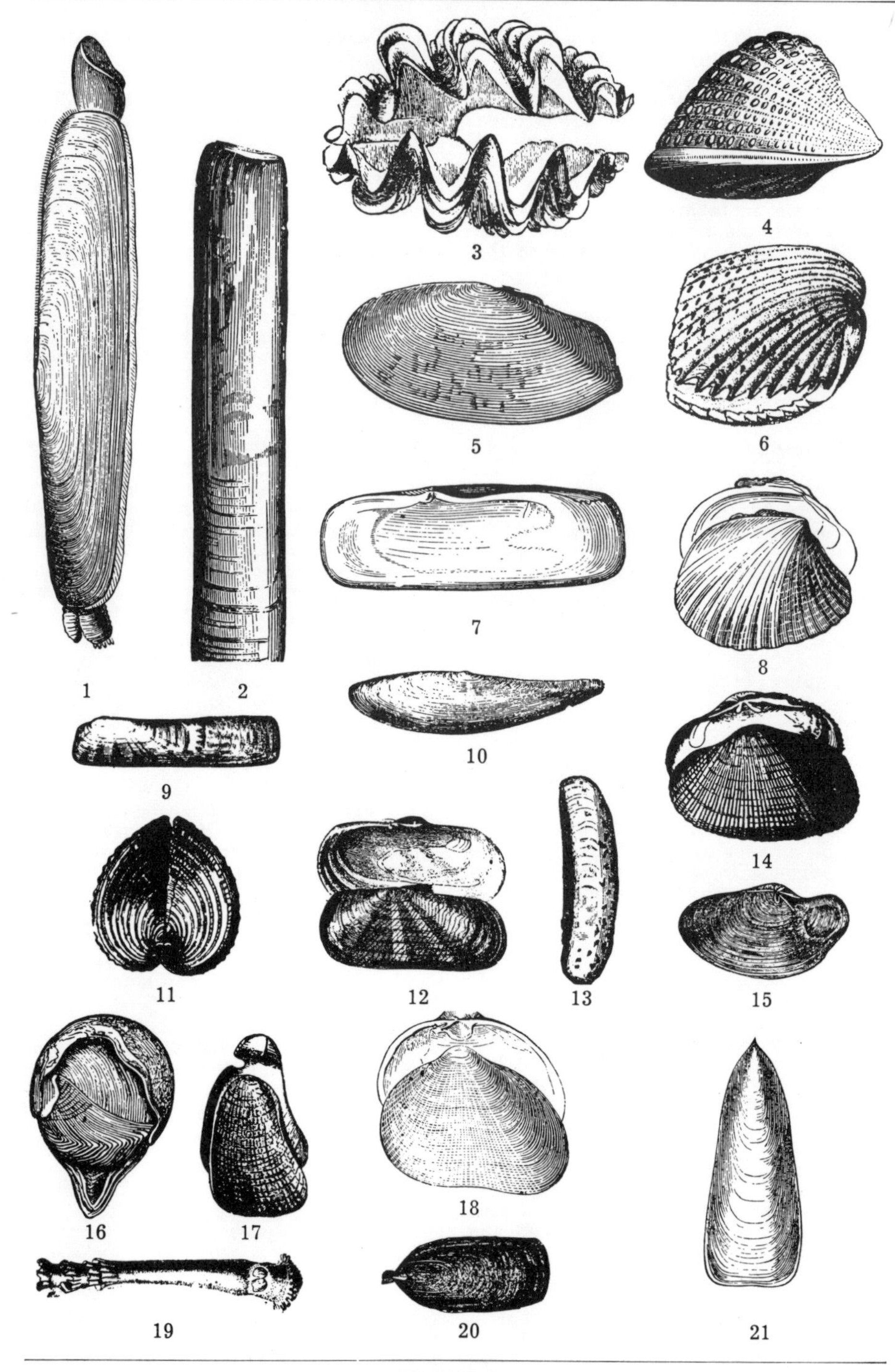

PLATE 95

1. **Pharus legumen,** L. Britain. A long slender, horn-colored shell that belongs to the Solen family and much resembles some of the razor shells. There is only the one species in the genus. 4 to 5″.

2. **Solen vagina,** L. England. A long yellowish-white shell which is quite straight and is called the Grooved Razor Shell. They burrow in the sand and as one end is apt to be slightly above the surface of the sand, they are usually avoided as much as possible by bathers. 5″.

3. **Tridacna gigas,** L. Philippines and Tropical East Indies. The Giant Clam stands supreme as the largest shell in the world and the largest Bivalve. It attains 3 to 4 ft. and weighs from 300 to 500 pounds. It is in a class by itself. The shell is white and the mollusk that protrudes around the fringe of the aperture is a brilliant purple. The power of the two valves in closing is beyond belief. Anything caught in its grasp is killed or drowned, hence it has few enemies. Boring shells attack it in great quantities, but seldom reach the mollusk itself. I have had wonderful specimens of Ring Shell Money sent me by the natives of the British Solomon Islands, that was made entirely from the hinge of these mollusks. This barter money, made centuries ago is 4 to 6″ in diameter and now very rare and in much demand.

4. **Cardium hemicardium,** L. West Australia. These diverse shaped Cardiums are always of great interest to collectors. You can hardly call this a Heart Shell and yet it belongs to that great family. It is white and covered with ridges and pimples. 2″.

5. **Gara ferroensis,** Chem. Faroe Islands. A very neat thin shell well covered with fine lines and mottled surface. The Gara used to be called Psammobia and there are about 80 known species in the genus, all rather attractive shells. See also Plate 76.

6. **Cardium fragrum,** L. China. A very neat white shell much the shape of **unedo**. It is covered with scalloped edges ornamented with red dots. 1½″.

7. **Novaculina gangetica,** Bens. Calcutta, India. A flat shell of a drab appearance but the base of the shell is white. It burrows in the mud like all of its near relatives. There are four known species of which this one is most easily available. 2¼″.

8. **Adacna laevuiscula,** Esch. Caspian Sea. In this great sea there are very few shells, owing to the low salinity of the water. The few that have been collected appear like degenerate forms of common Mediterranean Sea species. This shell much resembles the small Cardium edule. The marine bivalves in this sea comprise 3 genera of one or two species each. 1½″.

9. **Solen Sloanei,** Gray. Australia. A trim little narrow species, which is mottled with light brown in the form of cross bars. 3″.

10. **Soletellina elongata,** Lam. Moluccas. A narrow thin elongated shell of a purplish shade. Not common. 2½″.

11. **Cardium inversum,** Lam. Nicobar Islands. A small white shell in which the outer edges of the two valves roll over at right angles to the main body of the species. A remarkable form. 1½″.

12. **Solecurtis (Macha) strigillatus,** L. Italy. An attractive shell usually colored pink. It has very fine smooth ridges and the two valves are open at each end, only meeting in the middle. Lives buried in mud flats. 3″.

13. **Cultellus cultellus,** L. Australia. A fine mottled shell which is not real common. The 21 species in the genus are scattered over the whole world. Some forms are rare. 2″.

14. **Asaphis tahitensis,** Rve. Tahiti. It is not as thick as some of the other forms and only lightly and drably marked. Only occasional specimens have been seen. 2″.

15. **Anatina truncata,** Lam. A rather thin shell usually found in the mud in shallow bays. There are 45 species in the world mostly from Pacific. 2 to 3″.

16. **Jouannetia cumingii,** Sow. New Caledonia. A very curious shell, hard to describe. It is round and composed of different pieces, not being a true bivalve in the sense usually understood. It is a borer and only two or three species known, all rare. Comes after the Pholadidea. 1¼″.

17. **Pedum spondyloidon,** Gmel. Seychelles but also in Philippines. The single species in the genus lives usually in sponges. It is a very strange shape shell as cut indicates. The lower valve has an edge that fits over the upper valve. Cut above the hinge is for the siphon of the mollusk. 2½″.

18. **Lutraria arcuata,** Desh. Japan. There are 30 or more species of this genus scattered sparingly all over the world. They range from 1½ to 4″. Some of them are entirely different in form from this cut.

19. **Brechites (Aspergillum) vaginiferum,** L. Philippines. It is called the watering pot shell. It starts as a bivalve and ends up an a univalve of one piece. See the tiny embryo bivales on cut near the top of shell. The head is filled with tiny holes. 6 to 8″. Scarce.

20. **Lingula anatina,** L. Philippines. A broad flat thin Brachiapod that is not real common. They often live in clusters, attached to a stone or rock by a small thread. There are about 200 living species of Brachiapods and thousands of fossil forms. 1½″.

21. **Lingula hiaus,** Swain. Tonkin, China. One of those very strange Brachiapods that is of a greenish color and as thin as tissue paper. A splendid species, of which there are very few living forms in the world today. 1¼″.

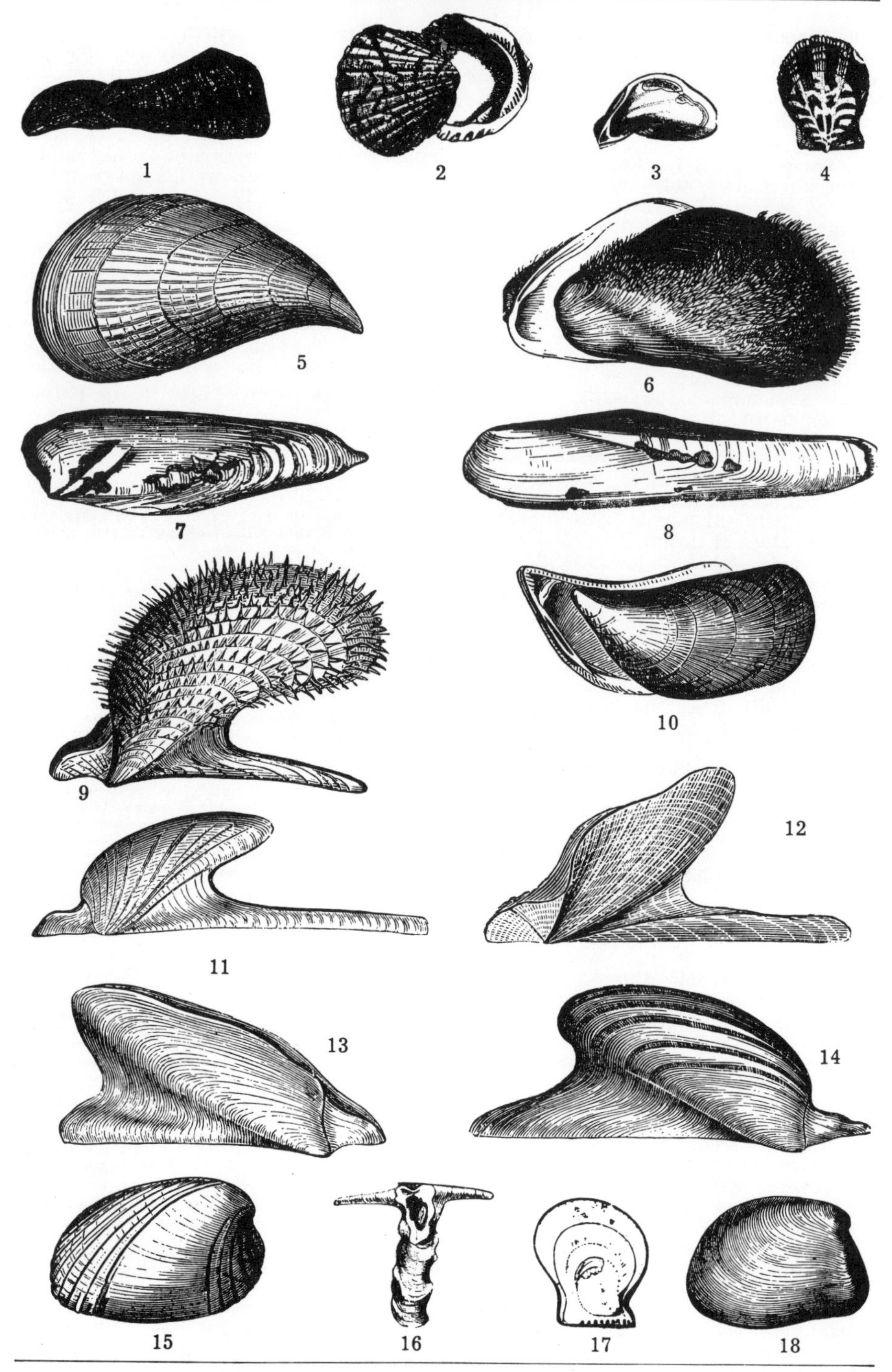

PLATE 96

1. **Arca tortuosa,** L. Australia. A very strange shaped shell, usually drab or white, the two valves are always twisted. 3″.

2. **Pectenculus laticostatus,** Q&G. New Zealand. One of the neat medium sized species from the southern hemisphere. It is deeply ridged and covered with irregular dark markings. They are called Dog Cockles in the old country where they are much admired. 2″.

3. **Dreissenia polymorpha,** Pallas. England. While this form of shell approaches the Mytilis or Mussels, they are classed as coming close to the fresh water clams. They inhabit the canals and small creeks where the water is often slightly saline. Many of the species vary enormously in shape and our friends across the sea are fond of making extensive collections of these curious forms. 1 to 1¼″.

4. **Hinnites sinuosus,** Gmel. Britain. A small genus of only 3 known species. This shell with crumpled valves is of fairly good color. Not common. 2″.

5. **Mytilus decussatus,** Dkr. Patagonia. A fine smooth shell of medium dark color, fairly common to this southern latitude. It is similar in form to many other species. 2½″.

6. **Modiola australis,** Gray. South Australia. A fine shell of russet color with darker shadings, under a very hairy periostracum. Fairly common. 2½″.

7. **Lithophaga cumingiana,** Dkr. Central America. A sharp pointed species which can bore as fine and smooth a hole in wood or rock as could be done by man. Collectors who live near the seashore have many chances to study the habits of such interesting species. The shell is fairly light colored. 2″.

8. **Lithophaga attenuata,** Desh. Hawaii. An elongated borer quite common in this territory. There are 65 known forms scattered over the world and they are a fascinating lot. Light colored. 2½″.

9. **Pteria tarentiana,** Lam. Mediterranean Sea. A splendid shell with bristly hairs and interior of brilliant dark pearl. Of unusual shape, it is fairly common but seems slow to come on our market. 3 to 4″.

10. **Septifer bilocularis,** L. (nicobaricus). Philippines. This genus contains about 16 species all of which are closely allied to the Mytilus but are usually much smaller than the specimens of that genus. They are very common where found. A dark shell. 1¼″.

11. **Pteria crocea,** Chem. Mauritius. There are some 70 species of these fine odd shells. All have a pearly interior and the shells usually have a strong periostracum. As this outer covering becomes warm and contracts, it will break the shells to bits. Most shells seen in collections are broken up in this manner. They can be preserved if coated with a solution of white shellac but this must be done soon after they are collected. That preserves the shell coating from action of the air. This little yellowish fellow has an unusual slender hinge. 3″.

12. **Pteria signata,** Rve. Moluccas. One of the remarkable forms of this genus which may be common but not often seen in collections. It is of a dark shade of color and interior is iridescent pearl. 3″.

13. **Pteria iridescens,** Rve. Moluccas. It is of a yellowish color splashed with shades of brown. Very thin, hinge slight. 2½ to 3″.

14. **Pteria heteroptera,** Lam. Australia. A brownish shell with light stripes which adds to its beauty. Thin with a bright pearly interior. 2 to 3″.

15. **Modiolaria trapezia,** Lam. Falkland Islands. There are about 8 species in this part of the world. They are round, fairly thin, smooth, with lines on two sections of the shell. 1½″.

16. **Malleus vulgaris,** L. Hammerhead Oyster. Philippines. An odd shell with very wide hinge and long wide base. The Oyster proper only lives in a small part of the shell. They attain 6 to 8″. There are 15 varieties in the Pacific. The wide hinge suggests the Hammerhead Shark, hence the common name.

17. **Melina (Perna) attenuata,** Rve. New Caledonia. There are 50 species of this genus and they are a very strange lot. The shell is pearl, usually covered with a shelly substance of a scaly nature. Where the shells are thick they will take a high polish. This form is about 3″.

18. **Modiolaria discors,** L. Australia. Most of the species of this genus are around 1″ covered with a periostracum and they often live in clusters. As many as a dozen are often found in one bunch. This species is brown. 1″.

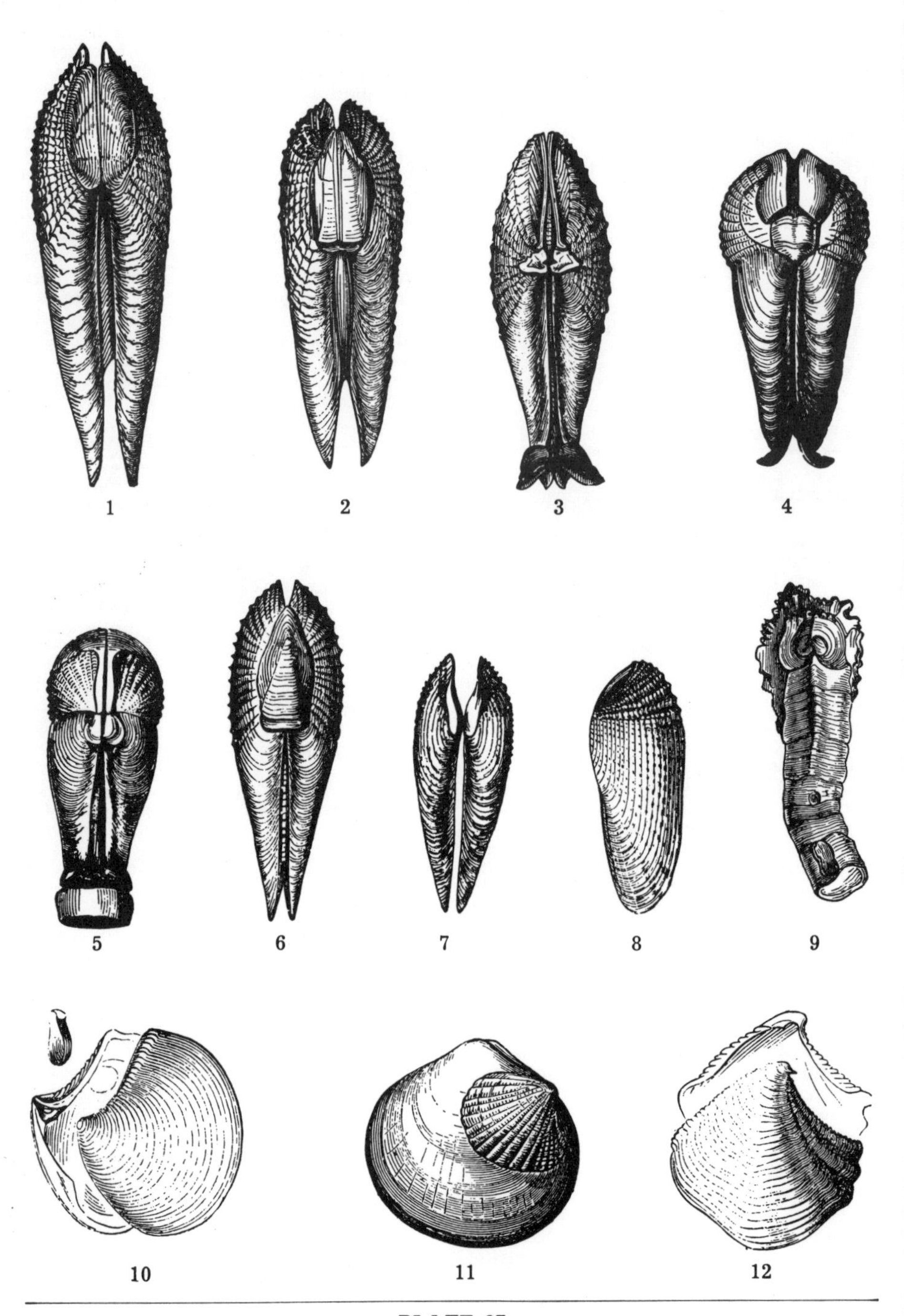

PLATE 97

1. **Pholas dactylus,** L. Central America. This beautiful white shell is somewhat similar to the Angel Wing of Florida. It burrows in the mud as do many of the Pholas. Shells are not so easy to obtain as shell collectors are very scarce in that region. 4 to 5″.

2. **Pholas chiloensis,** King. Peru. This species is fairly common but shells are scarce, as so few are ever collected. It burrows in the sand and lives in colonies. The first thing to do is to find the colony which is not always very easy. 3 to 4″.

3. **Talona explanata,** Speng. (Clausa Gray). West Africa. There is only the one species in the genus and it is somewhat similar to closely related forms. The top of the shell has fringe as shown. It used to be called T. Clausa and is so labeled in old collections. 3 to 4″.

4. **Parapholas concamerita,** Desh. Philippines. Fairly common on the rocky coasts of the various islands. It is a borer and specimens have been rare of late, as collectors do not seem to find them easily. There are only a half dozen species of this genus. 3″.

5. **Pholadidea melanura,** Sow. Panama. They are very neat white borers and there are a dozen or more species scattered over the world. It is quite different in shape from others of its class. The round knobby top of the shell is distinct. 3″.

6. **Pholas orientalis,** Gmel. India. A beautiful white pointed shell common along the coast of India and other localities. Many of the Pholas hold their valves together with a basal plate so that good pairs are easy to preserve in the cabinet if they are well prepared when collected. 3 to 4″.

7. **Barnea manillensis,** Phil. Manila. A fine shell which lives in burrows along the coast of various islands. Quite similar to the dozen other species in the genus. 2½″.

8. **Barnea candida,** L. Ireland. It is called the White Piddock in the Emerald Isle and has a thick convex shell with 25 to 30 rows of prickles. It is white with shades of brown. Lives in a hole in rock and is very sensitive to danger, immediately retiring into its shell. 2½″.

9. **Brechites (Aspergillum) strangei,** A.Ads. Victoria, Australia. One of the most remarkable of all bivalve shells or is it a univalve? See the two embryonic shells near top of shell. It started that way and as it grew developed an elongated one piece shell. When it is about a third of an inch in diameter it starts to grow in all directions but with a definite pattern in view. At the top it forms a series of plates with tubes in them. Rather scarce. 3″.

10. **Myodora striata, Quoy.** New Zealand. It is pure white, lower valve perfectly flat and upper valve convex. Has fine circular striations. Usually dredged. 1½″.

11. **Myochama keppeliana,** Ads. Bass Strait, Australia. A very strange shell for a bivalve, as it lives on the back of other shells. The specimen I have before me is ¾″ of a deep yellow color and was living on a valve of Pectenculus. The lower valve as in the genus Anomia, has assumed the shape of the shell of its host, and upper valve is convex. There are 8 known species.

12. **Myodora brevis,** Stutch. Tasmania. It is pure white, lower valve slightly concave and upper convex. Hinge about 1″. There are 26 known species. 1½″.

The Atlantidae consists of about 20 species of Atlanta and 4 species of Oxygyrus. The shells are minute, glassy, compressed and keeled. They are sprightly little mollusk, probing every object within reach, twisting the body about and swimming with the shell downward with sudden jerks. There are almost endless species of very small marine mollusks which you will find just as fascinating to study as the larger forms.

✦

The Semelidae consist of the small shells of Cumingie, but mainly of the large genus of Semele (Used to be called Amphidesma). There are likely 100 species and 30 or more fossil forms from the Eocene on. On many ocean shores they are the most common shell seen, specially the single valves. Most of them are white or drab color.

✦

The Pholadidae include the genera of Pholus, Barnea, Zirfaea, Navea, Talona Pholadidea, Jounnetia, Xylophaga, Parapholus, Martensia, Teredo, etc. All of the many genera are borers in some form.

APPENDIX

PLATE 98

1. **Turbo setonis,** Gmel. S.W. Australia. The shell is completely covered with verticle striations, also vertical zigzag light brown markings. Operculum covered with small nodules. 3″.

2. **Tonna (Dolium) caniculata,** L. Philippines. A round light brown smooth shell with faint circular white stripes. Apical whorls lighter color. 3″.

3. **Thais textilosa,** Lam. New Zealand. The shell has 4½ whorls. Over three-fourths of its size is last whorl. Has seven circular ridges. Light horn color. 3″.

4. **Tectarias pagoda,** L. Pacific generally. A fine conical shell the largest of the genus. Has two rows of pointed nodules on last whorl and usually one on upper part. Gray color, usually flesh inside. 2 to 3″.

5. **Pleurotomaria hirasei,** Pils. Japan. A fine conical shell, usually white, completely shaded with red diagonal stripes on each whorl. The notch is 1 to 1½″. Interior of a white pearly color. Very rare. Found up to 500 fathoms.

6. **Ovulum ovum,** L. Pacific generally. Often called the Egg Shell as it is of a smooth glistening white, oval, each end of aperture is curved into almost a ring. The name of this genus was changed some years ago to Amphiperas but collectors like to stick to Ovulum. 3″.

7. **Xenophora pallidula,** Rve. Japan and Philippines. The shell is conical, trochiform, whorls flattened carrying shells, corals, stones arranged and attached anywhere on the exterior surface, which completely disguises the shell. Lower surface is free, of a pale yellowish color. Most of the many forms inhabit deep water and are numerous in the Java and China seas. Some species are of huge size. 3 to 4″.

8. **Argonauta hians,** Sol. Japan. Each whorl has vertical wrinkles, the top has two rows of dark nodules and the whole shell is of a medium dark brown color. Usual specimens 2 to 3″.

9. **Thatcheri mirabilis,** Angas. Japan. The shell is angularly pyriform, solid, spire prominent, shorter than the aperture, whorls flattened above, strongly keeled at the periphery contracted below. Aperture with a broad incurved sinus between the extremity of the last keel and the junction of the body whorl. Basal canal wide and opened. Columella smooth. Outer lip simple. One of the most remarkable deep water shells in the world and usually rare.

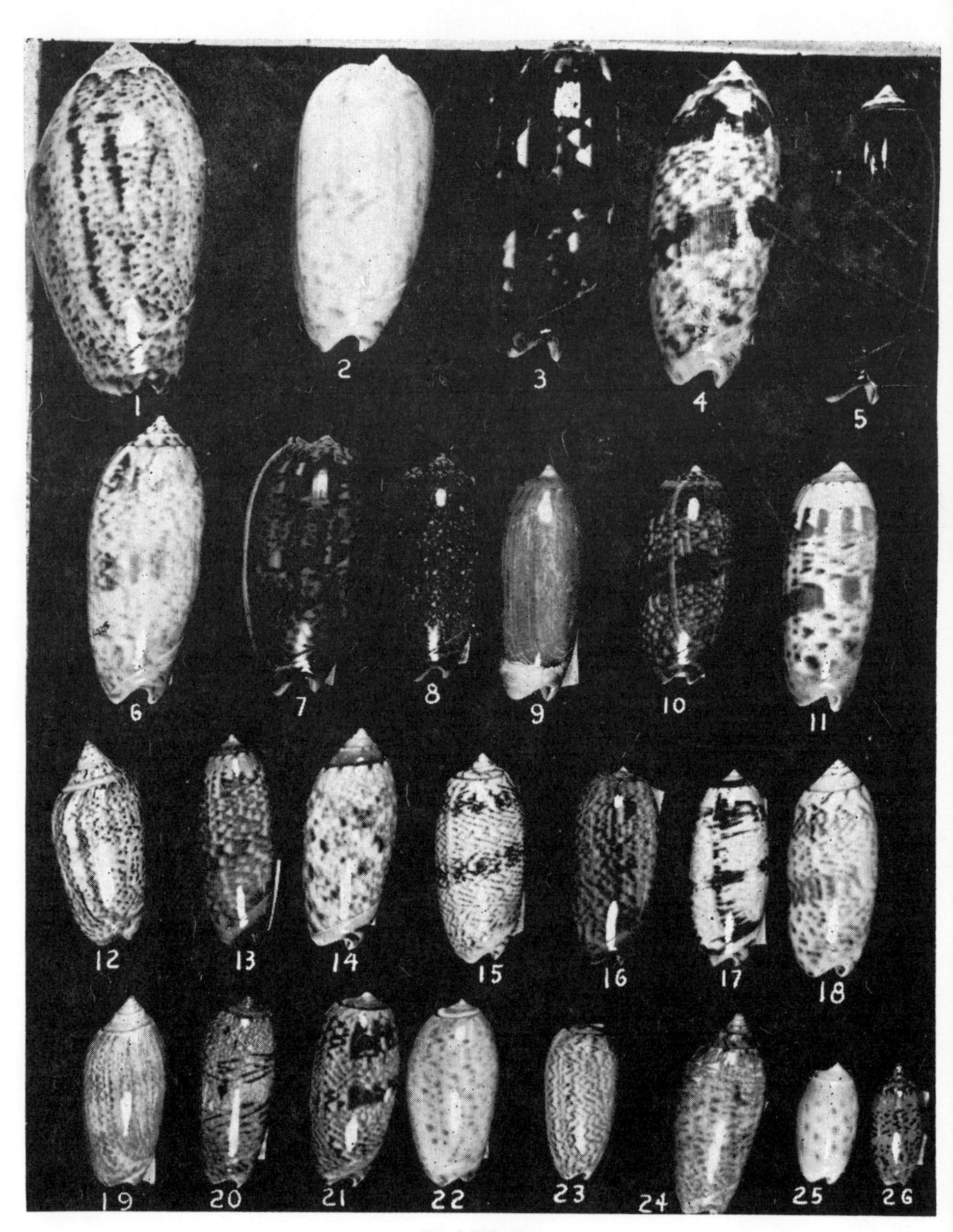

PLATE 99

1. **Oliva angulata,** Sol. Gulf of California. One of the thickest and strongest shells of the genus. It is angular, mottled with light brown and usually has perpendicular darker stripes. Lip beveled, interior often flesh colored when fresh. 3″.

2. **Oliva sericea,** Bolt. Philippines. A rather strong thick light colored shell with faint markings. Interior white. Usually lives below the tide line, hence rather scarce. 2¼″.

3. and 4. is **Oliva erythrostoma,** Lam. Philippines. Years ago scientists changed this old name to Minacea by Bolten. I never have agreed. It is a most striking shell, brilliantly mottled with black, brown and white. Aperture deep orange-yellow. 2″ to 3″.

5. **Black Olive.** Any oliva may be found either black or yellow which are simply nacre put on over the usual pattern. I get black ones that are evidently erythrostoma with the orange aperture and others have white aperture which must be some other shell. 2½″.

6. **Oliva erythrostoma tremulina,** Lam. Philippines. A finely mottled shell. Some markings being zigzag. 2½″.

7. **Oliva mauritiana,** Mart. Philippines. You see this shell labeled Oliva oliva, Bolt. just another of Boltens nonsensical changes. It is a richly mottled dark shell with blotches of black. Edge of lip often flesh colored . 2″.

8. **Oliva sanguinolenta,** Lam. Philippines. A dark species with fine dots of white, basal section rich red, which usually identifies the shell. Almost 2″ Philip.

9. This cut is simply a rich yellow form of some unknown species.

10. **Oliva tricolor,** Lam. Philippines. The entire shell is richly adorned with zigzag markings of yellow, green and white. 1¾″.

11. **Oliva circinata,** Mart. Philippines. Mottled much like tremulina but the apex is usually completely filled with nacre. 2″.

12. **Oliva venulata,** Lam. Gulf of California. The shell is somewhat angular completely covered with brown dots on a lighter background. 1¾″.

13. **Olivia porhyretica,** Mart. Philippines. It is mottled with tent-like marks but may be a variety. Aperture purplish. 1¾″.

14. **Oliva annulata,** Gmel. Philippines. The shell when flesh has an unusually glistening polish. Has numerous dark spots, and apex is salmon. Nearly 2″.

15. **Oliva evania,** Duc. Philippines. This shell has all the markings of **sanguinolenta,** but of a lighter shade and seems to be uniform. Base has a lighter shade of reddish. 1¾″.

16. **Oliva elegans,** Lam. Philippines. Has dark zigzag markings on a lighter background and some specimens are close to tricolor. Aperture flesh color. 1¾″.

17. **Oliva ornata,** Marrat. Australia. A brilliant light colored shell with various markings of brown. 1¾″.

18. **Oliva caerulea,** Bolt. Philippines. It has dots and zigzag markings of yellow and brown. Aperture purple. 1¾″.

19. **Oliva pindarina,** Duc. Gulf of California. The color pattern is vertical faint stripes and dots of brownish. 1½″.

20. **Oliva spicata melcheri,** Mke. Panama. Spire only slightly elevated less so than type. Body color brownish in spots, and blotches. 1½″.

21. **Oliva fumosa,** Mart. Philippines. The body is covered with zigzag and other shaped markings of a blackish color. 1½″.

22. **Oliva tigrina,** Lam. Philippines. Has the form of bulbosa in being more rounded and is covered with dots. 1½″.

23. **Oliva mustellina,** Lam. Philippines. Apex is almost flat, body completely covered with zigzag markings of a shade of brown. 1¼″.

24. **Oliva spicata,** Bolt. Old name was arenosa, Lam. which I like better. The shell is slightly pyriform, spire elevated and body is completely covered with light brown spots. 1¾″.

25. **Oliva tessellata,** Lam. Australia. Always a small shell with few botches regularly spaced and aperture is a deep lavender. About 1″.

26. **Oliva kaleontana,** Duc. Gulf of California. Looks like a small form of spicata but the several hundred specimens seem to be about uniform in size of 1″.

PLATE 100

1. **Terebra dimidiata,** L. Philippines. A tall slender shell of reddish color with faint diagonal stripes on the whorls. 4/5″.

2. **Terebra muscaria,** Lam. Japan. Tall and slender like the preceding but each whorl has two rows of square spots of brown. Usually light color and smooth. 4/5″.

3. **Terebra nebulosa,** Sow. Zanzibar. A medium size shell with two rows of squarish light brown spots on each whorl. 2½″.

4. **Terebra triseriata,** Gray. Japan. One of the most slender forms of the genus with around 25 whorls. Each whorl is separated by a raised spiral. Uncolored. 2½ to 3″.

5. **Terebra variegata,** Gray. Gulf of California. A finely mottled small shell with oblong row brownish spots at base of each whorl and similar smaller spots above. 2″.

6. **Terebra torquata,** A&R. Japan. A small slender shell. Each whorl has perpendicular stripes set close together. Shade of brown. ″2.

7. **Turris virgo,** Wood. Gulf of Mexico. Usually at 20 fathoms. A pure white shell with circular ridges. Some specimens have a tinge of yellow. Leathery operculum fits aperture. 3 to 4″.

8. **Cerithium vertagus,** L. Philippines. The shell is usually whitish but some are light brownish. There are perpendicular ridges on the whorls. 2¼″.

9. **Columbrarium pagoda,** Less. Okinawa. Here is a real Pagoda shell. The whorls have a row of spines on top edge. Lower spire up to 1½″. Whole length usually 2½ to 3″.

10. **Cymatium (Triton) exaratum,** Rve. Queensland. A neat form with ridges and nodules, aperture flat and nobbed. Light color, some reddish. 2¼″.

11. **Fasciolaria australasiae,** Perry. So. Australia. A neat round light brownish shell with black operculum that perfectly fits aperture. 3″.

12. **Fasciolaria trapezium,** L. Nicaragua. Of typical form with row of nodules on top of each whorl. Usually has light brownish skin. Dark operculum 3 to 5″.

13. **Potamides ebeninus,** Brug. A dark colored shell with small nobs on each whorl. Aperture flaring white and dark operculum. 3″.

14. **Fasciolaria trapeziem audoni,** Jonas. Philippines. This variety is a richer color than the type often showing much red and fine dark lines. It attains 4 to 6″. The specimen illustrated is a small 2″ dark specimen.

15. **Cymatium pyrum,** L. Philippines. The shell is of a reddish color and covered with irregular rows of knobs. Lip thick and lower canal usually bent backwards. 2 to 3½″.

16. **Murex capucinus,** Lam. Philippines. The shell is usually of a rich black color throughout but some specimens are lighter with a tinge of red. Varices are prominent and the circular rings or lines make it an attractive shell. 2½″.

17. **Vasum rhinoceros,** Gmel. East Africa. The shell is of a light brown color with circular lines of white. One row of nobs, and spire is elevated. Aperture toothed and white. 2½″.

18. **Cymatium chlorostoma,** Lam. Cooks Ids., South Pacific. The shell is a light grayish color and strongly reticulated throughout. Outer lip wide, aperture yellowish red. 2½″.

19. **Megalotractus proboscidifera,** Lam. Australia. The figure shown is a young yellow specimen of a shell that grows to two feet and with mollusks weighs 25 pounds or over, one of the two largest marine univalves in the world. The baby at the side was taken from an egg that contained a half dozen of same size about 1″. I had never seen a figure of the young before is reason I illustrate it.

20. A baby shell from egg.

1
2
3
4
5
6
7
8
9
10
11
12
13
14
15
16
17
18
19
20
21
22
23
24
25
26
27
28
29
30
31
32
33

PLATE 101

1. **Haliotis cyclobates,** Peron. Queensland. A really attractive small shell when first taken from water. The last whorl consists of most all of the shell, which is adorned with waves of white and brown. Interior iridescent. 2 to 2½″.

2. **Haliotis emmae,** Gray. So. Australia. The shell is wrinkled, thin, and the round knobs that adorn the main whorl extend around the shell, only 6 usually being open. Interior is wrinkled and iridescent. 2½″.

3. **Haliotis glabra,** Chem. Philippines. A small richly colored shell showing shades of cream, greenish and other colors on different specimens. Has 6 holes and interior is smooth pearl. About 2″.

4. **Patella transmerica,** Sow. Queensland. The color of interior is a pearly bronze the 22 perpendicular stripes show through. Center of interior is grayish, shape of owl. 1½°.

5. **Cassis japonica,** Rve. Japan. The shell is grayish with 4 interrupted bands of square blotches. Lip reflexed and showing brown transverse marks. 2″.

6. **Cassis abbreviata,** Lam. Gulf of California. The shell has circular tiny ribs but near crown there are two rows of small knobs. Lip is reflexed with 4 brown patches at back. 2″.

7. **Vermetus nigra,** Lam. Philippines. A black solid shell which has to be broken off rock. It usually consists of only one huge coil, the whole shell is usually 2″.

8. **Polinices bicolor,** Phil. Queensland. The shell is of a horn color, rather flat for its size, umbilicus open partly covered by prominent callous. 2″.

9. **Latirus prismaticus,** Mart. Philippines. The shell is white with vertical interrupted bands of red. Aperture white. Nearly 2″.

10. **Mitra plicata,** Kien. Philippines. A plicated white shell with vertical brown bands. Base aperture toothed. 1½″.

11. **Mitra sanguisuge,** L. Queensland. A small shell with vertical plications of rich dark color, one white circular band in middle of whorl, aperture dark. All Mitra are beautiful and this one specially so. 1½″.

12. **Mitra hanleyana,** Sow. Queensland. The shell is smooth shining of uniform shades of brown. Lower half of last whorl mostly white. 1½″.

13. **Mitra filaris,** Lam. Queensland. The shell is adorned with circular brown ridges throughout. Upper part of whorl is white. Little over 1″.

14. **Mitra glabra,** Swain. So. Australia. The color is light yellowish-brown, smooth, base of aperture tooth. 2½″.

15. **Cymatium kleineri,** Sow. South Africa. A small rugged shell. The picture is not quite typical, being shorter. 1″.

16. **Conus brazieri,** Sow. Queensland. A small smooth yellowish shell with one white band in middle of last whorl. Tip of shell shows pink. 1½″.

17. **Olivancillaria auricularia,** Lam. Uruguay. A short stubby shell of about two whorls and usually uncolored. Aperture white. 1¼″.

18. **Olivancillaria braziliana,** Lam. Brazil. The top of the shell is almost flat and lighter colored than rest of shell. Color gray. ¼″.

19. **Oliva peruviana,** Lam. Peru. The shell is slightly angular, apex elevated. Color pattern spotted with shades of brown but there are many uncolored forms of dainty shades. Very variable. 1¼″.

20. **Olivia ispidula,** L. Philippines. An extremely variable shell running from white to black and all shades in between. Usually 1 to 1½″.

21. **Olivancillaria cauta,** More. The shell is almost entirely one whorl, wide aperture, grayish color. 1¼″.

22. **Olivancillaria acuminata,** Lam. A slender yellowish-brown shell from So. Australia, but it may be other shades. 1½″.

23. **Olicancillraia steeriae,** Rve. India. A slender strong shell with brilliant markings of brown shading. 1½″.

24. **Olivancillaria subulata,** Lam. The specimen figured is of a smooth gray color, the apical whorl being tiny. 1½″.

25. **Olivancillaria gibbosa,** Born. Brazil. A very variable shell. The specimen figured is mottled gray, apical whorls darker, and there is a wide band near base of brown blotches on yellow. Other shells may be differently colored throughout. 1½″.

26-27. **Olive bulbosa,** Bolt. East Africa. This shell has many patterns, the ones illustrated being the lighter shades. Other specimens may be richly mottled with black. 1¼″.

28-29. **Olive ispidula,** L. Philippines. Another extremely variable variety the two figures showing extreme types. Some are entirely dark, others mottled and many have a rich lavender inside. It is very common on some of the islands.

30. **Mitra plicaria,** L. Queensland. This small reddish and white shell has two prominent dark bands. Base of shell is pink. Slightly over 1″.

31. **Conus kiiensis,** Kuroda. Japan. Upper whorls are carinated, main color pattern light brown with two darker bands. Lip thin, sharp. Slightly over 1″ scarce.

32. **Siphonalia trochula,** Rve. Japan. The shell is light brownish, with many spiral lines, aperture strong and there are interior spiral lines. 1¼″.

33. **Siphonalia cassidaeformis,** Rve. Japan. The shell has a row of nodules on top of last whorl, six circular brown lines, lip strong, interior white. 1½″.

PLATE 102

1. **Melo flammea,** Gmel. Queensland. A very handsome "Baler" Shell, richly mottled with shades of brown and white. 3 to 5".

2. **Turris unedo,** Val. Japan. One of the fine large forms of this wonderful family that contains many hundred species. It has prominent row of knobs on upper part of whorl and entire shell faintly dotted with brown dots. 3".

3. **Turris kaderlyi,** Lisch. Japan. The shell differs from preceding form in being usually smaller, with prominent brownish ridges on each whorl. 2½".

4. **Strombus epidromus,** L. Philippines. It is easy to always remember this fine white shell by the prominent flaring up. 3".

5. **Terebra chlorata,** Lam. Philippines. The fine species has only come on the market in recent years. It has two rows of square spots on each whorl. 3".

6. **Turris crispa,** Lam. Philippines. One of the finest of the genus being of good size and brilliantly marked with rows of squarish blotches of black on a white background. 3½".

7. **Verconella pyrulata,** Lam. Tasmania. This fine whitist shell is covered with prominent ridges and resembles a Fasciolaria. It belongs to the great Family of siphonalia. 3½".

8. **Fusus dupetithouarsi,** Kien. Gulf of California. A fine white shell usually covered with a fuzzy yellowish epidermis. Runs 4 to 5".

9. **Turritella tigrina,** Glen. Gulf of California. A handsome shell each whorl being covered with brownish zigzag markings. 3".

10. **Ancilla albocallosa,** Lisch. Japan. This shell is one of the finest of the genus, being smooth and richly adorned with brown above and below, the main body whorl being flesh color. 2¼".

11. **Ancilla 'urasia.** Lisch. Japan. This fine shell looks like a small edition of the above species and much resembles it in every way. 1½".

12. **Latirus turritus,** Gmel. Philippines. A smallish shell, completely covered with brilliant reddish-brown markings. 1½".

13. **Vermetus sipho,** Lam. Australia. A unique example of this remarkable genus which assumes all sorts of forms in its natural growth. You never see two twisted alike. 2".

14. **Ranella lampas,** L. Philippines. A young very knobby shell often with rich red aperture, unknown on larger shells. 2".

15. **Latirus castaneus,** Rve. Panama. A handsome shell with perpendicular ridges, the whole body being a yellow color. All of this genus are fine colored shells. 2".

16. **Turris granosa,** Helb. Japan. The shell has fine circular ridges and row small nodules on upper whorl. Uncolored. 2".

17. **Strombus marginatus,** L. Philippines. A shell with always distinct characteristics. Pyriform, small rows of nodules on top of main whorl, a rich shading of light brown throughout, flaring lip. 2".

18. **Cassis inornata,** Pils. Japan. A small round shell with faint brown markings tiny tubercles on top of main whorl, outer lip edged with brown spots. 1¾".

19. **Mitra chrysalis,** Rve. Philippines. The shell has faint circular ridges and body mottled with reddish-yellow. 1¾".

20. **Mitra intermedia,** Kien. Australia. The upper whorls are ornamented with broken line of black, lower whorls with wider lines and widest in middle of last whorl. Perpendicular ridges throughout. 2".

21. **Tenegodus weldi,** Woods. Australia. A small shell irregularly coiled with opening as usual in top of each whorl. 1".

22. **Terebratella rubicunda,** Sow. Australia. A smallish triangular reddish Brachiapod, usually found attached to and living with other shells. The Brachiapods are not true mollusca, as they have an internal structure. Only about 200 species living in the world and thousands of fossil forms.

23. **Delphinula atacta,** Rve. Japan. A small form with flat top, whorls ornamented with ridges and a few spines, usually stubby. Of a reddish color and fairly rare. 1½".

24. **Turbo gruneri,** Lam. South Australia. A small fairly smooth shell with circular ridges and of a reddish-brown color. 1¼".

25. **Cancellaria obesa,** Lam. Gulf of California. A small shell which much resembles our reticulata of Florida, except that it is more smooth, has two brownish bands and thin lip. 1".

26. **Cuma coronata,** Lam. East Africa. A small uncolored shell completely covered with pointed tubercles. The cuma are a branch of Thais. 1¼".

1
2
3
4
5
6
7
8
9
10
11
12
13
14
15
16
17
18
19
20
21
22
23
24
25
26
27
28
29
30
31
32
33

PLATE 103

1. **Conus omaria,** Brug. Philippines. A rather slender rounded shell of brownish shade, completely covered with tent-like markings. 2½″.

2. **Conus quercinus,** Hwass. Philippines. A handsome yellowish pyriform shell completely covered with fine striae. 2-2½″.

3. **Conus pennaceus,** Born. Japan. Only slightly pyriform completely covered with tent-like markings. 1¾″.

4. **Conus regularis,** Sow. Gulf of California. A handsome pyriform shell with zigzag brownish markings. Spire elevated to point. 2″.

5. **Conus planorbis.** Born. Philippines. Uniformly reddish-yellow with bands of lighter color. Spire nearly flat with irregular markings. 1″.

6. **Conus anemone,** Lam. Australia. Slightly pyriform, rather light and thin with light brownish markings. 1½″.

7. **Conus retifer,** Mke. Okinawa. A short pyramidal shell of reddish-chestnut with few white tent-like markings on a russet background. 1¼″.

8. **Conus eburneus,** Hwass. Indian Ocean. The shell has a white background covered with squarish blotches of black. 1¼″.

9. **Conus cancellatus,** Hwass. Japan. A white shell with elevated spire and fine circular ridges. 1½″.

10. **Conus figulinus,** L. Philippines. Uniformly light glossy brown with fine circular lines. Very distinct in color. 2 to 2½″.

11. **Conus interruptus,** Brod. Gulf of California. A pyriform shell with elevated spire, completely covered with dots and sometimes darker shades. There is a very dark form called mahoganyi. 1½″.

12. **Conus orbignyi,** Aud. Okinawa. A thin and slender shell very unusual in this genus. Spire elongated. There are wavy marks of brown and circular lines. 2″.

13. **Cypraca arenosa,** Gray. South Seas. Base is white, sides very light brown with four distinct brownish bands. 1½″.

14. **Cypraea walkeri,** Gray. Queensland. Upper surface shade of light brown, few spots along lower sides, wide brownish band. Rounded below, teeth strong, and color about same as top. 1¼″. Rare.

15. **Cypraea dicipiens,** Smith. West Australia. Base and sides are a rich shining black. Top mottled with white and light black. A brilliant rare shell that always used to sell for $10. 1¾″.

16. **Cypraea ventriculus,** Lam. South Seas. Base shining, sides reddish-black, top white and russet color. A rich colored distinct shell. 1¾ to 2″.

17. **Cypraea vitellus,** L. Philippines. It is of a light shade of brown with white spots. Base white. 1½ to 2″.

18. **Cypraea hungerfordi,** Sow. Japan. Base is creamy-white also lower sides. Just above is a fringe of black and top is mottled with dots and markings of russet. A deep water shell and always rare. 1¼″.

19. **Cypraea xanthodon,** Gray. Queensland. Lower sides have a few brown spots, base flesh color, teeth brownish, top is reddish-chestnut with two faint circular bands. Little over 1″. Rare.

20. **Cypraea subviridis,** Rve. Queensland. Top with chestnut markings, usually has a broad darker band. Base flesh color. Over 1″.

21. **Cypraea carneola,** L. Pacific generally. Base flesh color, side grayish, top with yellowish bands. Shell figured is 1″ but they are found to 3″ and frequently are confused with the rare aurantia.

22. **Murex phlorator,** Ad. & Rve. Japan. A small winged form, almost triangular. Only faint shadings of brown. 1¾″.

23. **Strombus gibberulus,** L. East Africa. The top whorls are usually distorted and body has circular brownish band of different widths. Aperture lavender. 1″½.

24. **Strombus floridus,** Lam. Pacific generally. A small stubby shell with russet markings, white band, thick lip. Very common. Imported by ton for novelty trade. Over 1″.

25. **Cymatium vespaceum,** Lam. Australia. Shell usually white with perpendicular ridges and knobs, lips wide. Over 1″.

26. **Cymatium rebecula,** Ch. Philippines. A small typical Triton completely covered with nodules and very variable shades of yellow, reddish and brown. 1½″.

27. **Cymatium weigmanni,** Ant. Panama. Row of knobs on top of whorl, circular brownish lines, each whorl flattened on top. 2 to 3″.

28. **Turbo porphyrites,** Mart. Philippines. An odd rounded triangular shell mottled with dark color. 1″.

29. **Turbo coronatus,** Gmel. East Africa. Top flat, two rows of nodules on last whorl, color usually whitish. Over 1″.

30. **Latiaxis japonica,** Dkr. Japan. Usually white, each whorl is adorned with a crown of flattened spines, clear to apex. Resembles a Chinese Pagoda. 1¾″.

31. **Drupa horrida,** Lam. Philippines. Base of shell is flat, aperture deep lavender, balance of shell covered with nodules and usually whitish. 1½″.

32. **Siphonalia pallida,** B&S. Gulf of California. The shell is entirely pure white, whorls have sharp points and are angular. 1½″.

33. **Cantharis gemmata,** Rve. Panama. A glistening white shell with perpendicular ridges and entire shell has circular ridges. 1¾″.

PLATE 104

1. **Thais bufo,** Lam. Queensland. Spiral lines on last whorl with two rows of small knobs. Outer lip edged with brown, interior flesh color. An attractive shell of this genus. 2″.

2. **Latirus belcheri,** Rve. Japan. Has a row of pointed nodules on each whorl. Ground color whitish with longitudinal wide stripes of black usually somewhat zigzag. An attractive shell. 2″.

3. **Siphonalia filosus,** Kuruda. Japan. An elongated shell with slight nodules in middle of each whorl and longitudinal markings of light brown. About 8 whorls. Aperture reflexed backward. 2½″.

4. **Ranella pulchra,** Gray. Japan. The Winged Ranella is a good common name. The five whorls have perpendicular nodules and the two wings encircle the aperture. 2″.

5. **Turritella flammulata,** Kien. Gulf of California. The shell is elongated as are all of this genus, with a ridge separating each of the 12 whorls. Of a grayish color and slightly wrinkled. 2½″.

6. **Murex eurypteron,** Rve. Japan. This shell is very closely allied to **aduncus** and it is very hard to separate them. Of about the same size and number of winged varices. A remarkable species of little over 2″.

7. **Conus ione,** Fult. Japan. The shell has only slight markings of brown and very closely resembles **sieboldi.** The apex is concave between each whorl. 2″.

8. **Siphonalia spadicea,** Rve. Japan. Similar in form to No. 3. The last whorl has small circular ridges and upper whorls have vertical ridges. Uncolored. 2″.

9. **Ranella corrugata,** Perry. Japan. The shell is completely covered with small nodules, usually a smaller row between, one higher. Aperture moludated, white. 1½″.

10. **Turris cosmoi,** Sykes. Japan. Each whorl has a row of small nodules in the center and between the whorls is a low row of pointed nodules of brown. A very dainty uncolored shell of 2″.

11. **Turris fascialis,** Lam. Japan. Each whorl has a row of small nodules and the lower part of the last whorl three rows. The nodules are light brown. 2½″.

12. **Turbo stenogyrus,** Fisch. Japan. A small beautifully mottled shell. Longitudinal alternate stripes of white, and brownish, granular but smooth surface. 1½″.

13. **Aporrhais pes-pelicani,** L. East Indies. A very unique and common shell now used in vast quantities in the novelty manufacturing business. It is not unusual to see a whole barrel full in a wholesale house. The peculiar fingers do not appear on the young shell. 1¾″.

14. **Siphonalia fusoides,** Rve. Japan. Typical form of many of the smaller shells of this genus, spiral lines, base turned back. 1¾″.

15. **Latiaxis lischkeana,** Dkr. Japan. This beautiful shell much resembles **Japonicus,** is white, but the row of flat spines at edge of each whorl, curve upwards. The body whorl is covered with rows of very tiny spines. 1½″.

16. **Latiaxis pagoda,** A. Ads. Japan. This little fellow of 1″ has rows of upturned spines on each whorl. The spines are flat. Practically uncolored.

17. **Murex modesta,** Fult. Japan. A fine small shell with three thin wings, drab color and almost round aperture. 1¼″.

18. **Murex penchinati,** Crs. Japan. A small shell of the form of our Florida **rufus** and **salleanus.** Some are reddish and some black. Usual specimens 1½″.

19. **Pecten singaporensis,** Sow. Queensland. The shell has 18 prominent rounded ridges and is mottled with reddish, brownish and white. Usual specimens 1½″.

20. **Ishnochiton proteus,** Rev. New South Wales. A new little species with a light streak down the back and lighter on edge. 1¼″.

21. **Ovula volva,** L. Japan. Often called the Weaver Shuttle Shell, as each end is elongated like the shuttle used in weaving cloth. Aperture is elongated and oval on edge. Usually flesh colored in fresh specimens and ranges from 2 to 4″.

22. **Turris coffea,** Smith. Philippines. A dark shell, upper part of each whorl is black and lower part lighter colored, with circular ridges throughout. 2½″.

23. **Ishnochiton virgatus,** Rve. Victoria, Australia. The back is light colored and mottled, with darker edge. 1¼″.

24. **Dentaliem vernedei,** Hanley. Japan. This is one of the largest varieties of the genus, which consists of several hundred species throughout the world. The East and West Coast of North America have a large number of species, mostly 1 to 2″, many highly polished and all have a hole in each end. This species is white and ranges from 4 to 5″ in length.

PLATE 105

1. **Voluta daviesi,** Fult. Japan. A tall slender shell with vertical striae and squarish brown blotches on each whorl. Always deep water. 4″.

2. **Voluta mentiens,** Fult. Japan. Somewhat similar to preceding species but with circular striae as well as vertical riblets. The brown shadings are more suffused. 4/5″.

3. **Voluta cancellata,** Kuroda. Japan. The shell is practically uncolored with vertical striae and ridges. 3″.

4. **Voluta delicata,** Fult. Japan. A small uncolored shell with vertical ridges more prominent on the upper whorls. 1¾″.

5. **Fusus laticostata,** Desh. Philippines. A typical shell in form, with circular ridges all shaded brown. 2½″.

6. **Voluta cumingii,** Brod. Gulf of California. One of the smaller forms of this genus which seldom attains over 1″. It has vertical ridges of brown, lip is wide at edge and aperture narrow.

7. **Fuscosurcula mirabilis,** Sow. Japan. A handsome member of the Turris Family with curved vertical stripes of brown. 2½″.

8. **Alipurpurea centrifuga,** Hinds. Gulf of California. This fine little white triangular shell was formerly a Murex and looks like one. Each of the three ridges has a central point. Rare. 1½″.

9. **Cymathium dunkeri,** Lisch. Japan. This is a fine member of the great Triton group. It has vertical ridges or knobs, aperture strong and rounded, tail at bottom curved backward. Aperture orange. Main color pattern, shades of brown. 3½″.

10. **Cymatium gutturnium,** Koch. Philippines. The shell is mostly uncolored, knobby, tail curves backwards, aperture rounded, and all well covered with nacre. Aperture may be orange. 2¼″.

11. **Tudicle spinosa,** H&Ads. Queensland. I had never seen this shell until summer of 1950. My other books only illustrate one species. I have an idea this shell is rather rare as owner said he had only taken 10 specimens in years. May be deep water. The main shell is well rounded with row of points on top of whorl, which extend to about top. The extended canal has a few points also. Generally of a white color with faint brownish markings. Only a very few forms in the genus.

12. **Voluta pulchra,** Sow. Barrier Reefs, where it has only been found to any extent on one island but likely lives on many when that vast territory is better known. The shell is covered with white on a light reddish background. As its name indicates it is one of the grand shells of the genus. 2″.

13. **Voluta caroli,** Ired. Barrier Reefs, Australia. A comparatively rare shell anywhere. It is a shiney flesh color with three rows of brown markings. Some specimens are uncolored. 2 to 3″.

14. **Murex denudatus,** Perry. Queensland. This shell is of a light pinkish color to white and much resembles our salleanus of Florida. 2″.

15. **Murex motacilla,** Ch. Barbadoes. A neat little chap with tail curved backward, prominent ribs and knobs, few sharp short spines. 1½″.

16. **Murex recurvirostris,** Brod. Gulf of California. The three prominent rounded ridges on this shell are marked with dark brown and white, few short spines and tail is straight. 1¾″.

17. **Murex triformis,** Rve. Tasmania. There are two prominent upright ridges on body whorl, aperture is flat and flaring. Almost uncolored. 2½″.

18. **Harpa nobilis,** Bolt. Philippines. I suspect this shell is found everywhere in Pacific and Indian Oceans. It is a small form but very richly colored with red, brown and white. 2″.

19. **Harpa minor,** Lam. Philippines. One of the small forms of the genus. Fresh specimens are richly mottled with black, brown and white. 1½″.

PLATE 106

1. **Thais pica,** Blv. Philippines. The shell has two rows of pointed knobs on last whorl and one row above. Ground color white, richly adorned with jet black markings. Nearly 2".

2. **Thais rudolphi,** Lam. Philippines. This is a handsome shell if you can get a specimen that cleans up well. It has two rows of faint knobs on last whorl. Entire shell has circular tiny white lines. Aperture edged with brown, columellar part flesh colored. 2".

3. **Thais haustrum,** Mart. Australia. The remarkable thing about this shell is its aperture which in specimen figured measured 2" and the entire shell 2½". The general color is brown, edge of aperture well marked, columellar section white.

4. **Thais succincta,** Mart. New Zealand. The shell is uncolored, has vertical striae and circular ridges. Aperture white. 2".

5. **Thais mancinella,** L. Philippines. The whole shell is ornamented with flat pointed knobs. Ground color mostly white. Aperture a rich yellow. 2".

6. **Cassis crumena,** Brug. Gulf of California. The shell much resembles the Cameo Shell with rows of small knobs of russet color, aperture white, upper part narrow and wider below which is one of its main distinctive features. I suspect it lives below tide lines as it is not very common. Average size 2".

7. **Strombus mauritianus,** Lam. East Africa. A narrow elongated shell covered with russet markings, usually arranged in circular rows. Lip notch is prominent. Always used to be considered fairly scarce but since war has appeared on the market in fair quantity. Nearly 2".

8. **Strombus melanostomus,** Sow. Philippines and Pacific generally. A very attractive shell. Back has row of knobs in upper section and less below. Upper part of aperture terminates in a spire of a half inch. Lower section reflexed backwards with prominent notch. The whole aperture richly adorned with black and yellow markings. 2½".

9. **Strombus japonicus,** Rve. Japan. An elongated pointed shell upper whorls uncolored, last whorl broadens out, shaded brown with two faint circular rows of white. The upper part of the lip extends upward on the two previous whorls. Aperture white. 2½".

10. **Strombus isabella,** Lam. Philippines. A fat chunked rounded shell of light brownish color with irregular markings throughout. Upper part of lip thick and rounded, thin below at notch. 2".

11. **Lischkeia argenteonitens,** Lisch. Japan. A thin deep water shell with row of small knobs on each whorl, fading out entirely in apex. Aperture thin. The whole shell has a brilliant sheen, common in very deep water shells. 2".

12. **Turcicia coreanis,** Lam. Japan. Somewhat similar to preceding shell to which it is closely allied. It has a tiny row of knobs, another at sutures, which continue to apex. Base of shell flat with tiny circular lines. It is a very deep water shell and shows some sheen. 1½".

13. **Turcicia crumpii,** Lisch. Japan. Smaller than preceding species, has only faint knobs, but covered with spiral lines, with base same. 1½".

14. **Calliostoma meyeri,** Phil. Tasmania. A handsome conical shell completely adorned above and below with circular ridges made up of fine beading. Apex bluish. 1½".

15. **Solarium maximum,** Phil. Japan. This genus is also called Architectonica. One of the largest of the genus. It has rows of square spots above and below each whorl at suture. It has no central aixs, and umbilicus can be clearly seen to top, each whorl ringed with serrated edge. In the older days was called the Staircase Shell. 2½".

16. **Eugyrina subdistorta,** Lam. Tasmania. This genus name used to be a subgenus of Tritons or Cymatium. The shell has characteristics of both a Triton and Distorsio. It is of light color with shadings of brown, faint circular rows of knobs. 2½".

17. **Calliostoma haliarchus,** Mke. Japan. The shell forms a perfect shiney cone terminating at a point of apex. The whorls have faint circular lines and base the same. The naturally smooth shiney surface makes it a very attractive shell. 1¾".

18. **Cassia areola,** L. Pacific generally. Shell figured is from Australia. It is generally smooth and completely covered with circular rows of square patches of brown. Lip strong reflexed and has usually inner row of small sharp teeth. 2 to 3".

PLATE 107

1. **Cassis japonica,** Rve. Japan. Also figured on Plate 101.

2. **Cassis abbreviata,** Lam. Gulf of California. Also figured on Plate 101.

3. **Cassis pila,** Rve. Japan. The shell much resembles japonica, but spots are more faint or entirely lacking. 2½".

4. **Cassis achatina,** Lam. Cape Verde Ids. A small round form with faint flesh color. Rather shiney. 1¼".

5. **Melongena galeodes,** Lam. Philippines. Usually dark color, row of short spines on top of whorl, another similar row near base. Body lines with circular rows of brown. 2".

6. **Turbo petholatus,** L. Philippines. With the green operculum which boys during war called "Cat eyes." But many of them are white. The shell is very smooth and shiney and brilliantly mottled, hardly any two of the same pattern. 2".

7. **Ranella ranelloides,** Rve. Japan. A typical small shell of this genus, covered with rows of tubercles. Light brown in color. 1¾".

8. **Strombus auris-dianae,** L. Philippines. The back of the shell may be rough or smooth, aperture a deep reddish-orange, with a short spire on top of lip. 2".

9. **Distorsio constrictus,** Brod. Panama. The aperture is distorted as usual similar to the Florida form. Color white. 2".

10. **Distorsio anus,** L. Philippines. The most remarkable shell of the genus, the whorls being distorted, covered with knobs, and the face of the aperture is broad, covered with rich nacre. 2".

11. **Latirus cingulata,** L. Panama. An attractive shell covered with circular rings of black on a white background, but none can be seen in life as they are covered with a brown periostracum. There is a sharp tooth at base of aperture. 1¾".

12. **Latirus polygonus,** Geml. Philippines. A handsome shell with notched perpendicular ridges of black color, either white or flesh colored between. 1¾".

13. **Cantharus insignis,** Rve. Gulf of California. This shell is one of the large forms of the genus and is perpendicularly marked with stripes of blackish-brown. 2".

14. **Nessarius magnifica,** Lisch. Japan. Some years ago they changed the name of the genus to Hindsia, and you may see the shells with either name. This form is one of the finest, with vertical ridges on each whorl, only faintly marked. 1¾".

15. **Cancellaria laticostata,** Kob. Japan. This shell is closely allied to reeveana, Crs. It is covered with fine ridges and only faintly marked with brown. 1¾".

16. **Cancellaria cassidaeformis,** Rve. Panama. An unusual form reticulated, a row of points on top of whorl, interior of aperture ridged. 1½".

17. **Cantharus erythrostoma,** Lam. West Australia. A fine small shell with vertical ridges, tinged with brown. Ground color yellowish. 1¼".

18. **Siphonalia nodosa,** Mart. New Zealand. The shell is roundish with rows of small tubercules and vertical bands of brown. 1¾".

19. **Vasum cornigerum,** Lam. Philippines. The shell is pyriform, covered with tubercles well marked with white and black color. 2".

20. **Latirus smaragdula,** Lam. Philippines. A round bulbous shell with very fine circular lines, usually a reddish-brown or blackish color. 1¾".

21. **Thais tuberculata,** Lam. Philippines. There are very few shells of this genus so richly adorned with pointed spines, and black color. 1¾".

22. **Crucibulum imbricata.** Sow. Gulf of California. There are Cup and Saucer shells all over the world and this is one of the best. Always irregular in shape, the inside cup is pure white. 1½".

23. **Astraea olivaceum,** Wood. Gulf of California. A low pyramidal shell covered with lines only faintly colored. Below most of the shell is flesh color with reddish at center. 1½".

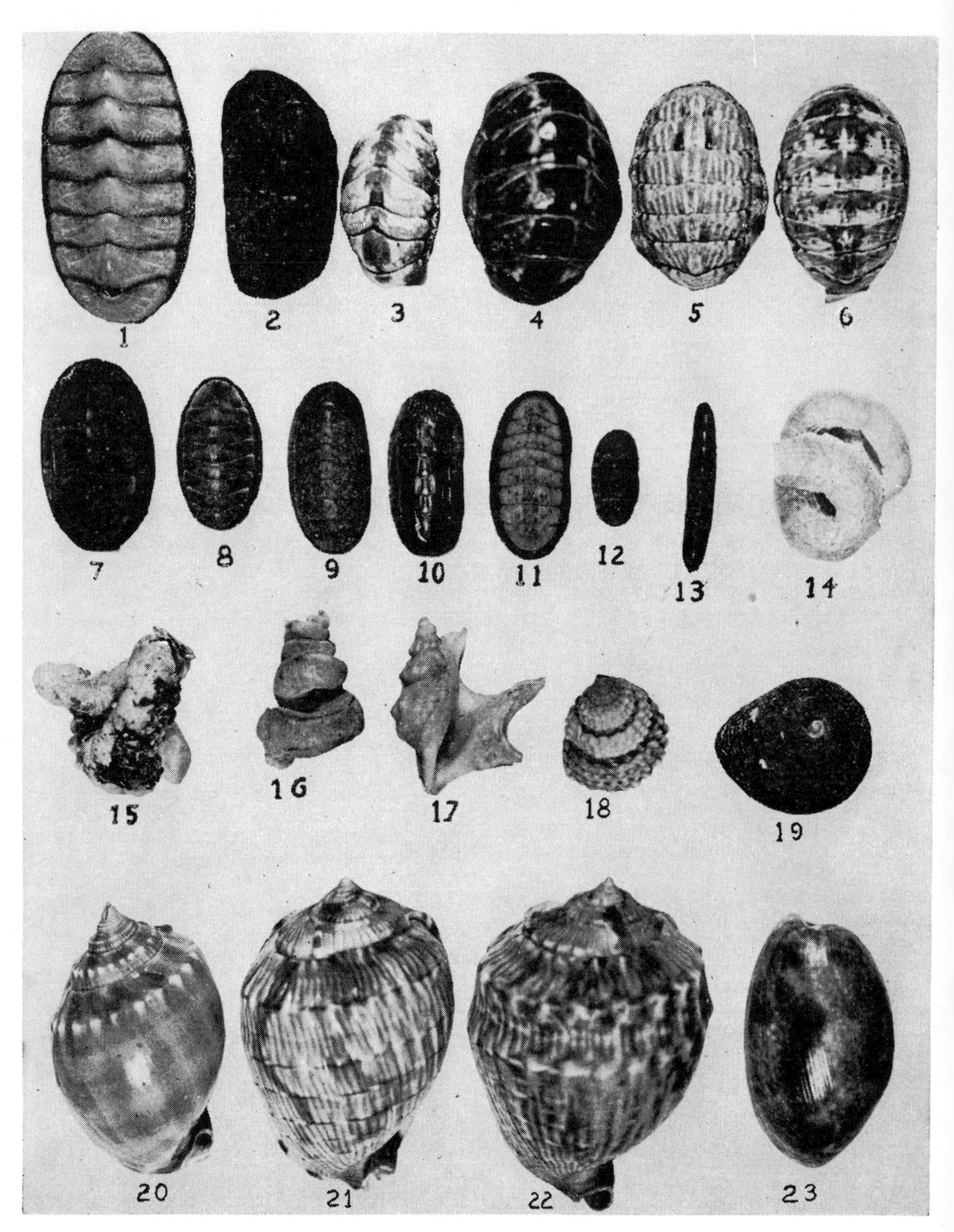

PLATE 108

1. **Chiton tuberculatus,** L. Panama. A handsome grayish form, with snakeskin mantle. Interior is sea green. You will see them clinging to the rocks where the surf is heavy. 2½″.

2. **Callistochiton pulchellus,** Gray. Gulf of California. The shell is dark colored, plates are richly lined in fan shape. Interior greenish. 2″.

3. **Chiton sqamosus,** L. West Indies. A rugged chap, the plates above usually covered with incrutations or bryozoa. Interior is pale greenish. One of most common forms. 2″.

4. This is a specimen of the same shell, buffed down to the hard shell. You would never recognize the rich black color bordered with white and brown. Interior black and green.

5. **Chiton marmoratus,** Gmel. Bahamas. In its natural state the back is adorned with faint stripes on a gray background. Mantle like snakeskin. Interior rich green. 2″.

6. This is the same shell richly buffed. The back is now a rich green-white and some faint stripes. A real beauty.

7. **Ishnochiton contractus,** Rve. Also shown in Figure 11 this plate.

8. **Ishnochiton lineolatus,** Blv. South Australia. A handsome grayish-brown shell with very fine lines. Mantle slightly darker brown. 1¼″.

9. **Ishnochiton cariosus,** Pils. South Australia. The patterns is somewhat like preceding species, slightly narrower, the mantle light brown. 1¼″.

10. **Ishnochiton torrei,** Iredale. South Australia. Top two shades of brown, mantle still lighter, some specimens show red color. 1¼″.

11. **Ishnochiton contracturs,** Rve. So. Australia. The main body is light grey with faint markings and mantle is a darker brown. 1½″.

12. **Ishnochiton tricostalis,** Pils. So. Australia. This little fellow is hard to describe. The back is richly adorned with lines in form of triangle, and the narrow mantle is barred. 1″.

13. **Stenochiton longicymba,** Blv. So. Australia. One of the dainty forms of this great family, very narrow with glossy back. Looks more like a worm than a sea shell. Interior is dark. 1½″.

14. **Vermetus filosus,** Rve. Japan. It is just naturaly grows flat on some smooth rock or shell. It frequently uses the host as part of its body whorls and when removed shows many holes. Uncolored. 1½″.

15. **Vermetus novahollandiae,** Rouss. Australia. It just naturally grows in a twisted mass, twining its coils around one another. Uncolored. 2″.

16. **Tenegodus (Siliquaria) anguina,** L. Philippines. One of the finest of the genus but seldom two alike in form. The top of the whorls have an opening the entire length. Uncolored. 2″.

17. **Aporrhais pes-pelicani,** L. Also figured on Plate 104.

18. **Echinella coronaria,** Lam. Philippines. All of the shells of this genus resemble Tectarias, being trochiform, nobby, and often richly adorned with color, as is this species. 1″.

19. **Calliostoma meyeri,** Phil. Tasmania. Also figured on Plate 106.

20. **Cassis bandatum,** Ire. Australia. This shell much resembles glauca, which is found all over the Pacific area. The specimens sent me come from Australia, are faintly spotted, smooth, shiney, lip flesh colored. Top of each whorl has a row of small tubercles. 2½ to 3″.

21. **Cassis plicata,** L. Australia. A very handsome glossy shell with vertical flat ridges, five faint circular stripes of brown and wider irregular vertical stripes same. Top of shell has one row of very small knobs. Lip has five narrow bands. The shell is not very common in collections in this country. 3″.

22. **Cassis bicarinata,** Jon. Australia. Superficially this shell resembles the previous species but there are prominent distinctions. It has two rows of knobs on upper part of last whorl. The vertical lines are less pronounced and there are wavy brownish markings throughout, Specimen figured was 3″.

23. **Bulla tenuissima,** Sow. South Australia. One of the largest species of the genus I have seen. Usually 2″ or more. The back has wavy dark markings and the top has open umbilicus.

PLATE 109

1. **Pectenculus maculata,** Baird. Gulf of California. One of the finest marked shells of the genus being completely covered with zigzag markings of deep brown. 2½″.

2. **Crassatella kingicola,** Lam. Tasmania. The shell is covered with a dark periostracum, which if removed would show flesh color. This genus has shells that range from 1 to 4″. 2½″.

3. **Mactra obesa,** Desh. Queensland. The shell is mostly uncolored, some red patches on the umbones. 2¼″.

4. **Lucina exasperata,** Rve. Philippines. The circular white shell is finely reticulated, interior edge is often lined with pink. 2″.

5. **Callista erycina,** L. Australia. The handsome shell is richly lined and shaded with purple, interior also shows some of that color. 2½″.

6. **Callista aurantiaca,** Sow. Gulf of California. A shell of natural brilliant polish, as are most all of this genus. Main color yellowish-brown. Some are mostly yellow. 2½″.

7. **Cytherea chemnitzi,** Hanley. West Australia. A very beautiful shell with thick upraised ridges, thickly scalloped. Only faint shadings of brown. 2½″.

8. **Cardium flavum,** L. Queensland. A rather flat elongated shell with many fine rounded ribs, shaded with yellowish-brown. Interior partly yellow. 2″.

9. **Cardium rusticum,** L. Malta. The shell is finely ridged with shadings of brown throughout. 1½″.

10. **Cardium setosum,** Redfield. Australia. A very dainty shell, with faint ridges and pinkish-yellow color. 2″.

11. **Cardium procerum,** Sow. Gulf of California. A solid shell with prominent rounded vertical ridges, which are faintly marked with brown. 2½″.

12. **Cardium elatum,** Sow. Gulf of California. A fine smooth yellow shell with faint vertical lines throughout. 2½ to 3″.

PLATE 110

1. **Vola alba,** Lam. So. Australia. The shell has flat ridges shaded with light purple. Interior is lined with purplish brown. 2½″.

2. **Vola laqueatus,** Sow. Japan. The shell has eight ridges on the flat valve and more on convex. Color pattern shade of brown. 2½″.

3. **Pecten bifrons,** Lam. Australia. The shell is rich purple and even deeper color inside. 2 to 3″.

4. **Pecten fulvicostatus,** A&R. West Australia. The shell has 7 prominent ribs and is covered with fine vertical lines. Of a brownish shade of color, the interior is lighter. 2″.

5. **Pecten radula,** L. Philippines. The shell has about 12 vertical ribs. Upper valve usually dotted with black and lower uncolored. 2 to 3″.

6. **Pecten circularis,** Sow. Gulf of California. The type color is dark, but the specimen figured is what I call the reddish variety, being blotched with red and yellow, lower valve may be all yellow. Very variable, some shells showing much more red than others. 2″.

7. **Pecten squamatus,** Gmel. Japan. The shell ranges thru various colors, the back often has flat spiney surface. 1½″.

8. **Tellina elegans,** Gray. Sicily. The specimen figured is most all white with some shade of yellow. Very thin. 3″.

9. **Tellina salmonea,** Rvs. New Zealand. The umbones are red shading lighter to edge. 2¼″.

10. **Tellina albinella,** Lam. So. Australia. The shell is thin and all red, even darker inside. 1¾″.

11. **Callista kingi,** Sow. Tasmania. A richly colored shell, with stripes of white and light black. Interior white. 1¾″.

12. **Callista planatella,** Lam. West Australia. One of the finest of the genus with natural polished ridges, shadings of brown and white. 2″.

13. **Paphia sulculosa,** Phil. Queensland. A richly marked shell with natural polish, adorned with black and flesh colors. Interior umbone section yellow. 2″.

14. **Cardium biagulatum,** Sow. Gulf of California. A small shell of about 1″, with prominent vertical ribs and one end flat.

15. **Corbis Sowerbyi,** Rve. West Australia and Philippines. I call it one of the most beautiful of all bivalves. The circular elevated ridges are tipped with pink stripes. There are other bivalves that approach it, but never exceed. 2″.

16. **Paphia turgida,** Lam. Queensland. The shell is richly marked with zigzag markings, of shades of brown. Interior has blotch of same. 1¾″.

17. **Tellina capsoides,** Lam. West Australia. The shell is almost entirely white, one end flattened, interior white. 1½″.

18. **Waldhemia flavescens,** Lam. Tasmania. A typical roundish Brachiapod with faint ridges on lower half. 1½″.

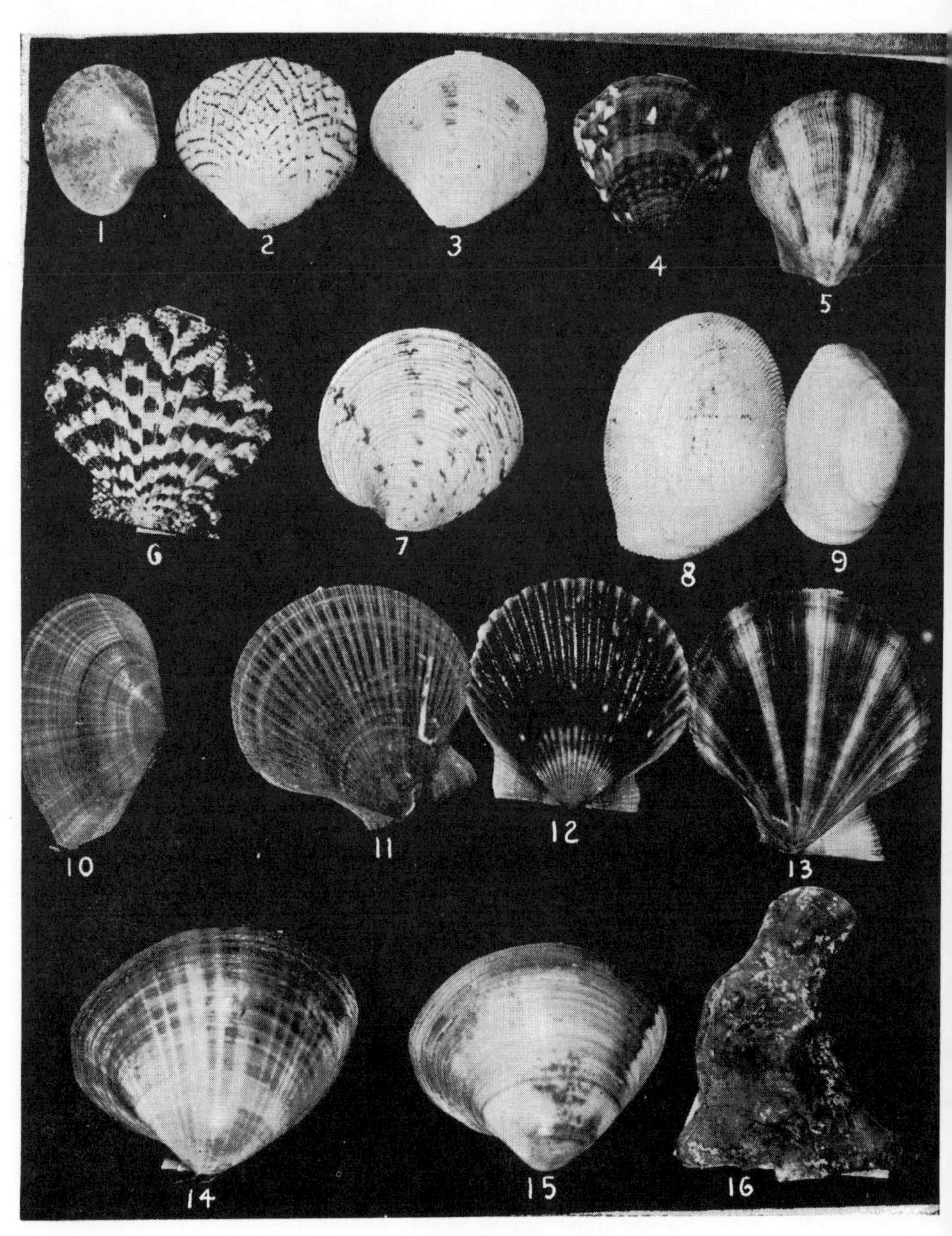

PLATE 111

1. **Sunetta excavata,** Hanley. West Australia. The shell is very shiny and only faintly colored. Interior white. 1¼″.

2. **Circe scripta,** Lam. About same size as preceding shell, both valves are covered with zigzag markings of brown on white. Interior only faintly colored. 1½″.

3. **Circe rivularis,** Brug. Philippines. An almost circular shell with ringed lines and few faint blotches of color. Shell is depressed at umbones. Interior partly covered with pinkish-lavender. 1½″.

4. **Cardita crassicostata,** Sow. Gulf of California. It has curved vertical ridges. Main color reddish, white blotches lower edge. Umbones curved. 2″.

5. **Spondylus acanthus,** Mawe. Japan. One of the small forms of this great genus the spines being mere pricks. Shell has 3 vertical bands of reddish color on white background. 2″.

6. **Pecten pallium,** L. Philippines. One of the most beautifully colored shells of the genus, with often zigzag stripes of reddish-brown and white. 2″.

7. **Dosinia victoriae,** Gat. and Gab. Australia. The shell is finely ringed with lines and has 4 vertical rows of brownish spots. 1¾″.

8. **Tellina scobinata,** L. Philippines. The shell is completely covered with lines and chevron markings. Mostly uncolored. 2¼″.

9. **Tellina staurella,** Lam. Japan. A neat small form that ranges from pure white, to vertical reddish stripes. 2″.

10. **Tellina virgata,** L. Philippines. One of the finest colored shells of the genus being well striped with reddish and white. 2″.

11. **Pecten hindsii novarchus,** Dall. Puget Sound. The true hindsii is pale pink and this variety is lavender and very much rarer. 2″.

12. **Pecten gibbus amphicostatus,** Dall. Gulf of Mexico. Upper whorl is brownish-black, lower, yellow. 2″.

13. **Pecten hericeus,** Gld. Puget Sound. A handsome finely ridged shell of a rich shade of pink color. 2″.

14. **Mactra eximia,** Desh. Queensland. A fine shell with stripes of faint brownish color. 2½″.

15. **Mactra trigonella,** Lam. So. Australia. A triangular shell mostly uncolored, with just trace of brown on umbones. 2″.

16. **Malleus albus,** Lam. So. Australia. These shells much resemble flat oysters with wings, but some forms have no wings. This shell has short wings and is uncolored. 3″.

PLATE 112

1. **Lingula anatina,** Lam. Philippines. One of the unique brachiapods of the world. I found it fairly common living attached to other shells, with a cord often most 2″ long. It is of a greenish color. 1½″.

2. **Soletellina violacea,** Lam. Philippines. The thin elongated shell is well marked with violet color. Likely a variety of what is called elongata. 1½″.

3. **Venus flexuosa,** L. Australia. A small triangular form with prominent ridges. About 1″.

4. **Pecten vesicularis,** Dkr. Japan. A small well-marked shell which is variable, seldom two alike. Interior has color also. 1″.

5. **Paphia variegata,** Sow. Broome, N. W. Australia. A small finely mottled shell, which is variable and no two exactly alike. 1¼″.

6. **Gari lineolata,** Gray. New Zealand. A rather small thin shell with faint brown markings. Interior also shows some color. 1¼″.

7. **Anomalocardia subrugosa,** Anton. Gulf of California. The shell has transverse ridges with two dark bands across same. Interior white. A splendid variety of the Venus complex. 1½″

8. **Cardium rachetti,** Don. Tasmania. The shell is almost pure white, smooth except upper part of each valves has prominent striae. 1½″.

9. **Cardium lyratum,** Sow. Philippines. A gorgeous shell with fine striae and brilliantly mottled with coral red. Interior flesh-white. Rather scarce.

10. **Pitaria vulnerata,** Brod. Gulf of California. A handsome mottled smooth shell much like the Callistas of that region. 1¼″.

11. **Pitaria japonica,** Lam. Japan. The fine shell is a flesh-white with similar shading inside. 1¼″.

12. **Pecten transquebarica,** Gmel. Philippines. The dark colored shell is completely covered with ridges in the form of minute knobs. 1½″.

13. **Mactra dissimilis,** Desh. So. Australia. A triangular shiney shell with rich dark coloring both inside and out. 1½″.

14. **Circe gibba,** Lam. Philippines. A thick bulbous shell covered with striae and shaded with brown. 1¾″.

15. **Venus scalarina,** Lam. South Australia. A small oblong shell with both circular and vertical dots and zigzag marks. Attractive. 1¼″.

16. **Mactra pura,** Rve. Australia. The color is mostly white with some thin periostracum. General shape triangular. 2″.

17. **Pitaria citrina,** Lam. Philippines. A bulbous almost round shell with yellowish markings. Interior has some shading of dark color. 2″.

18. **Chione matadon,** P&L. Gulf of California. A real Venus shell with circular striae and shadings of brown. Interior has a rim of dark color. 2″.

19. **Venus columbiensis,** Sow. Gulf of California. The shell is almost round covered with stariae and markings of various shades of brown. 2″.

20. **Pectenculus reevei,** Mayer. Japan. An almost round dark colored form with curved striae throughout. 2″.

21. **Gari stangeri,** Gray. Auckland, N. Z. A shiny smooth shell with faint shadings, much like some of the Tellinas. 2″.

22. **Paphia litterata,** L. Philippines. The shell is thin and beautifully marked with brown hieroglyphic tracings. 2 to 2½″.

23. **Crassatella gibbsi,** Sow. Gulf of Mexico, 20 fathoms. As taken from the sea it has a dark periostracum, but cleaned up shows fine dots of brown. The finest species in this territory. 2″.

24. **Cardita laticostata,** Sow. Panama. Fine vertical ridges and two bands of white. 1⅛″.

25. **Arca antiquata,** L. Philippines. It has the usual striae and is covered with dark periostracum. 2″.

26. **Arca granosa,** L. Philippines. The shell is pure white when cleaned, with nodulated striae throughout. 2 to 3″.

HOW TO PRONOUNCE THE LATIN NAMES

They are easy to learn when you become familiar with them. Take it slow on the start and you will learn them from day to day. In a short time they will be just as easy to pronounce as the names of your own family.

Every branch of Natural Science uses names of this character to designate species, genera, families, etc. There is a certain style that follows through each branch of nature. When you are familiar with the style of shell names, you will instantly recognize them as belonging to conchology.

Here are a list of names taken at random from the front of this book, and their proper pronunciation. It is believed with this brief aid, most students will be able to figure out the rest.

Voluta junonia	Volú-ta ju-nó-ni-a
Cassis tuberosa	Caś-sis-tu-ber-ó-sa
Argonauta argo	Ar-go-náu-ta ár-go
Vermicularia spirata	Ver-mic-u-lá-ria spi-rá-ta
Terebra dislocata	Ter-é-bra dis-lo-cá-ta
Thais patula	Thá-is pét-u-la
Cymatium nobilis	Cym-á-ti-um nó-bil-is
Buccinum undatum	Buc-ciń-um undá-tum
Leucozonia cingulifera	Leu-co-zó-nia cin-gu-li-fé-ra
Melogena corona	Me-lón-ge-na co-ró-na
Tonna perdix	Tón-na per-dix
Xenophora conchyliophorus	Xe-nóph-o-ra con-chyl-i-óph-o-rus
Murex rufus	Mú-rex rú-fus
Busycon perversum	Bus-ý-con per-vér-sum
Neptunea decemcostata	Nep-tu-néa dĕ-cem-cos-tá-tá
Strombus gigas	Stróm-bus gi-gas (hard g)
Livonia pica	Li-vó-nia pí-ca
Conus proteus	Có-nus pró-te-us
Fasciolaria gigantea	Fas-ci-o-lá-ria gi-gán-tea
Voluta virescens	Vo-lú-ta vi-rés-cens
Cypraea cervus	Cyp-rá-ea cer-vus (c like s)
Astraea longispina	As-traé-a lon-gi-spí-na
Bullia striata	Búl-li-a stri-á-ta
Spondylus americana	Spón-dy-lus a-mer-i-cá-na
Chama macerophyla	Chám-a mac-er-o-phýl-a
Pinna serrata	Piń-na ser-rá-ta
Pedalion elata	Pe-dal'-i-on a-lá-ta
Tellina radiata	Tel-lí-na ra-diá-ta
Cyrtodora siliqua	Cyr-to-dó-ra sil-í-qua
Cardium muricatum	Cár-di-um mu-ri-cá-tum

CONCHOLOGICAL ABBREVIATIONS

After the latin names of species, you will see an abbreviation of the author's name, who described said species. The following list is not intended to be complete, but covers most of the world authors of molluscan names.

Abbreviation	*Name*	*Nationality or Country*
Ad. A.	Adams, Arthur	English
Ad. H.	Adams, Henry	English
Ad. C. B.	Adams, Charles B.	American
Adan.	Adanson, M.	French
Alb.	Alberts, Johann Christ	German
Ald.	Adler, Joshua	English
Anc.	Ancey, C. F.	French
Ang.	Angas, Geo. F.	Australia
Ant.	Anton, H. E.	German
Anth.	Anthony, J. G.	American
Archer	Archer, A. F.	American
Auct.	Auctores (authors)	
B. & D.	Bavay & Dautzenberg	French
Bednl.	Bednall, William T.	Australian
Bartsch.	Bartsch, Paul	American
Baker	Baker, Frank C.	American
Baker, H. B.	Baker, H. Burrington	American
Bailey	Bailey, Joshua L. Jr.	American
Bens.	Benson, W. H.	English
Berry	Berry, S. Stillman	American
Berth.	Berthelot, Sabin	French
Bielz	Bielz, E. A.	German
Bgt.	Bourguignat, M. J. R.	French
Blainv.	Blainville, H. M. de	French
Bld.	Bland, Thomas	American
Blf.	Blanford, W. T.	English
Boettg.	Boettger, Dr. Oscar	German
Bourg.	Bourguignat, J. R.	French
Braz.	Brazier, John	Australian
Brod.	Broderip, W. J.	English
Brooks	Brooks, S. T.	American
Brug.	Bruguiere, J. G.	French
Binn.	Binney, Amos	American
Burnp.	Burnup, Henry C.	English
Bush	Bush, Katherine J.	American
C & F	Crosse & Fischer	French
Calc.	Calcara, P.	Italian
Call.	Call, R. E.	American
Caill.	Cailliaud, Frederick	French
Cantr.	Cantraine, F.	French
Carp.	Carpenter, P. P.	English
Caz.	Caziot, Le Commandant	French
Chase	Chase, E. P. & M. E.	American
Chemn.	Chemnitz, J. H.	German
Ckll.	Cockrell, T. D. A.	American
Conn.	Connolly, Maj. M.	English
Clench	Clench, W. J.	American
Conr.	Conrad, T. A.	American
Coop.	Cooper, Dr. J. G.	American
Cooke, Jr.	Cooke, Jr. T. Montague	Hawaii
Cooke	Cooke, A. H.	English
Couth.	Couthouy, J. P.	Australia
Cox	Cox, Dr. James C.	American
Crist	Christofori, G.	French
Crosse	Crosse, H.	French
Dall	Dall, W. H.	American
Da. Cost.	DaCosta, E. M.	English
Dautz.	Dautzenberg, E. M.	French

Abbreviation	*Name*	*Nationality or Country*
Deb.	Debeaux, O.	French
DeCamp	DeCamp, W. H.	American
Desh.	Deshayes, G. P.	French
Desm.	Desmoulius, Chas.	French
Dillw.	Dillwyn, Lewis W.	English
Don.	Donovan, Edward	English
D'Orb.	D'Orbigny, A.	French
Dohrn.	Dohrn, Dr. H.	French
Drap.	Draparnaud, J.	French
Dkr.	Dunker, Dr. W.	German
Ducl.	Duclos, M.	French
Dup.	Dupuy, D.	French
Dupois	Dupois, Com. Paul	Belgian
Eichw.	Eichwald, E. von	German
Esch.	Eschcholtz, Dr.	German
Ehrenb.	Ehrenberg, Dr.	German
Fabr.	Fabricus, O.	Swedish
Fag.	Fagot, Paul	French
Fbs.	Forbes, Edward	English
Fer.	Ferussac, J. B. L. D. de	French
Finl.	Finlay, H. J.	New Zealand
Fisch.	Fischer, Paul	French
Frier.	Frierson, L. S.	American
Fult.	Fulton, Hugh	English
Garr.	Garrett, Andrew J.	American
Gabb	Gabb, Wm. M.	American
Gass.	Gassies, J. B.	French
Gld.	Gould, Dr. A. A.	American
Gmel.	Gmelin, J. F.	German
G.´A.	Godwin-Austen, Lt. Col. H. H.	English
Goodr.	Goodrich, Calvin	American
Gregg	Gregg, Wendell O.	American
Gray, A. F.	Gray, A. F.	American
Gray	Gray, Dr. J. E.	English
Grat.	Grateloup, J. P. D. de	French
Greg.	Gregorio, Antonio de	Italian
Guild.	Guilding, Lansdowne	English
Gude	Gude, G. K.	English
Guppy	Guppy, R. L. L.	English
Gut.	Guttierrez, Dr.	Cuban
H. & J.	Hombron & Jacquinot	French
Hald.	Halderman, S. S.	American
Hanl.	Hanley,Sylvanus	English
Hanna	Hanna, G. Dallas	American
Hann.	Hannibal, Harold	American
Hartm.	Hartman, Dr. D. W.	American
Hedl.	Hedley, Chas.	Australian
Hemp.	Hemphill, Henry	American
Hend.	Henderson, Junius	American
Heyn.	Heynemann, D. F.	German
Hde.	Heude, R. P.	French
Hds.	Hinds, R. B.	English
Hid.	Hidalgo, Dr. J. G.	Spanish
Hink.	Hinkley, A. A.	American
Hombr.	Hombron, M.	French
Humph.	Humphreys, J. D.	English
Hull	Hull, Arthur F. B.	Tasmania
Hutt.	Hutton, F. W.	New Zealand
Ired.	Iredale, Tom	Australian
Issel	Issel, Arthur	Italy
Jacq.	Jacquinot, H.	French
Jeffr.	Jeffreys, J. Gwyn	English
Jick.	Jickeli, C. F.	German
Jouss.	Jousseaume, Dr.	French
John.	Johnson, C. W.	American

Abbreviation	*Name*	*Nationality or Country*
Kien.	Kiener, L. C.	French
Kob.	Kobelt, Dr. W.	German
Kust.	Kuster, H. C.	German
Lam.	Lamarck, J. B.	French
Lea.	Lea, Isaac	American
Lewis	Lewis, Dr. James	American
Less.	Lesson, R. P.	French
Lindstr.	Lindstrom, G.	Danish
Lisch	Lischke, Dr. C. E.	German
Linn. or L.	Linne (Linnaeus) Carl von	Swedish
Loc.	Locard, Arnold	French
Lowe	Lowe, H. N.	American
M. & S.	Melville & Standen	English
Mab.	Mabille, M. P.	French
Macglv.	Macgillivrayi, W.	English
Marie	Marie, E.	French
Marsh.	Marshall, Wm. B.	American
Mart.	Martyn, Thomas	English
Martini	Martini, Fr. H. W.	French
Mayn.	Maynard, Chas. J.	American
Melv.	Melvill, J. Cosmo	English
Midd.	Middendorf, A. T. von	Russian
Mich.	Michaud, A. L. G.	French
Migh.	Mighels, J. W.	American
Mke.	Menke, C. T.	German
Mlldff.	Mollendorf, Otto F. von	German
Mtg.	Montagu, George	English
Mlldff.	Montfort, Denys de	French
Montr.	Montrouzier, M.	French
Monts.	Monterosatha, Marquis, di	Italian
Moq. Tand.	Moquin-Tandon, M. G.	French
Morch.	Morch, Otto A. L.	German
More.	Morelet, M. Arthur	French
Morse	Morse, E. S.	American
Mouss.	Mousson, Albert	French
Mozl.	Mozley, Alan	American
Mull.	Muller, Otto F.	German
Nev.	Nevill, Geoffrey	English
Newc.	Newcomb, Wesley	American
Nutt.	Nuttall, Thomas	English
Odhnr.	Odhner, Nils H.	Swedish
Oldr.	Oldroyd, Mrs. Ida S.	American
Artm.	Oatman, A. E.	American
Pall.	Pallary, Paul	French
Payr.	Payraudeau, P. C.	French
Parr.	Parreyss, Dr.	German
Pfr.	Pfeiffer, Dr. Louis	German
Peile.	Peile, A. J.	English
Palad.	Paladilke, M.	French
Pett.	Pettard, W. F.	Tasmania
Phil.	Philippi, Dr. R. A.	German
Pils.	Pilsbry, Dr. Henry A.	American
Pse.	Pease, Wm. Harper	American
Pons.	Ponsonby, J. H.	English
Prest.	Preston, Henry B.	English
Prash.	Prashad, B.	India
Poli.	Poli, Xavier	Italian
Prm.	Prime, Temple	American
Q. & G.	Quoy & Gaimard	French
Raf.	Rafinesque, C. S.	French-American
Recl.	Recluz, C. A.	French
Redf.	Redfield, John H.	American
Roberts	Roberts, S. Raymond	American
Risso	Risso, A.	French
Roch.	Rochesbrunne, A. T.	French

Abbreviation	*Name*	*Nationality or Country*
Romer	Romer, Dr. Edward	French
Rehdr.	Rehder, Harold A.	American
Rossm.	Rossmassler, E. A.	German
Rve.	Reeve, Lovell	English
Sars.	Sars, G. O.	Norwegian
Saint-Simon	Saint Simon, Dr. Alfred de	French
Schum.	Schumacher, C. J.	German
Serv.	Servain, Georges	French
Shutt.	Shuttleworth, R. J.	German
Simp.	Simpson, Chas. T.	American
Shimek.	Shimek, B.	American
Souv.	Souverbie, Dr.	French
Sow.	Sowerby, G. B.	English
Sol.	Solander, D.	English
Smith, E. A.	Smith, E. A.	English
Smith, M.	Smith, Maxwell	American
Speng.	Spengler, L.	German
Stimp.	Stimpson, William	American
Strki.	Sterki, Dr. V.	American
Stoss.	Stossick, Adolph	Italian
Strg.	Strong, A. M.	American
Strns.	Stearns, R. E. C.	American
Spix	Spix, J. B.	Brazil
Sut.	Suter, Henry	New Zealand
Strob.	Strobel, H.	German
Stab.	Stabille, J.	Italian
Swain.	Swainson, William	English
Tap-Can.	Tapparone-Canefrio, C. M.	Italian
Theob.	Theobald, W. Jr.	English
Trosch.	Troschel, F. H.	German
Tate	Tate, Ralph	Australian
Thiel.	Thiele, Prof. Johannes	German
Tomlin	Tomlin, J. R. LeB.	English
Torre	Torres, Carlos de la	Cuban
Tryon	Tryon, Geo. W.	American
Turt.	Turton, W. H.	English
Torr	Torr, Dr. W. G.	Australian
Val.	Valenciennes, M. A.	French
Vanatta	Vanatta, E. G.	American
Vands.	Vander Schalie, Henry	American
Walker	Walker, Bryant	American
Van. Mart.	Van Martens, Dr. Edward	German
Wein.	Weinkauff, H. C.	German
Westr.	Westerlund, Dr. C. A.	German
Welch	Welch, D'Alte A.	Cuban
Wilt.	Willett, G.	American
Woll.	Wollaston, T. V.	English
W. G. B.	Binney, W. G.	American
Wrt. B. H.	Wright, Berlin H.	American
Wrt. S. Hart	Wright, S. Hart	American
Wink.	Winkworth, R.	English
Ziegl.	Ziegler, Dr.	German
Zet.	Zetek, James	Panama

CLEANING SHELLS

MARINE SHELLS. If your shells have no periostracum, or epidermis and you wish to remove the sea growths, or loosen them, so they will come off easily with a knife, simply grasp them with tweezers and immerse in a crock of muriatic acid. Hold them for a moment, and then rinse in water. If not enough, try it again, and again.

Shells that have no periostracum can be dipped in weak acid for a moment and it will brighten them up. If your shells only need a little cleaning to brighten them up, use Clorox which you can now secure at most any good grocery store.

This solution is fine to brighten up shells and it is used for bleaching corals, to make them white. In fact it seems to be unusually useful in cleaning many kinds of shells. Do not leave them in the solution for more than a few hours, or over night. Then wash the shell with a brush and water.

If your shells have a periostracum or epidermis as many call it, they require different treatment. This epidermis is animal matter whereas the shell is lime. If there are sea growths on the epidermis they must be removed first with the muriatic acid. For generations collectors have used chloride of lime or concentrated lye to eat off this epidermis and it took from 3 to 6 days to do it. The most modern methods is to use caustic soda. One pound to a gallon of water will remove or loosen the epidermis in 24 hours so that it slips off easily. The soda comes in tin cans and should not cost over 10 or 15c a pound. In fact in 100 pound cans it is as low as 5c. This does not harm the shell at all and works quickly.

If you wish to put a high polish on your shells, take a porcelain kettle of fair size and place it in a solution of 1/3 muriatic acid and 2/3 water. Bring to a boiling point. Grasp your shell with tweezers as before, and immerse in this boiling solution for just a moment and then in ice water. This polish will last forever. You will not have real success with shells of soft texture. It works better with shells that are very hard, as most of them are.

Practice in the art of cleaning makes perfect work. We call it putting them through the "beauty shop," and you can obtain really wonderful results if you persevere.

LAND SHELLS. Most shells that live on land, only require washing. If they are simply soiled or shells from an old collection, use a little sapolio and small hand brush. A good small syringe is very useful in cleaning the interior whorls.

FRESH WATER SHELLS. Treat same as land shells, except if they are covered with algea or greenish mould, it is well to immerse them in oxalic acid for an hour or more, and then wash with a soft brush and water. Do not leave them in the acid too long or they may be ruined. Very minute, shells can be nicely cleaned by putting them in a vial with sand and water and shaking vigorously.

TO REMOVE SOFT PARTS. Simply immerse the shell in boiling water for 2 to 5 minutes, then place in cold water so they can be handled, and use a sharp wire hook. Attach to foot of mollusk and by pulling gently the entire soft parts will come out easily. If the soft parts should break off, it is then best to place the shells in alcohol for a few days. Then lay in shade to dry and there will be no unpleasant odor. Very small and minute shells can be placed in alcohol when first collected. Leave a few days and then spread out to dry.

SHELL CABINETS

MOST collectors keep their shells of moderate size and small forms in cabinets of drawers. I give herewith a style I have found very useful.

The cabinet is 60 inches high, 28 inches across the front and 24 inches deep. This style and size just makes 20 drawers as follows: 6 drawers 1⅜ inches, 6 drawers 2 inches deep. Then a ⅞th inch strip across front to strengthen. 6 drawers below the strip 2½ inches deep and 2 drawers 4 inches deep. Have made of white-wood, well kiln dried. The drawers will be light to handle and will forever move freely.

Large shell over 4 inches diameter it is best to place on shelves in a glass cabinet or built in walls. If specimens collect dust as they surely will, they can be washed as often as necessary with warm water and if unusually dirty, use sapolio which will bring back their original fine condition. I have had many types of cabinets but the above, arranged in tiers, I found were best for a private home. In museums, the problem is different, and they often use steel dustproof receptacles, which are of course more expensive.

Suitable trays for specimens I have found to be of the following sizes: 1½ by 2 inches; 2 by 3 inches; 3 by 4 inches; 4 by 6 inches. Two of one size equals the next, and they fit nicely into any size drawer. The depth should be uniformly ¾ inch. They can, when made, be covered with any colored paper desired, glazed or plain. Small or minute shells can be mounted in glass-topped boxes that are round or oblong, using very dark blue cotton. Another way which takes much less rooms, is shell vials, which can be made by almost any glass manufacturing concern. I have used 3 sizes only, all 50 mm. long, round bottom and they are universally called shell vials. Smallest size 8 mm. thick, next size 12 mm., third size 20 mm. These three sizes will hold almost anything you will want to put in vials. The labels should always be placed inside. I have always used cotton in place of corks, saves room in cabinets and protects shells.

The main advantage of vials, is they take up so little room. There are many genera of small shells of which you will only have five to a dozen species, perhaps all in the smallest vials. They will all go in a 1½ inch tray. There may be a genus you will have 200 to 500 vials, and they can be conveniently arranged in trays alphabetically or nearly so.

A tier of cabinets described above in Mr. Webb's Conchological den. There is another similar row to left.

INDEX OF SCIENTIFIC NAMES

INDEX OF COMMON NAMES AND GENERAL INFORMATION